P9-DWV-561

$18.00

Plumbers Handbook

Revised Edition

By

Howard C. Massey

Craftsman Book Company
6058 Corte del Cedro P.O. Box 6500 Carlsbad, CA 92008

Acknowledgements

The author expresses his sincere thanks and appreciation

- to Mr. Boyd A. Arp, assistant county manager of Metropolitan Dade County, Miami, Florida, for granting persmission to use excerpts from the plumbing sections of the South Florida Building Code to authenticate the information in this book.

- to Mr. A. T. Strother, Executive Director of the Plumbing Industry Program of Miami, Florida, for granting permission to use excerpts from his article "A History of Plumbing" which appeared in *The Florida Contractor Magazine.*

- to the Florida Energy Committee and the Environmental Information Center of the Florida Conservation Foundation, Inc. for pertinent information necessary to authenticate the thermosyphon and pumped solar water heating system.

- to Mr. David Leidel, Chief Plumbing Inspector, and Mr. Larry Glover, assistant plans examiner (plumbing), Metropolitan Dade County, Miami, Florida, for their expertise and assistance in interpreting the intent of certain sections of the plumbing code.

- to American Standard for providing the plumbing fixture roughing-in measurements (as illustrated in Chapter Sixteen).

- to Josam Manufacturing Company for providing the plumbing fixture carriers (as illustrated in Chapter Sixteen).

To my wife, Hilda, for her untiring assistance in editing and typing this manuscript.

Library of Congress Cataloging in Publication Data
Massey, Howard C.
Plumbers handbook.

Includes index.
1. Plumbing--Handbooks, manuals, etc.
I. Title.
TH6125.M37 1985 696'.1 85-9668
ISBN 0-910460-49-3
First Edition ©1978 Craftsman Book Co.
Second Edition ©1985 Craftsman Book Co.
Second Printing 1987

Illustrations by
Gretchen Egge, Bill Grote, and the author

Cover by
Ann B. Sweeney

Contents

Plumbing And The Plumber

If you have chosen plumbing as your profession, you should find it one of the most challenging and satisfying of all construction trades. The possible variations in design, layout, and installation methods in any building can present a stimulating challenge to any professional plumber. Yet, lack of knowledge and failure to understand the minimum requirements of modern plumbing codes have left many would-be plumbers discouraged, frustrated, and confused.

Learning plumbing from a code book can be a very difficult task. This book is intended to help anyone interested in doing professional quality plumbing work grasp the important design and installation principles recognized as essential to the trade. What you learn here should be applicable nearly anywhere in the U.S. regardless of the model code adopted by the jurisdiction where your work is done. Reading and understanding what is written here should be much easier than reading and understanding the code for anyone who is just learning the fundamentals of plumbing. However, this book is not the plumbing code and every plumber will have to refer to his local code from time to time. The minor variations in model plumbing codes are emphasized throughout this book and should be easily recognized as you read and compare sections of this book with your local code. *The basic principles of sanitation and safety remain the same, regardless of the geographical location.*

The art and science of plumbing came into being through the struggle of mankind against disease. The history of civilization is the history of plumbing. At the dawn of civilization, when two or three families gathered together to make a tribe, man drank from springs and streams. He made no provisions for the disposal of sewage and garbage. We can assume that when his place of living became fouled with kitchen refuse and human waste, he moved to a fresh camping ground. When disease killed members of his tribe because they neglected the laws of sanitation, he may have concluded that the gods were offended by the place where the tribe lived, and they moved to a new area. He did not understand that lack of cleanliness breeds disease.

Archeologists, while digging in various parts of the world, have confirmed that some ancient civilizations had developed plumbing systems for protecting health. At Nippur, in Babylon, archeologists have uncovered an aqueduct made of glazed clay brick that dates back to 4500 B.C. This aqueduct contained three lines of glazed clay pipe. Each section was eight inches in diameter and two feet long with a flanged mouth. Other excavations have revealed glazed clay pipe in jar patterns, concave and cone shapes and a sewage system complete with manholes.

On the island of Crete, at Broad Knossos, some of the palaces of ancient sea kings were equipped with extensive water supply and drainage systems. Evidence of plumbing fixtures constructed of hard clay were also discovered. The glazed clay pipe was found to be in perfect condition after 3500 years.

In ancient Greece further advances were made in cleanliness. Greek aqueducts took pure water from mountain streams into cities. Sewers, which exist to this day, carried away waste to the surrounding rivers.

While people may not have understood the causes of disease, and did not guard against pollution of the water supply, they did understand that bathing was a desirable habit. Greeks portrayed Hygeia, the goddess of health (from whose name we get the word "hygiene"), as supplying pure water to a serpent, the symbol of wisdom, signifying that wise people found health in a supply of pure water.

The ancient Egyptians realized the value of sanitation. Moses was acquainted with the sanitary science of the Egyptians and used it in the code of laws framed by him and found in the book of Leviticus.

The Romans in the time of Julius Caesar developed the principles of sanitation to a high art. Lead was unknown to the ancient Greeks and Egyptians, but it was a new substance to the Romans and was imported from the British Isles. The Romans called this substance *plumbum*. The words "plumbing" and "plumber" are derived from the Latin word for lead. The forefather of the modern plumber was called the "plumbarius," which meant a worker in lead. The uses to which lead was put by the Romans were much the same as today.

Two thousand years ago the city of Rome had an adequate water supply and sewage disposal system. Water was piped from hills and mountains 50 miles distant from the city. To bring this water into Rome, great overhead aqueducts and underground tunnels were built of masonry.

Branch lines from the aqueducts to the individual buildings required the use of pipe. Water was piped into the homes of the upper class for use in private bathrooms long before the development of the great public baths. The baths of Pompeii consisted of an entire room with the floor and walls of marble. Brass, bronze and silver were used in the fixtures in addition to lead.

With the advent of the public baths, bathing became an important function. The public bath became a form of family entertainment, an event much like going to the beach today. One of the most famous bath houses was a mile square and the basin could hold 3,200 people.

When Rome set out to conquer the world, they took their bathing habits with them. After taking what is now Great Britain from the Druids, the Romans built their own baths on the banks of the Avon River. As recently as 1875, in the city of Bath, archaeologists uncovered a Roman bath 110 feet long and 68 feet wide.

From as far back as 600 B.C. Rome had an elaborate drainage system called the *Cloaca Maxima*. This main was 13 feet in diameter and was joined by many laterals. It was constructed from three concentric rows of enormous stones piled one on the top of another without cement or mortar. It still exists and is used today in the drainage system of *modern* Rome.

In the 12th century, trade guilds were first organized in England. The first apprenticeship laws were passed in 1562 during the reign of Queen Elizabeth. These laws required an apprenticeship of seven years and made apprenticeship in all crafts compulsory. It was not until 1814 that the compulsory clause was removed and apprenticeship was made voluntary. The first known master plumbers' association was organized in England and incorporated in the College of Heralds of London.

With the discovery of the New World, man, like his ancient ancestors, sought to escape the dark and dirty cities of Europe for a fresh campground.

Although America has become a symbol of high standards in plumbing and sanitation, progress in the development of sanitation and plumbing was very slow. As the population of the early settlements increased, conditions deteriorated. Garbage and sewage dumped onto the ground and seepage from earth-pit privies polluted nearby wells.

Health conditions became so intolerable that eventually public sewers had to be installed underground and extended to each building. Although New York in 1782 installed the first sewer under the streets, Chicago is credited with having the first real city sewage system, constructed in 1855.

Plumbing as we know it today traces its roots back many centuries, but was not really perfected until the twentieth century. Even so, many of us were reared without the benefit of indoor plumbing. The open well, the pitcher pump, the outhouse, and the Saturday night romp in the old wooden tub on the back porch are still with us. The modern bathroom, city water, and the sewer system of today are taken for granted. Let us not forget — without today's "plumbarius" this would be a vastly different world in which to live. Truly, the plumber protects the health of our nation and the world.

Sanitary Drainage And Vent Systems

The private sanitary drainage system is the essential part of the plumber's work. It will be presented in this chapter as: (1) All the pipes installed within the wall line of a building and on private property for the purpose of receiving liquid waste or other waste substances (whether in suspension or in solution), (2) The pipes which convey this waste to a public sewer or other private, approved sewage disposal system, and (3) The vent system.

This whole drainage and vent system must be installed so that it is not a health or safety hazard to any individual or to the general public. Thus, most municipal authorities have adopted laws (codes) to protect the public health. Although sanitary drainage and vent arrangements are the heart of the plumbing system, most experts agree that this is the most complex, misunderstood, and misinterpreted section of the code. Professional engineers, master plumbers, and plumbing inspectors are frequently at odds on its intent and interpretation.

Interestingly enough, most of the questions asked and the isometric drawings required to complete the journeyman's and master's examinations are taken from this section of the code. It is worth noting also that the majority of cases coming before any Board of Rules and Appeals for clarification and resolution center on this same section of the code. Therefore it is important that you understand this section of the code.

Although details of plumbing installations vary, the basic principles of sanitation and safety remain the same, regardless of the code that has been adopted in your area. Any minor changes from the basic rules described here should be easily recognized and noted as you read and compare the sections of this book with the code used where you work.

Isometric Drawings

Before you can understand drainage and vent systems, you must be familiar with isometric drawings. Isometric drawings are the means of communication between plumbing professionals. They are used by the plumbing contractor to estimate the cost of new work and to show the job foreman how to *rough-in* a particular job. Anyone who deals with plumbing work must be able to make and interpret isometric drawings. You will quickly find that it is easy to read and make isometric illustrations.

Only three basic angles are needed to express the plumbing system: the horizontal pipe, the vertical pipe, and the forty-five degree angle pipe. Figure 2-1 can be a valuable guide in practicing these isometric angles. The only pieces of equipment necessary are a sharp No. 2 pencil and a 90-60 right triangle. By following the directions here and practicing the exercises in this chapter, you will be able to produce your own isometric projections or follow those made by others.

Directions For Practicing Isometric Drawings

First, draw a circle with a dot in the exact center. Place the letter *N* at the top of the page to designate the direction of north. See Figure 2-1. (Do not let the dotted line confuse you; it is used only to determine the proper angle necessary to illustrate the sanitary system as shown in unbroken solid lines.)

Using a 90-60 right triangle, square the short base with the right edge of your paper and draw line *A* through the center dot. (This represents *A*, the north-south horizontal pipe.)

Again, using the 90-60 right triangle, square the short base with the left edge of your paper and draw line *B* through the center dot. (This represents *B*, the east-west horizontal pipe.)

To arrive at the *C* vertical line, square the short base of the 90-60 right triangle with the lower edge of the paper and draw line *C*, using the long base for this vertical line connecting it with any of the horizontal lines, as desired.

To determine the placement of line *E*, divide the area equally between the horizontal line *B* and the dividing dotted line. Then draw line *E* east and west through the center dot. (This is necessary to depict the change in direction assumed by the 45-degree fittings of either a wye or 1/8 bend.) The same procedure will, of course, yield *D* north and south. The lower portion of Figure 2-1 shows a simple isometric drawing using all three basic angles that are used in designing rough plumbing for any building.

Fittings Within An Isometric Drawing

The lines on isometric drawings represent pipe and fittings. Symbols are used to show the location of fixtures. The figures that follow represent 16 common fittings used with no-hub pipe. The symbols are the same regardless of the type of pipe used.

Figures 2-2, 2-3, and 2-4 show typical isometric drawings. Each fitting in these drawings is numbered to correspond with a drawing of the same fitting shown below. Look at Figure 2-2. You will see that the "horizontal twin tap sanitary tee" (also known as an "owl fitting") is the same as fitting number 14 at the top of the page. The purpose of this fitting is to permit two similar fixtures to connect to the same waste and vent stack at the same level. In this case it connects two lavatories.

Before anyone can become proficient in the plumbing trade, he must learn how to construct isometric layouts correctly.

Plumbing Fixture Abbreviations

Common abbreviations are often used to identify various types of plumbing fixtures in isometric drawings as well as floor plans.

For example, some writers will use the letter *L* to designate a lavatory. Others may use "LAV." We have selected the following abbreviations to identify the plumbing fixtures used in isometric drawings and floor plans in this book.

Plumbing Fixtures	Abbreviated Symbols
Water closet	W. C.
Bathtub	B. T.
Shower	SH.
Lavatory	L.
Kitchen sink	K. S.
Clothes washing machine	C. W. M.

Typical plumbing fixture abbreviations

A more detailed and complete section on abbreviations, definitions and symbols is included in Chapter 18.

Definitions and Illustrations

One of the first steps in learning the plumbing process is identifying the basic piping arrangements as defined in your local code. Without this knowledge, you won't be able to design, lay out or install pipes and fittings. Figures 2-5, 2-6 and 2-7 show three sanitary isometric drawings which include all the major definitions of the basic drainage and vent systems. The important terms are shown in graphic symbol coding in the illustration, thus relating its use to other piping in the overall drainage and venting system.

Figure 2-5 is the sanitary isometric drawing of a typical two bath house, including a kitchen and utility room. Figure 2-5 is an installation on the *flat* connected to a public sewage system. Figure 2-6 is a typical one bath house, including a kitchen and utility room. This figure illustrates an installation on a *stack system* connected to a private sewage disposal unit (a septic tank). Figure 2-7 shows a typical battery of plumbing fixtures often found in a two story public building.

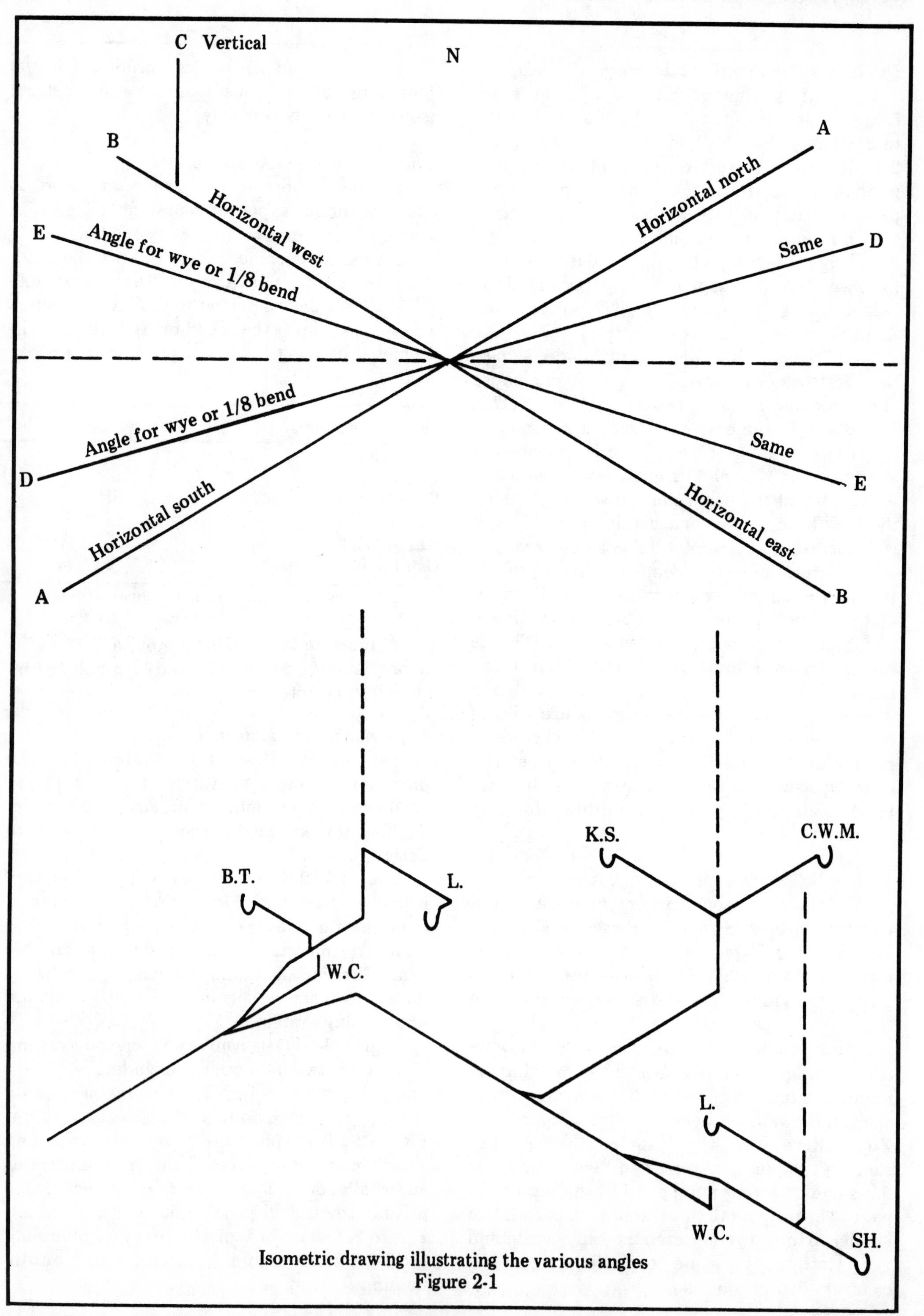

Isometric drawing illustrating the various angles
Figure 2-1

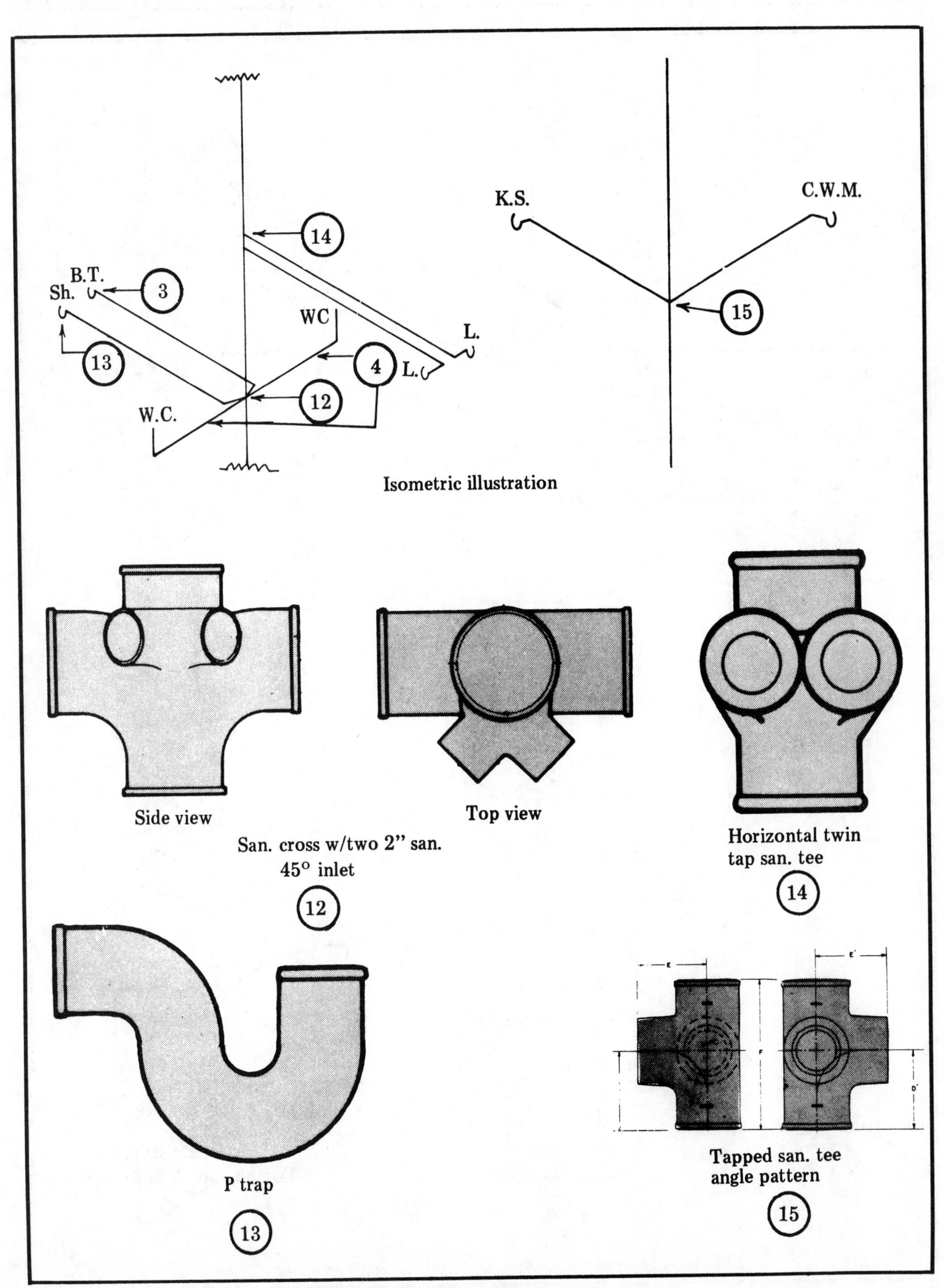

Fittings illustrated within isometric drawings
Figure 2-2

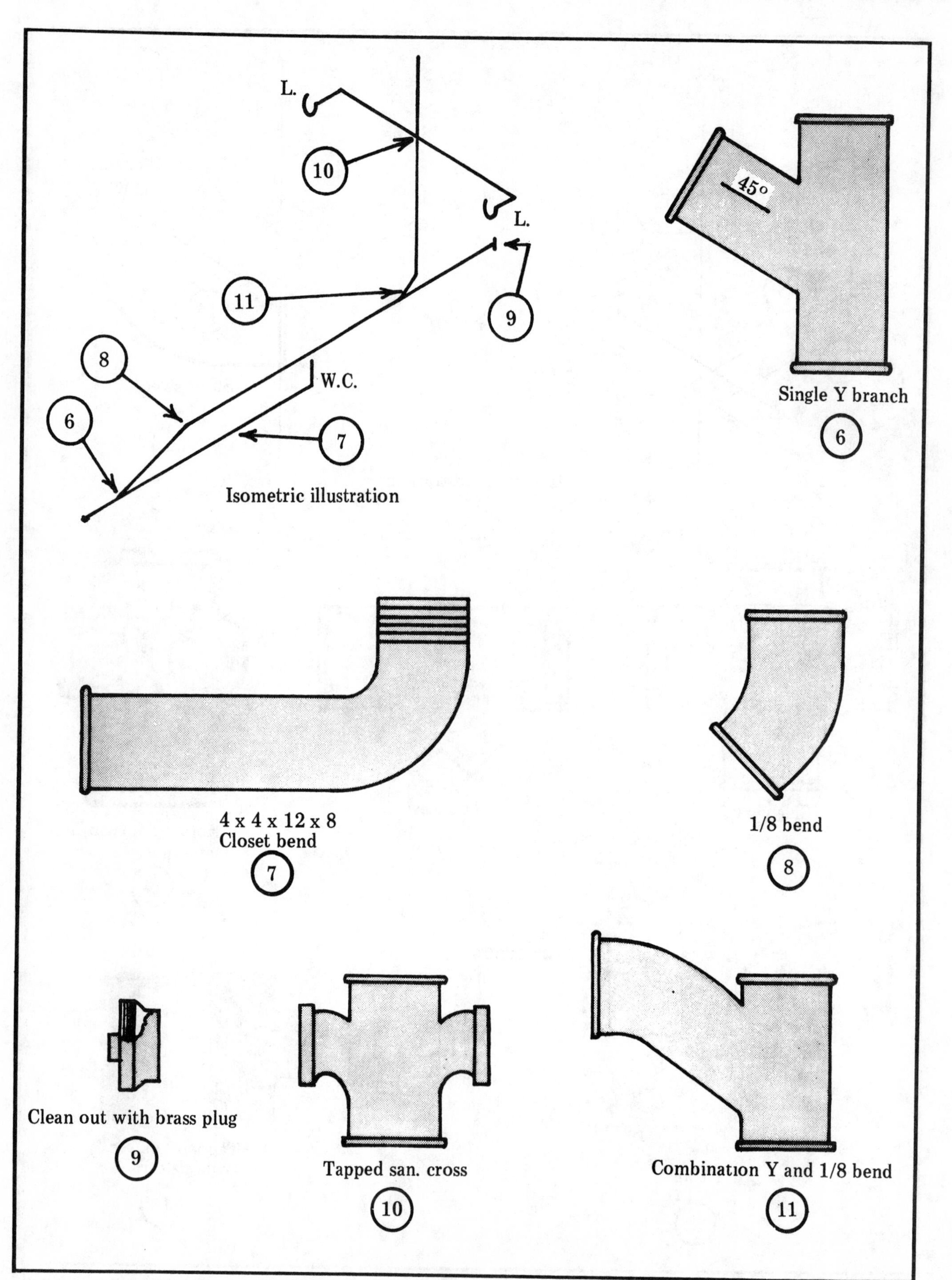

An isometric drawing and its fittings
Figure 2-3

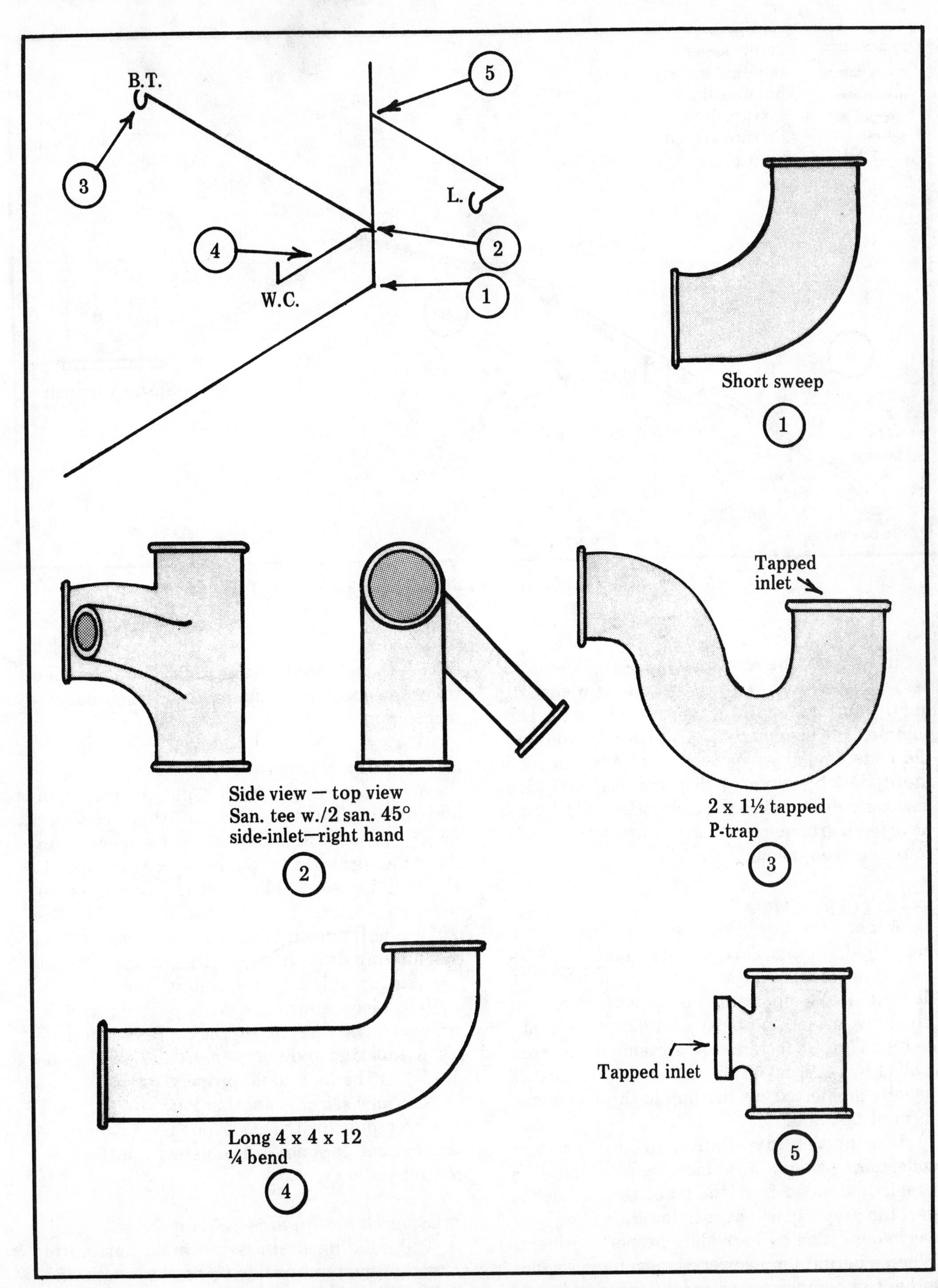

An isometric drawing and its fittings
Figure 2-4

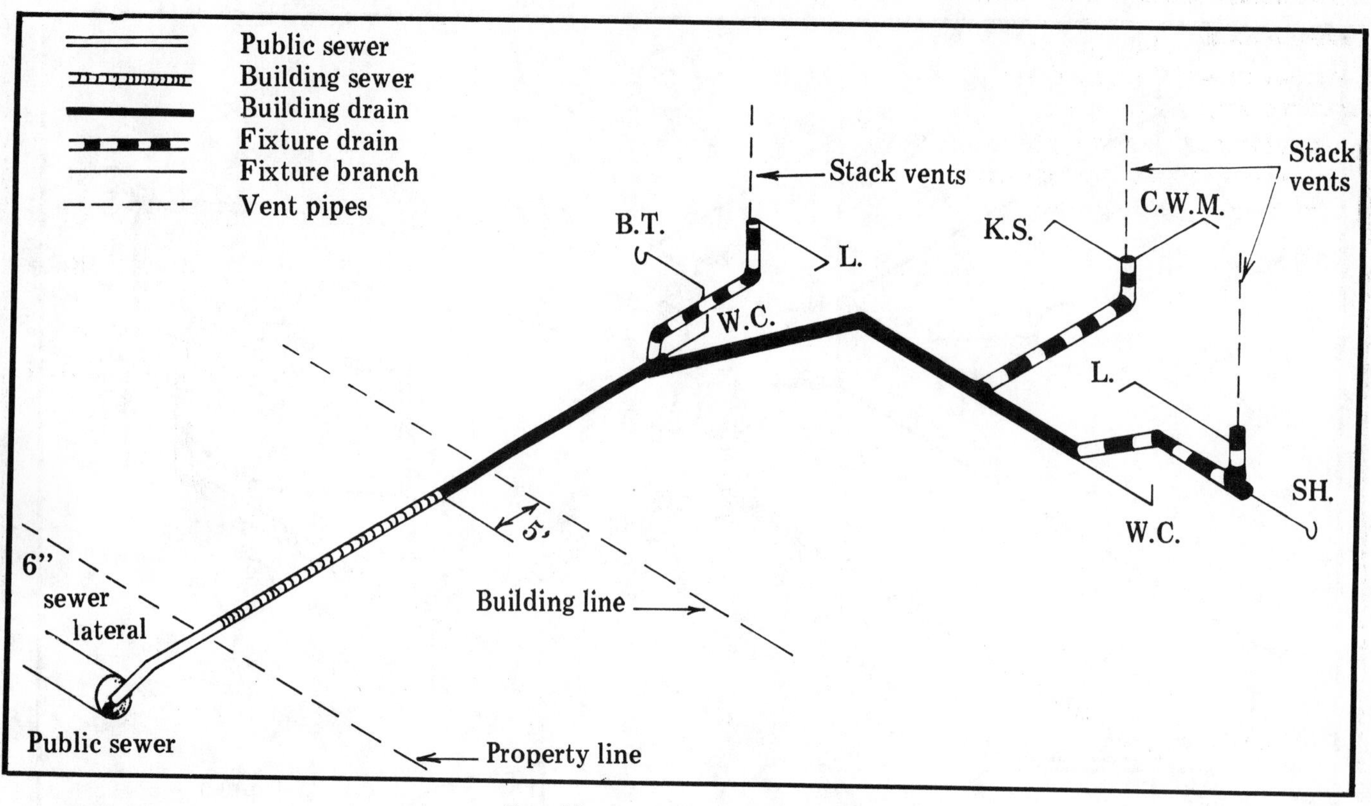

Graphic symbols of definitions
Figure 2-5

In defining the terms in Figures 2-5, 2-6 and 2-7, every effort has been made to use simple, easily understood language to clear up the complex and seemingly contradictory wording of the code. A particular section of pipe may be identified by several different terms with various definitions in the code. Here the terms have been grouped under a graphic symbol code to reduce confusion.

Public Sewer (Figure 2-5)

A *public sewer* may also be known as a *municipal sewer*. This sewage collection system is located in either a street, alley or a dedicated easement adjacent to each parcel of privately owned property. Public sewers are common pipes installed, maintained and controlled by the local public authorities. Its cost is usually supported by the public through some form of taxation.

During the installation of this sewage collection system, a 6 inch sewer lateral is usually extended from the main several inches past the property line of each lot. This allows for easy connection by individual property owners. When a permit for a sewer connection is issued, either to a homeowner or a plumbing contractor, the depth and location of the lateral are obtained from the local municipal engineering department.

Building Sewer (Figure 2-5)

A *building sewer* is that part of the main horizontal drainage system that conveys sewage or other liquid waste substances from the building drain to the public sewer lateral. The building sewer begins at its connection to the 6 inch lateral a few inches within the property line. It terminates at its connection to the building drain five feet (less in some codes) from the outside building wall or line.

It is also known as a *private sewer*, as it is not controlled directly by the public authority. Its installation and maintenance is the responsibility of the individual property owner.

Sanitary sewer is another name for this section of piping in the code book, as it carries sewage that does not contain storm, surface, or ground water.

Building Drain (Figures 2-5 through 2-7)

The building drain is the main horizontal collection system, exclusive of the waste and vent stacks and fixture drains, and is located

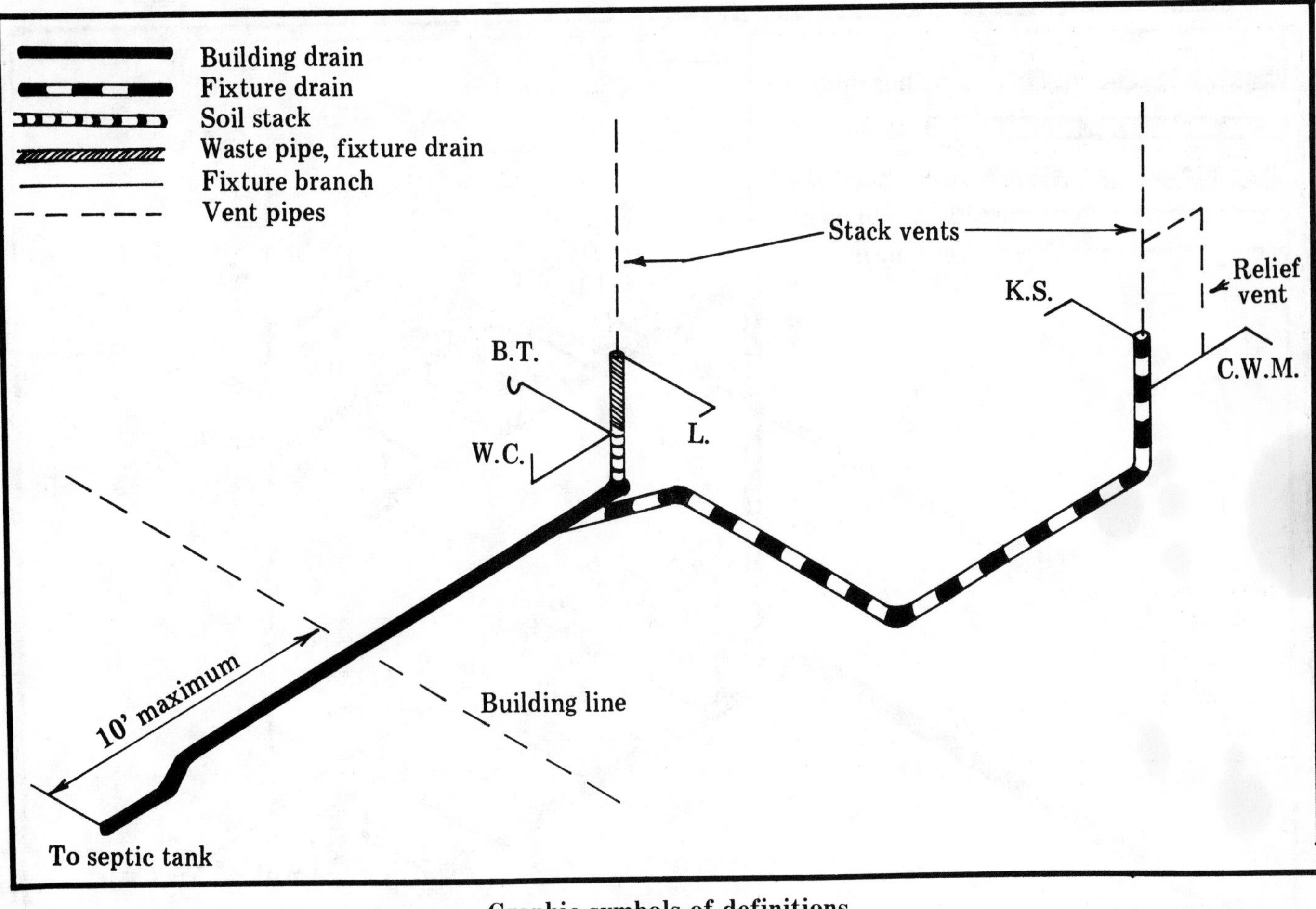

Graphic symbols of definitions
Figure 2-6

within the wall line of a building. It conveys all sewage and other liquid waste substances to the building sewer which begins five feet (less in some codes) outside the building wall or line.

It is also considered a *main*, as it acts as the principal artery to which other drainage branches of the sanitary system may be connected.

Fixture Branch (Figures 2-5 through 2-7)

The *fixture branch* is that portion of the drainage system that is composed of the pipe from the fixture trap to the vent serving that fixture. It may connect directly to a vertical vent stack above the floor or, in the case of a shower or bathtub, to the horizontal wet vent section beneath the floor. It is often referred to by plumbers as the "sink arm" or "lavatory arm."

Fixture Drain (Figures 2-5 and 2-6)

This is the drainage pipe that conveys liquid waste not containing body waste from a fixture branch to the junction of any other drain pipe. In the code it is also called a *waste pipe,* as it does not carry the waste from water closets, urinals, or bed pan washers.

Another term for this pipe is the *wet vent,* for it often conveys liquid waste from plumbing fixtures, excluding water closets, to the building drain and serves as a vent for these fixtures as well.

Soil Stack (Figures 2-6 and 2-7)

A *soil stack* is the vertical section of pipe of a plumbing system that receives the discharge of water closets, with or without the discharge from other fixtures, and conveys this waste substance, usually to the building drain.

The *branch interval* performs the same function as the soil stack and becomes an integral part of the soil stack. The only difference is in its vertical height. It usually corresponds to a story height but in no case can it be less than eight feet in length. Stacks also include any vertical pipe including the waste and vent piping of a plumbing system.

Horizontal Branch (Figure 2-7)

A *horizontal branch* is that portion of a drain

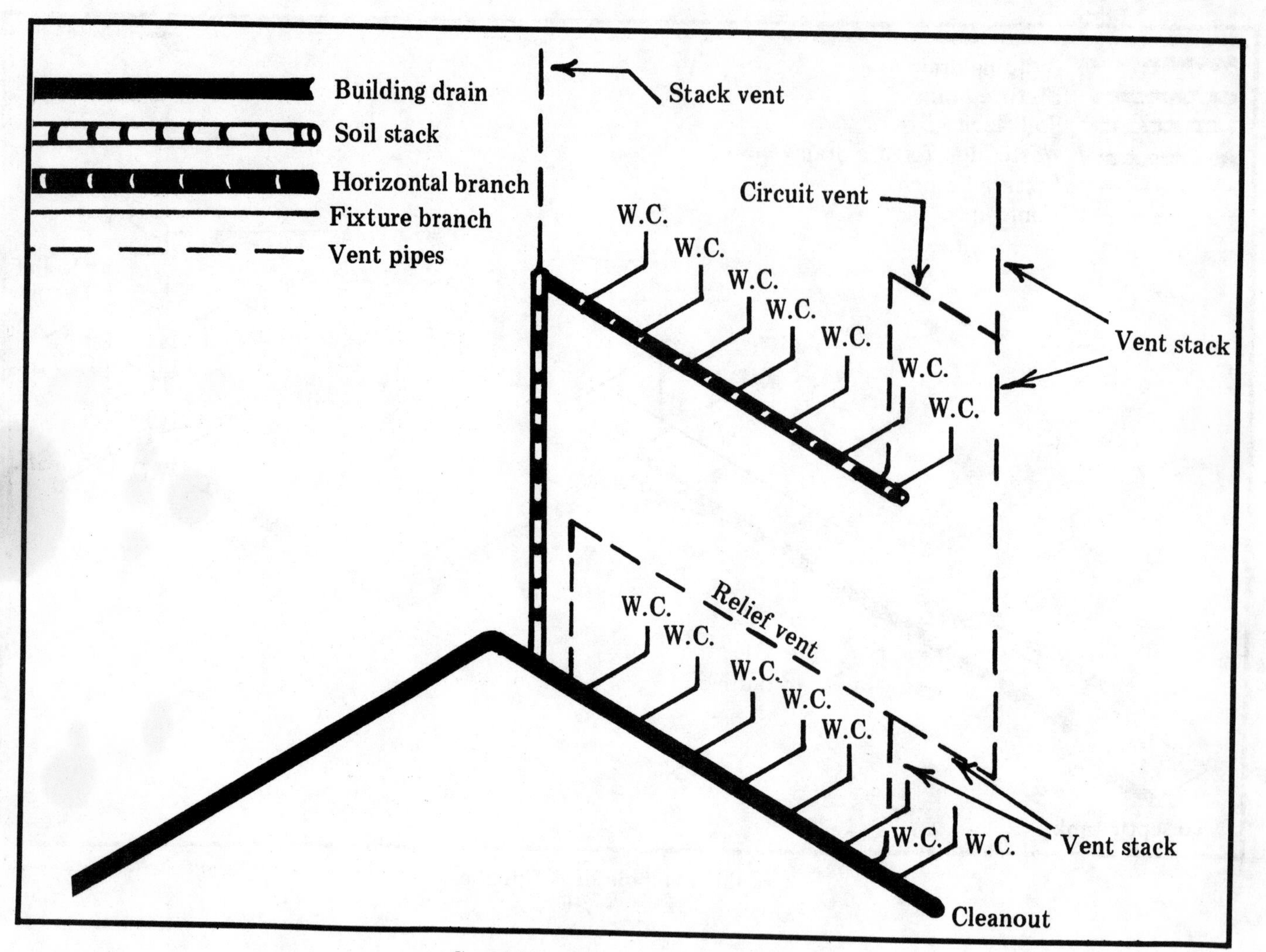

Graphic symbols of definitions
Figure 2-7

pipe extending laterally from a soil or waste stack that receives the discharge from one or more fixture drains. (See Figure 2-7.)

Common Vent (*Figure 2-5*)

A *common vent* is a vertical vent that serves two fixture branches that are installed at the same level. In Figure 2-5 this is the sink and clothes washing machine.

Continuous Vent (*Figures 2-32, 2-33, 2-38*)

A *continuous vent* is the vertical portion that is a continuation of the drain to which it is connected. It is also known as an *individual vent*.

Vent Header (*Figures 2-30 and 2-31*)

A *vent header* is a single pipe that receives the connection of two or more vent pipes and is then connected to the main vent stack or is extended to the atmosphere separately at one point.

Yoke Vent (*Figures 2-34 and 2-35*)

A *yoke vent* is a pipe connecting upward

Fixture Type	Fixture unit value as load factor	Minimum size of trap in inches
Bathtub (with or without overhead shower)	2 F.U.	1½
Shower stall, domestic	2 F.U.	2
Lavatory, domestic	1 F.U.	1¼
Water closet, tank operated	4 F.U.	3
Kitchen sink, domestic	2 F.U.	1½
Automatic clothes washer	3 F.U.	2

Note: Fixture units and trap sizes may vary from those listed above. Check local code requirements.

Fixture units per fixture
Table 2-8

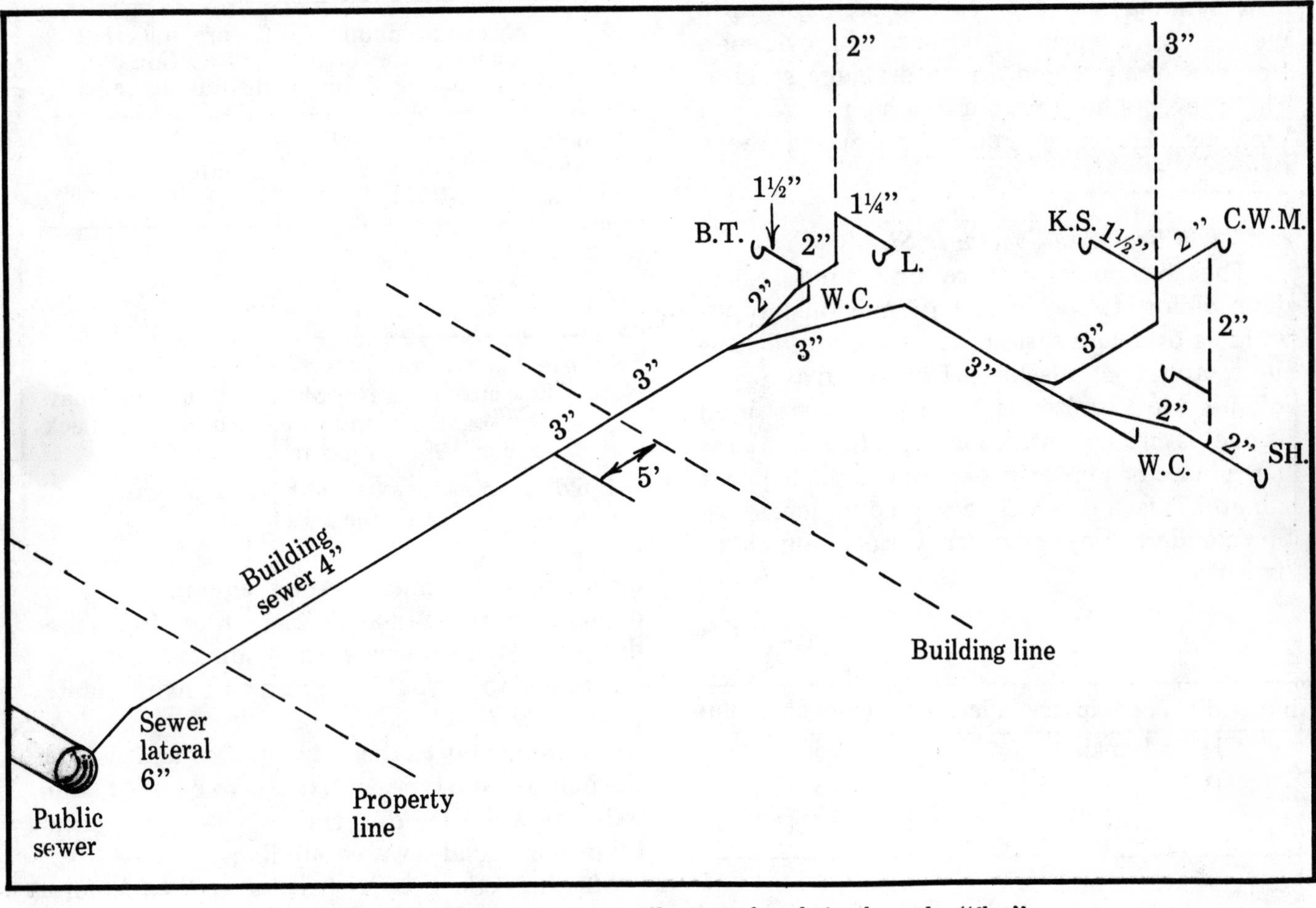

Plumbing drainage system illustrated and sized on the "flat"
Figure 2-9

from a soil or waste stack to a vent stack for the purpose of preventing pressure changes in the stack.

Main Vent (Figure 2-36)

The *main vent* is the principal pipe of a venting system to which vent branches may be connected. The main vent must connect at the base of a soil or waste stack below the lowest horizontal branch.

Stack Vent (Figures 2-5, 2-6, and 2-7)

A *stack vent* is nothing more than the extension of a soil or waste stack (dry section) up and through the roof of a building.

Vent Stack (Figure 2-7)

A *vent stack* is the vertical portion of a vent pipe. Its primary purpose is to provide circulation of air to and from all parts of a drainage system.

Relief Vent (Figures 2-6 and 2-7)

The *relief vent* is sometimes referred to in the trade as a "re-vent." Its primary function is to provide circulation of air between the drainage system and the vent of a plumbing system.

Circuit Vent (Figure 2-7)

A *circuit vent* functions similarly to a *branch vent*. Circuit vents serve two or more fixtures and rise vertically from between the last two fixture traps located on a horizontal branch drain. This vent must then connect to the vent stack.

Loop Vent (Figure 2-27)

A *loop vent* is different from a circuit vent only in that a loop vent loops back and connects to the stack vent, and not to the vent stack.

Vent System

A *vent system* consists of all the vent pipes

of a building and may be one or more pipes installed for the primary purpose of providing a free flow of air to and from a drainage system. This prevents back pressure or siphonage from breaking the water trap seals serving these fixtures.

How The Drainage System Is Sized

The maximum fixture unit load is the determining factor in sizing individual pipes within a drainage system. A fixture unit load is the total gallons discharged by the fixture per minute. Most lavatories, for instance, are rated as one fixture unit, which is equal to 7.5 gallons per minute or approximately one cubic foot per minute. This is generally accepted nationally as the standard flow rate for various plumbing fixtures.

Fixture drain or trap size in inches	Fixture unit value
1¼ and smaller	1 F.U.
1½	2 F.U.
2	3 F.U.

Special fixtures
Table 2-10

In your code book two tables list the various fixture load values, and from these tables the total fixture load for the building can be computed for any type of plumbing system.

Table 2-8 lists the most common plumbing fixtures and gives the fixture unit load value and trap size for each. Using this table the plumbing fixtures in Figure 2-9 can be tabulated to arrive at the total fixture unit load.

The second table (Table 2-10) does not usually carry a title, though it can be classified as a "Special Fixtures" table. To arrive at the fixture unit values for special fixtures, use Table 2-10. This will give you the fixture drain or trap size. The fixture unit values are usually determined by the drain or trap size recommended or supplied by the manufacturer.

To better understand how Table 2-10 may be used, we will use a special fixture as an illustration. A coffee urn is a special fixture and

	Maximum number of fixture units that may be connected to any portion of the building drain or the building sewer			
Diameter of pipe in inches	Fall per foot 1/16"	1/8"	1/4"	1/2"
2			21	26
3		20[2]	27[2]	36[2]
4		180	216	250

[2]Not over two water closets

Note: Pipe size, fall per foot and fixture units may vary slightly from those listed above. Check local code requirements.

Building drains, sewers and horizontal branches
Table 2-11

comes from the manufacturer equipped with a one-half or three-quarter inch drain from the drip pan. It has a waste drain smaller than 1¼ inches and thus would be rated as 1 fixture unit. (See Table 2-10.)

By referring to these two tables and listing the number and types of fixtures to be used, you will be able to determine the maximum fixture-unit load for any building.

In the code book there are also two more tables for sizing pipe based on fixture units in Tables 2-8 and 2-10.

The next table, entitled "Building Drains, Sewers and Horizontal Branches," lists the pipe size, fall (pitch) per foot, and maximum fixture units for each size. This table (Table 2-11) includes only the pipe sizes illustrated in Figure 2-9. Larger sizes are included in the code book.

The most generally accepted fall per foot is 1/8 inch. The 1/16 and 1/4 inch fall per foot is also acceptable, but is likely to be used for more difficult installations involving special conditions.

Table 2-12 appears next in most code books and is used for computing the size and permitted length of soil and waste stacks for multi-story buildings.

What does Table 2-12 really mean? Look at line three, for example. We have a waste pipe with a diameter of 2 inches. The maximum number of fixture units is 24. The last column shows that the total fixture units at any one floor or any one opening which carry waste into this 2 inch waste stack must not exceed 6.

Remember, this is a *waste stack* and not a *soil stack*. This means, according to our defini-

Diameter of pipe in inches	Maximum number of fixture units that may be connected to: Any horizontal fixture branch[1]	One stack of 3 stories in height or 3 intervals	More than 3 stories in height: Total for stack	More than 3 stories in height: Total at one story or branch interval
1¼	1	2	2	1
1½	3	4	8	2
2	6	10	24	6
2½	12	20	42	9
3	20[2]	30[3]	60[3]	16[2]

[1]Does not include branches of the building drain
[2]Not over two water closets
[3]Not over six water closets

Horizontal fixture branches, size and permitted length of soil and waste stacks
Table 2-12

tion, that this waste pipe — because of its size — conveys only liquid waste which does *not* contain body wastes.

Now turn to Table 2-8 and note the fixture units given for each type of plumbing fixture. Line one shows that, given these figures, you can install twelve bathtubs (total 24 F.U.) on this 2 inch stack at different floor levels, or three bathtubs (total 6 F.U.) on a branch drain connected to this 2 inch stack at any one story level. Or you could install 24 lavatories (total 24 F.U.) on this 2 inch stack at different floor levels, or six lavatories (total 6 F.U.) on a branch drain connected to this 2 inch stack at any one story level.

Fixtures with waste openings larger than 2 inches and fixtures that convey greasy waste are prohibited on a 2 inch waste stack.

Figure 2-13 illustrates how the maximum 24 F.U. can be used on a 2 inch waste stack for a five story building.

When water closets or similar fixtures are used, the waste stack is then referred to as a soil stack. The size is increased to 3 inches and is governed by the schedule opposite the 3 inch pipe in Table 2-12.

To find the cumulative fixture unit load for the two-bath house illustrated in Figure 2-9, we have listed the type of fixtures and fixture unit loads from Table 2-8, as follows:

Number and type of fixture	Total fixture units
2 tank operated water closets	8 F.U.
1 bathtub w/overhead shower	2 F.U.
1 shower stall	2 F.U.
2 lavatories	2 F.U.
1 kitchen sink	2 F.U.
1 automatic clothes washer	3 F.U.
	19 fixture units

Cumulative fixture unit load for Figure 2-9

We now have the total fixture units (19) for this particular building to compare and size the various drainage pipes included in Table 2-11.

At this point it would seem that the tabulation is complete. However, it is extremely important to be aware of all the possible restrictions, limitations, and exceptions imposed upon a drainage system, as these will supersede the established pipe sizes and fixture units listed in Table 2-11. Familiarize yourself with these subsections and footnotes, commonly scattered throughout the code book, before attempting to size drainage pipes for other buildings. Note, for example, that in Figure 2-9 the building sewer is sized as 4 inches, even though the total fixture units computed are only 19. According to Table 2-11, at 1/8'' fall per foot, a 3 inch pipe would certainly be large enough, as it is rated sufficient to safely carry 20 fixture units. The size of the building sewer, though, is controlled by one of the footnotes in the code mentioned earlier, which states that "the minimum size of a building sewer shall not be smaller than 4 inches."

In Figure 2-14 the building sewer shown is connected to a septic tank. Though this is a building sewer, it is not classified as such because its developed length does not exceed 10 feet. Therefore, that portion of the sewer pipe exceeding the 5 foot limit beyond a building exterior wall in this particular case may be considered part of the building drain and thus sized accordingly — 3 inches.

Another perplexing situation which could cause the young plumber confusion is illustrated in Figure 2-15. This is an addition to a building. (Figure 2-15 could also serve to illustrate a similar requirement of an accessory building on the same lot.) As is shown in the figure, an addition with a bathroom is being installed at the rear of an existing building. The plot plan shows the sewer from the new addition installed

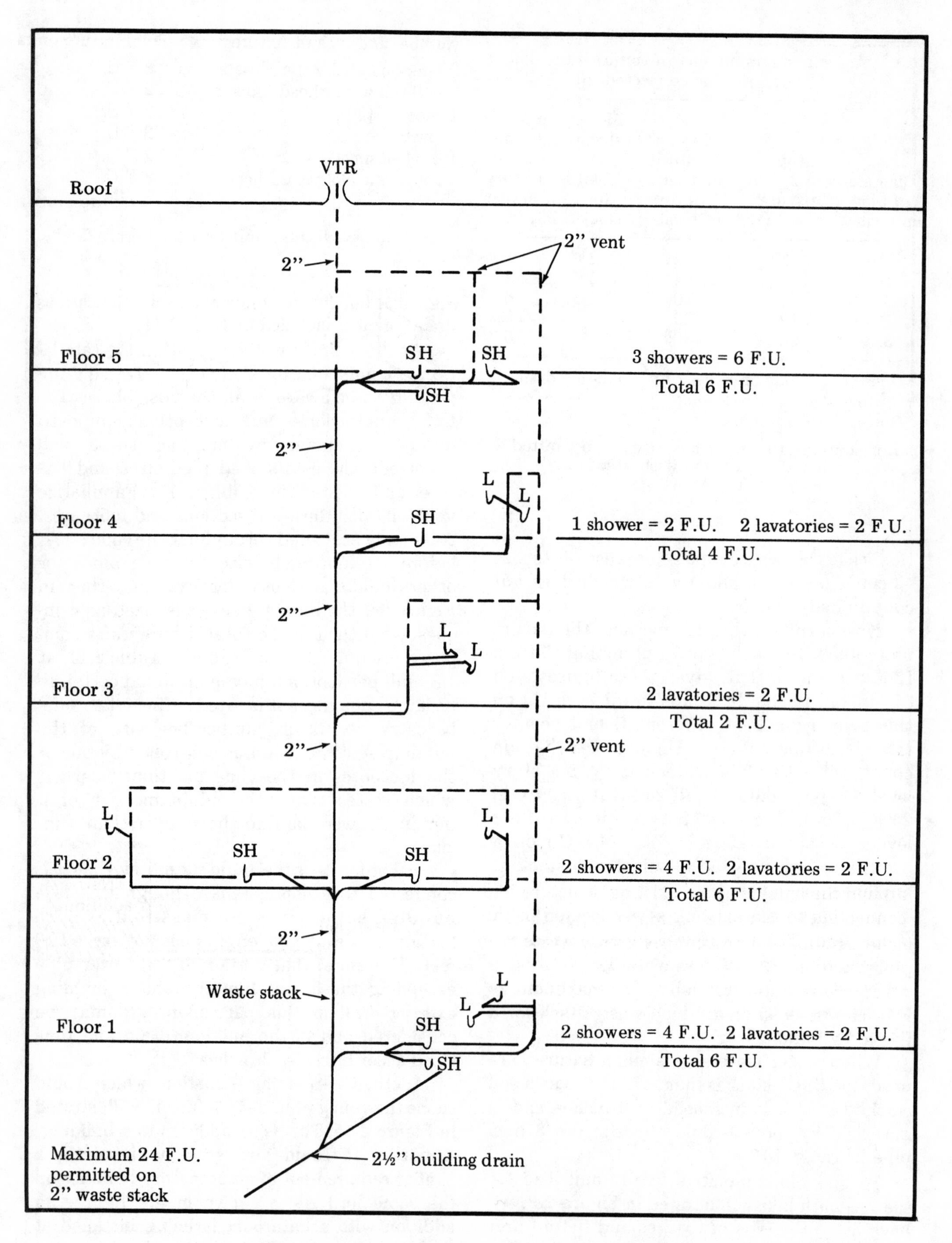

Plan showing maximum 24 F.U. on 2" waste stack
Figure 2-13

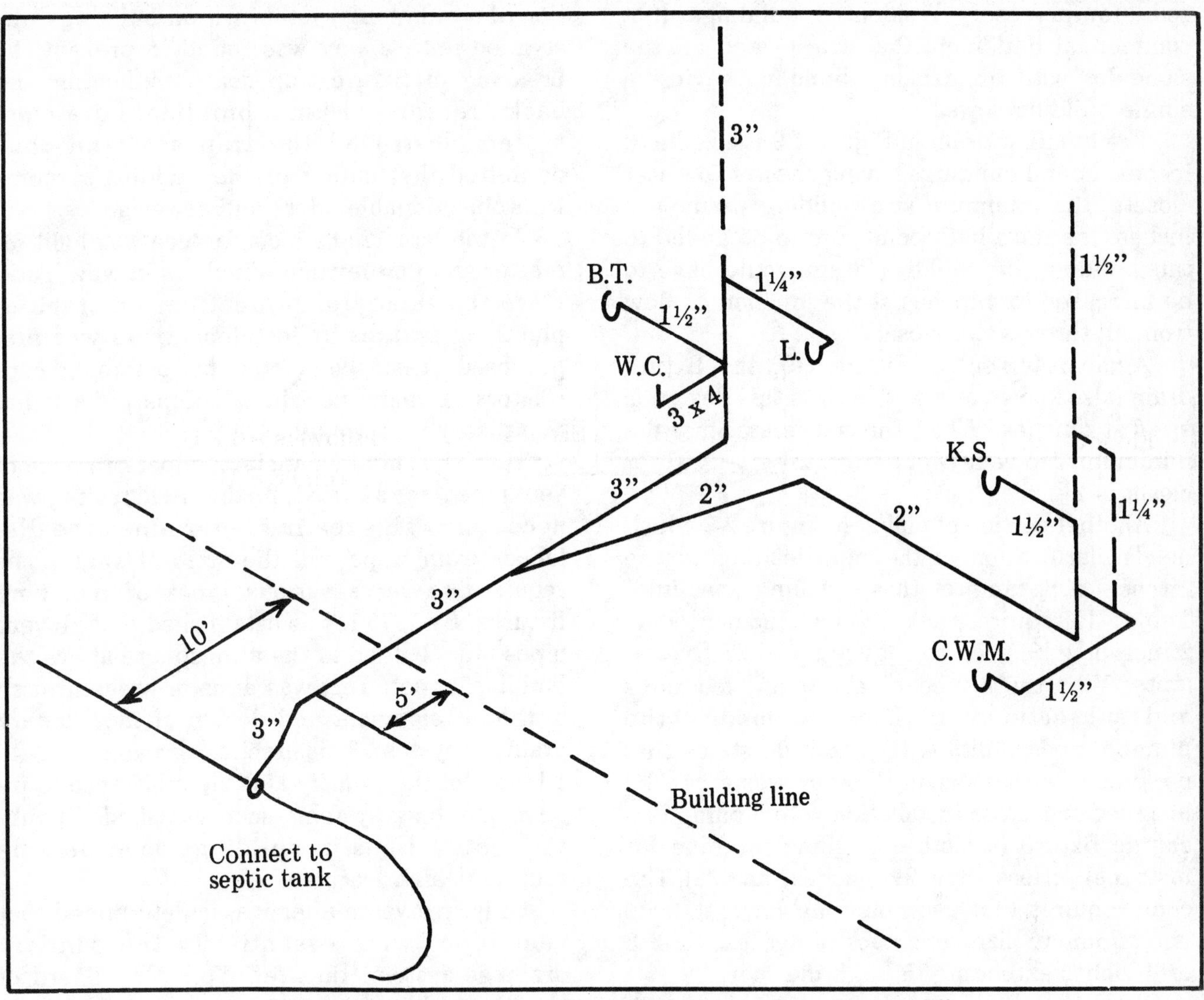

Plumbing drainage system illustrated and sized on the "stack"
Figure 2-14

around the outside of the existing structure and connecting to the existing sewer located in the front yard. The new sewer pipe is sized at 3 inches, when normally it would be required to be 4 inches. Why? The code has made an exception in this case and the interpretation of the code reflects this, as stated below:

> *For additions to residential buildings when soil and vent lines are inaccessible and it is necessary to install the sewer line outside and around an existing structure, such line shall be considered as a horizontal branch and installed in accordance with Table 2-11 in this book. This horizontal branch must connect into the existing building drain.*

This same allowance is permitted by the

Sewer installation for addition
Figure 2-15

code for accessory residential buildings (not commercial buildings) that are located on the same lot with an existing building having a single building sewer.

The building drain in Figure 2-9 is sized at 3 inches. For all buildings having one or two water closets, the minimum size building drain is 3 inches. If a third bathroom were to be added to this building, the building drain would have to be increased to 4 inches at the junction of flow from all three water closets.

Again referring to Figure 2-9, the fixture drain (also known as a wet vent) in this design is sized at 2 inches. Why? The code states that the minimum size vent (wet or dry) to serve a water closet is 2 inches.

Another section of piping in Figure 2-9 which needs clarification is the pipe leading to the kitchen sink and clothes washing machine. Table 2-11 requires that at ¼ inch fall per foot a 2 inch pipe be used to convey up to 21 fixture units. Why the 3 inch pipe? Again, footnotes and subsections in the code modify the plumbing possibilities. (1) The code states that no kitchen sink receiving greasy waste can be installed on a cross installation with a pump discharge fixture (a clothes washing machine for instance) of less than 2½ inches, and (2) The code requires that each building have at least one minimum size vent stack of not less than 3 or 4 inches extending through the roof.

Since both bathrooms have a 2 inch wet vent and since the code requires a 3 inch pipe up through the sanitary tap cross, it is economical to satisfy the code requirements by making this vent stack the main vent for this building.

Figure 2-14 shows another plumbing design in which the bathroom vent is used as the main or minimum size vent stack of 3 or 4 inches for the building. The kitchen sink and clothes washing machine are installed on a 2 inch waste pipe in this instance. Why is this legal? A 2 inch pipe can be used if the fixture connections to the waste and vent are at different levels and if a relief vent as shown in Figure 2-14 is used on the clothes washing machine fixture branch.

How The Use of Vent Pipes Emerged

Vent pipes are important to the successful functioning of plumbing fixtures and the sanitary drainage system. Although the first water closet was developed by an English inventor, Joseph Braman in 1778, the installation of a plumbing system within buildings was stymied until a way was found to prevent the breaking of fixture trap seals. Siphoning and back pressure within a building's drainage system destroyed the trap seal and thus permitted the fouling of the building's rooms with objectionable odors and sewer gases.

In the late 1800's a conference was held for master and journeyman plumbers in New York. Here the theory of protecting the traps of plumbing fixtures by installing vents was first proposed. Tests conducted by plumbing contractors on many new installations proved that this theory was workable.

However, before satisfactory performance of vent pipes was assured, further field testing was necessary. This testing determined the distances from traps and the sizes of vent pipes required to serve various types of plumbing fixtures. By 1875 it was established that all vent pipes must extend to the atmosphere above the building's roof. This was a major breakthrough in the development and design of the sanitary drainage system. This practice spread rapidly to all parts of the country and the resistance to indoor plumbing systems soon vanished. Plumbing installations in buildings soon became routine design features.

Only five years after it was determined that vent pipes were essential to the sanitary drainage system, the New York City Board of Health established minimum vent pipe sizes to serve the various size traps in existence at that time.

In the late 1800's many cities adopted and established separate plumbing codes to protect the health of people in densely populated areas. Plumbing practices varied considerably from municipality to municipality and the code requirements reflected these variations.

By 1921 a comprehensive effort to standardize the plumbing code was made by the U.S. Department of Commerce. Scientific experiments conducted by the National Bureau of Standards became the basis for many plumbing requirements.

States authorized the establishment of examining boards of qualified plumbers and other agencies to write plumbing regulations and amend these regulations when necessary to keep abreast of new materials and installation methods. Although plumbing practices still vary, the variations today are usually of minor

Diameter of soil or waste stack in inches	Maximum fixture units	Size and maximum length of vent piping in feet				
		1¼"	1½"	2"	2½"	3"
1¼	2	30	--	--	--	--
1½	8	50	150	--	--	--
1½	10	30	100	--	--	--
2	12	30	75	200	--	--
2	20	26	50	150	--	--
2½	42	--	30	300	--	--
3	10	--	30	100	200	600
3	30	--	--	60	200	500
3	60	--	--	50	80	400

Size and length of vent pipes
Table 2-16

importance.

How The Vent System Is Sized

Vent pipes are sized and arranged to relieve pressure that builds up as water is discharged into the sanitary drainage system by various types of plumbing fixtures. This free flow of air within the system keeps the back pressure or siphoning action from destroying the fixture trap seal.

In rural areas where inspections may not be required, or in municipalities where inspections may be lax, inadequate sizing and arrangements of vent pipes can cause the following problems to occur.

- Plumbing fixtures may drain slowly, as if a partial stoppage exists.
- Water closets may need several flushes to remove the contents from the bowl.
- Back pressure (known as positive pressure) within the drainage pipes, if strong enough, may force sewer gases up and through the liquid trap seals and into the building.
- Plumbing fixtures located a greater distance from a vent pipe than is permitted by code may, when the contents are released, siphon the liquid trap seal. (This action is known as negative pressure.)
- Because of the importance of the free flow of air within the drainage system, in very cold climates where the danger of frost forming on the inside of the vent pipes exists, increasers or other devices are used. (The frost is formed as the warm moist air flows up and out the vent pipes and makes contact with the cold frigid atmosphere.)

Figures 2-5, 2-6 and 2-7 reflect the importance and the close relationship between the various types and styles of roughing-in of vent pipes used with the drainage system.

Sizing of vent pipe, like sizing its cousin the soil and waste pipe, is determined by the maximum fixture unit load, its developed length, type of plumbing fixtures, and the diameter of the soil or waste stack it serves.

Table 2-16 (like Table 2-12) is used for computing the size and permitted length of vent stacks for multi-story buildings. Again, this table has been included, along with one example of its use, for future reference purposes. Larger sizes and lengths, of course, can be found in the code book table.

How is Table 2-16 interpreted? Look at lines 4 and 5, both involving a 2 inch waste stack. In line 4, the 2 inch waste stack serves 12 fixture units. If the height of the building does not exceed 30 feet, a 1¼ inch vent stack *may* be used. Should the building exceed 30 feet, but not 75 feet, a 1½ inch vent may be used.

Finally, if the building height exceeds 75 feet but does not exceed 200 feet, a 2 inch vent stack *must* be used. These are of course the minimum sizes permitted by the code. It is permissible to upgrade a plumbing system by using larger vent pipes, provided the vent pipe size does not exceed the diameter of the soil or waste stack it serves.

You will note in Table 2-16 that even though

the 2 inch waste pipe on line 5 is the same as line 4, as the maximum fixture unit load increases the height of the vent pipe must decrease. A larger load requires more air flow and air is restricted more and more as a pipe grows longer.

The basic principles of venting are summarized in most codes as follows:

- For each building having a single building sewer receiving the discharge of a water closet there must be at least 1 minimum size vent stack (of not less than 3 or 4 inches) extending through and above the building roof.
- Where there is an accessory building or buildings located on the same lot having one common building sewer, the minimum size vent stack or stacks to serve each accessory building may be sized in accordance with Table 2-16. Again, there is one exception. Should a water closet be located in the accessory building, the minimum size vent stack must be no smaller than 2 inches.
- No vent (wet or dry) for a water closet can be less than 2 inches in diameter.
- The diameter of the vent stack can not exceed the diameter of the soil or waste stack to which it connects.
- The diameter of an individual vent stack can not be less than 1¼ inches nor less than one-half the diameter of the drain to which it is connected. In other words, if a drain should be 3 inches, the minimum size vent is 1½ inches. If a drain is 4 inches, the minimum size vent is 2 inches.
- The pipe of the dry vent section of a loop or circuit vent can be one pipe size smaller than the diameter of the horizontal soil or waste drain it serves. As an example, if the horizontal drain is 3 inches, the minimum dry vent section of piping would be 2 inches.
- Plumbing fixtures which are separated by some distance from the vent serving other plumbing fixtures (and thereby requiring an individual or re-vent) must have a minimum size of 1¼ inches up to six fixture units (less any water closets or similar fixtures).
- Vent stacks in some cases can have a dual role. In certain types of installations they can be used to supply and remove the air from a drainage system and also serve as cleanouts for inserting a cleaning cable. See Figure 2-76 at the end of this chapter. These installations must meet the following requirements in order to qualify for this dual role: It is a one-story building with not more than one 90-degree change in direction in the drainage system; the vent stack is vertical throughout (without any offsets) and extends up through the roof; the vent stack is of the same size as the waste pipe it serves; the vent stack is not reduced beyond the following minimums.

  4'' must not reduce to less than 3''
  3'' must not reduce to less than 2''
  2'' must not reduce to less than 1½''

Wet Vents

This special method of venting is generally used in dwellings. It provides adequate protection for trap seals for permissible plumbing fixtures. Being a single piping system, it is economical and can serve to vent several adjoining plumbing fixtures when located on the same floor level. See Figure 2-17.

A wet vent may be vertical or horizontal as illustrated in Figure 2-5. It is represented in graphic symbol coding as a fixture drain or waste pipe. As a building drain it can convey only waste from fixtures with low unit ratings. This excludes water closets and similar fixtures. Because of the dual role of a wet vent, the plumbing code has placed a number of restrictions on its use. The restrictions, pipe sizes and maximum capacities for wet vents are as follows:

- Horizontal wet vents can not exceed 15 feet and can receive the discharge from fixture branches only.
- 2 inch wet vents can convey up to six fixture units but can not be used for urinals, pressure fixtures or for sinks with or without garbage disposers.
- 2½ inch wet vents can convey up to 10 fixture units. Water closets and fixtures requiring a waste opening greater than 2 inches can not use a wet vent.
- 3 inch wet vents can convey up to 16 fixture units. No water closets or other fixtures requiring a waste opening greater than 3 inches are permitted.

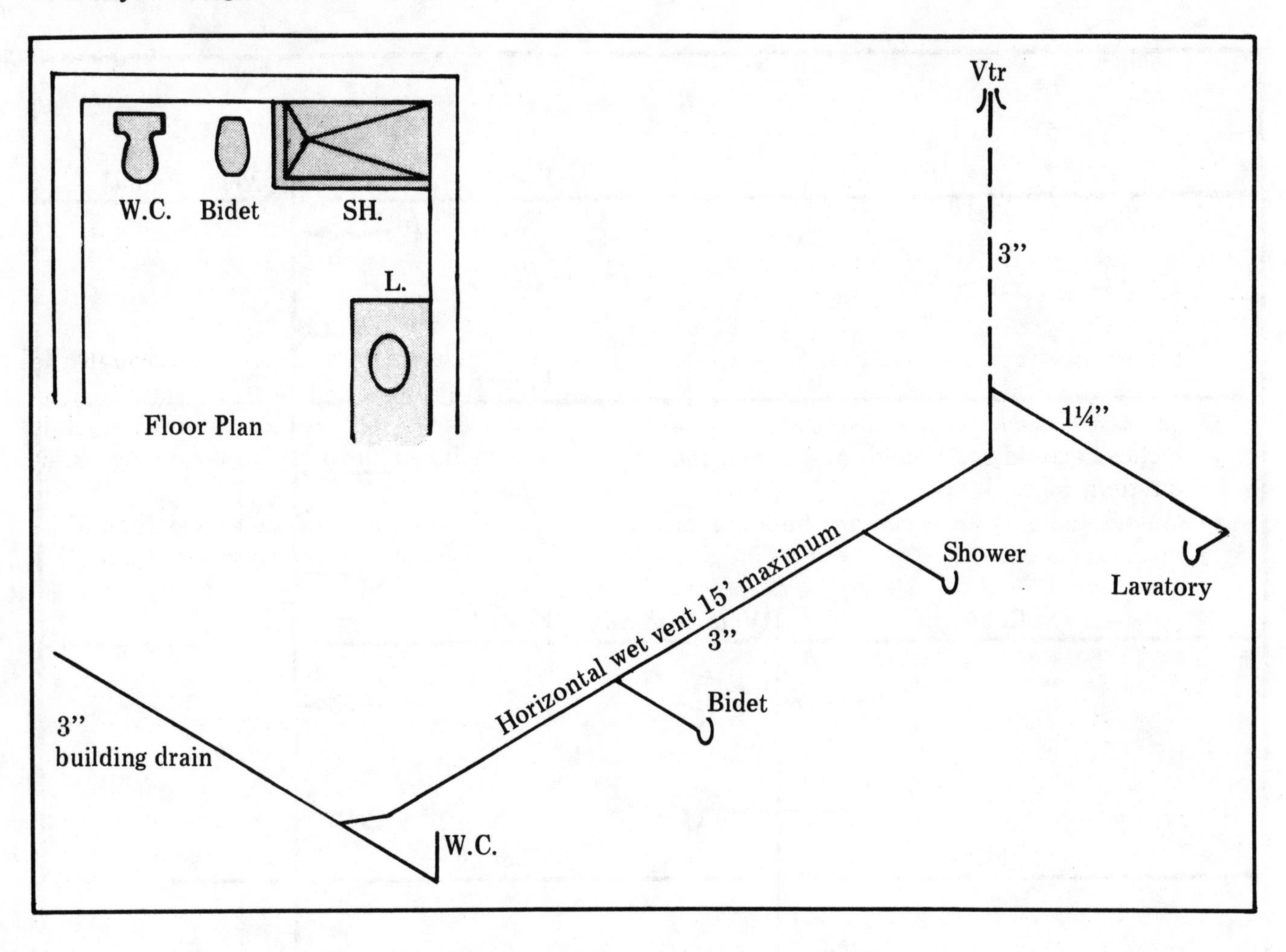

Horizontal isometric of wet vent
Figure 2-17

- 4 inch wet vents can convey up to 32 fixture units. No water closets or other fixtures requiring a waste opening greater than 4 inches are permitted.

Vertical Combination Waste and Vent

This special method of receiving waste and venting certain permissible plumbing fixtures is generally used in high-rise buildings. It can be used to advantage in any multi-story building where plumbing fixtures are located directly over each other on different floor levels. This is a single vertical pipe riser installation and thus is economical and practical to use.

The following partial Table 2-18 is used for computing the size and permitted length of combination waste and vent stacks.

Again we are limited by the code to the size of the stack, number of fixture units it may receive, type of plumbing fixtures and the total length of the combination waste and vent. Figures 2-19 and 2-20 illustrate the workings of a combination waste and vent pipe. Figure 2-19 depicts a four-story building having eight kitchen sinks, back to back, with a total of 16 fixture units. Figure 2-20 depicts a six story building having six kitchen sinks, one on each floor, with a total of 12 fixture units. Again the number of total fixture units and the maximum length (height) determine the diameter of the stack.

Diameter of stack in inches	Fixture units on stack	Maximum length in feet
2 (no kitchen sink)	4	30
3	24	50
4	50	100
5	75	200

Vertical combination waste and vent
Table 2-18

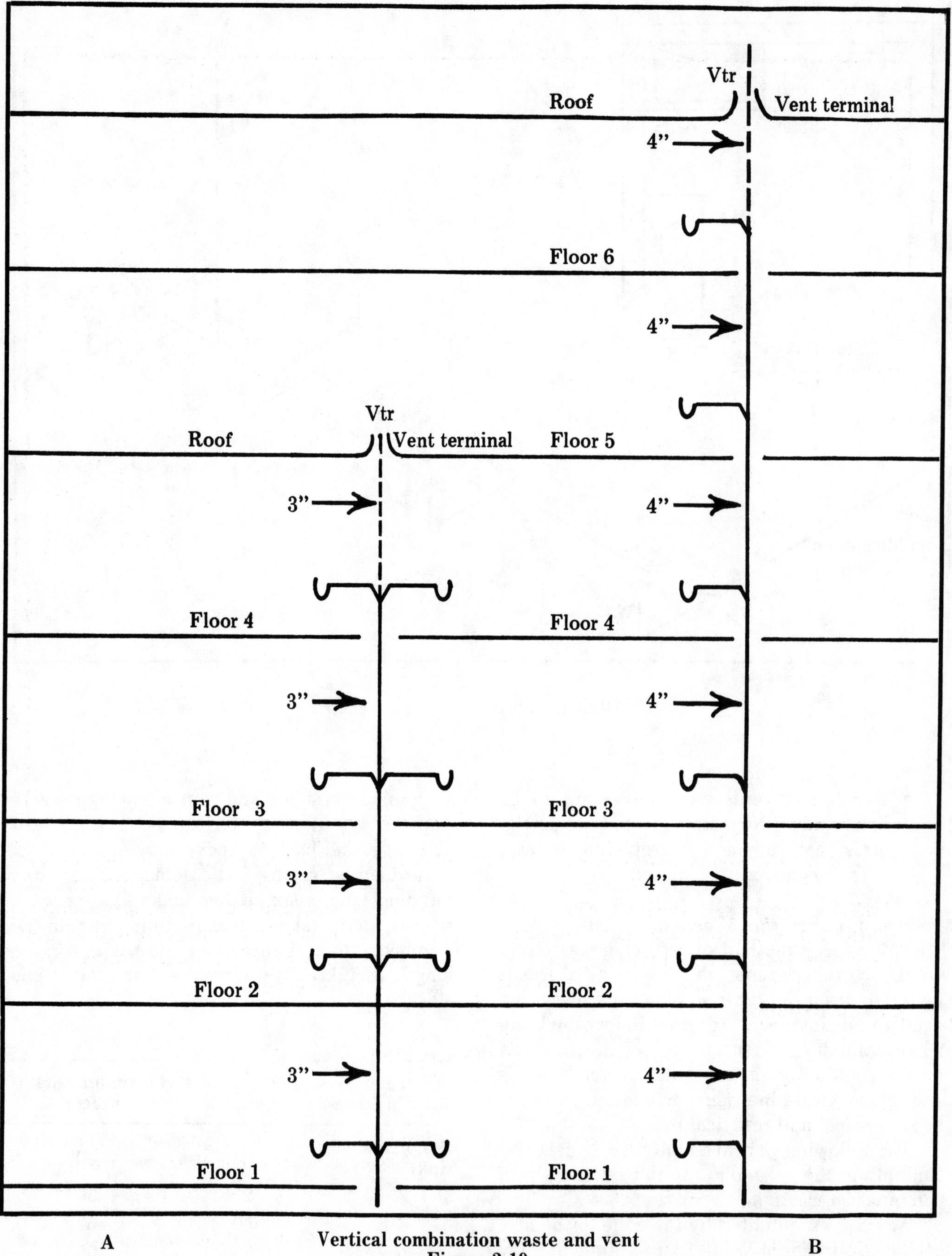

Vertical combination waste and vent
Figure 2-19

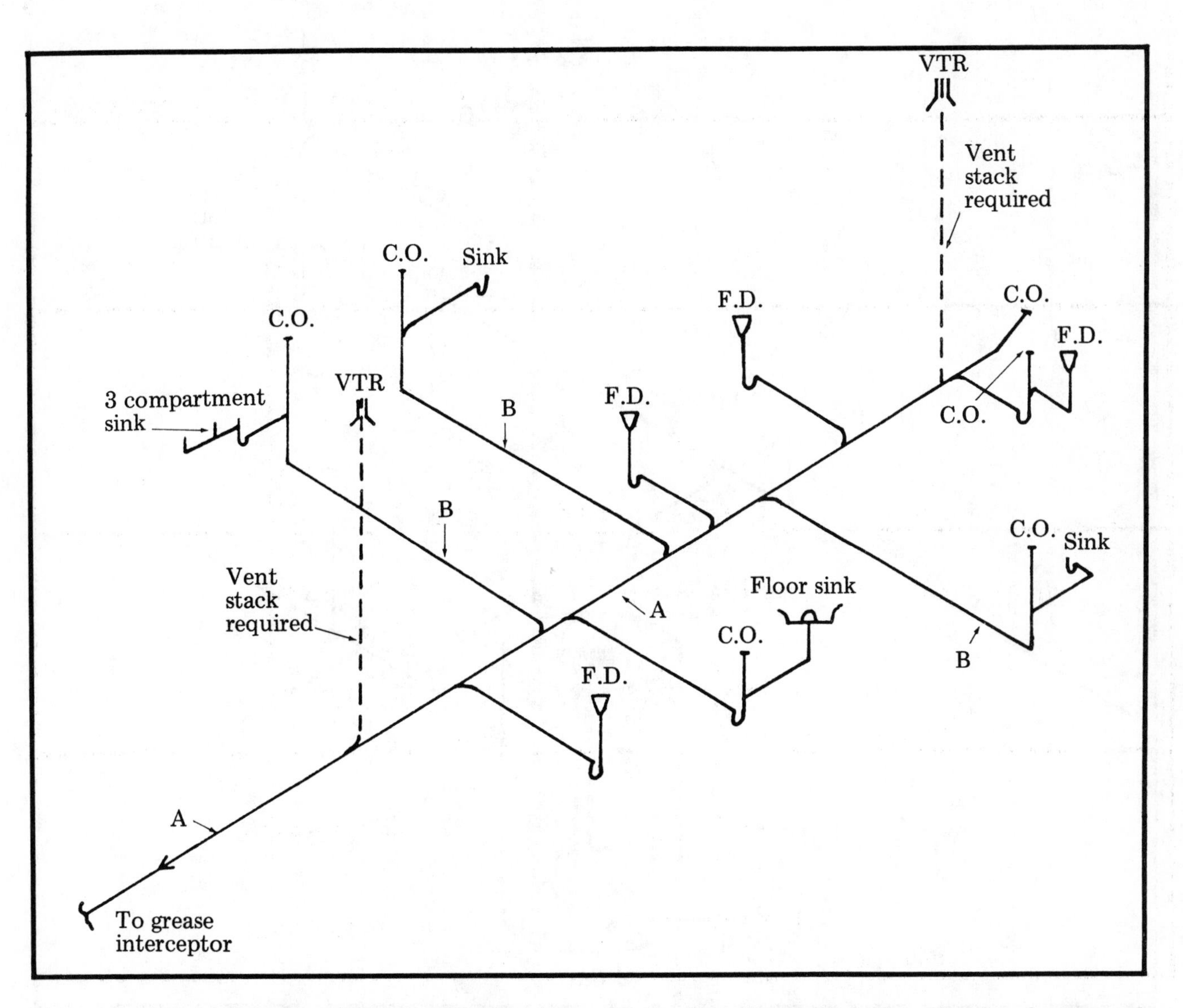

Note: A - Minimum size of main waste pipe (A) is 6", for a total of 96 F.U. For systems that exceed 96 F.U. the main waste pipe (A) must be 2 pipe sizes larger than the required sizes as listed in the code table for building drains and horizontal branch drains.

Note: B - Branch drains (B) for horizontal combination waste and vent system, as illustrated, must be 2 pipe sizes larger than the required sizes as listed in the code table for building drains and horizontal drains.

Horizontal combination waste and vent system
Figure 2-20

Referring to Table 2-18, the combination waste and vent in Figure 2-19A has to be 3 inches, thus accommodating the maximum number of fixture units permitted, 16. Since a four-story building is approximately forty feet high, the combination waste and vent pipe remains within the fifty foot limit shown in Table 2-18.

In Figure 2-19B it is important to note that the diameter of the combination waste and vent pipe must be 4 inches, in this instance, although the 12 fixture units are fewer than those used in Figure 2-19A, which had 16 fixture units.

Since a six-story building would exceed the maximum length permitted by the code for a

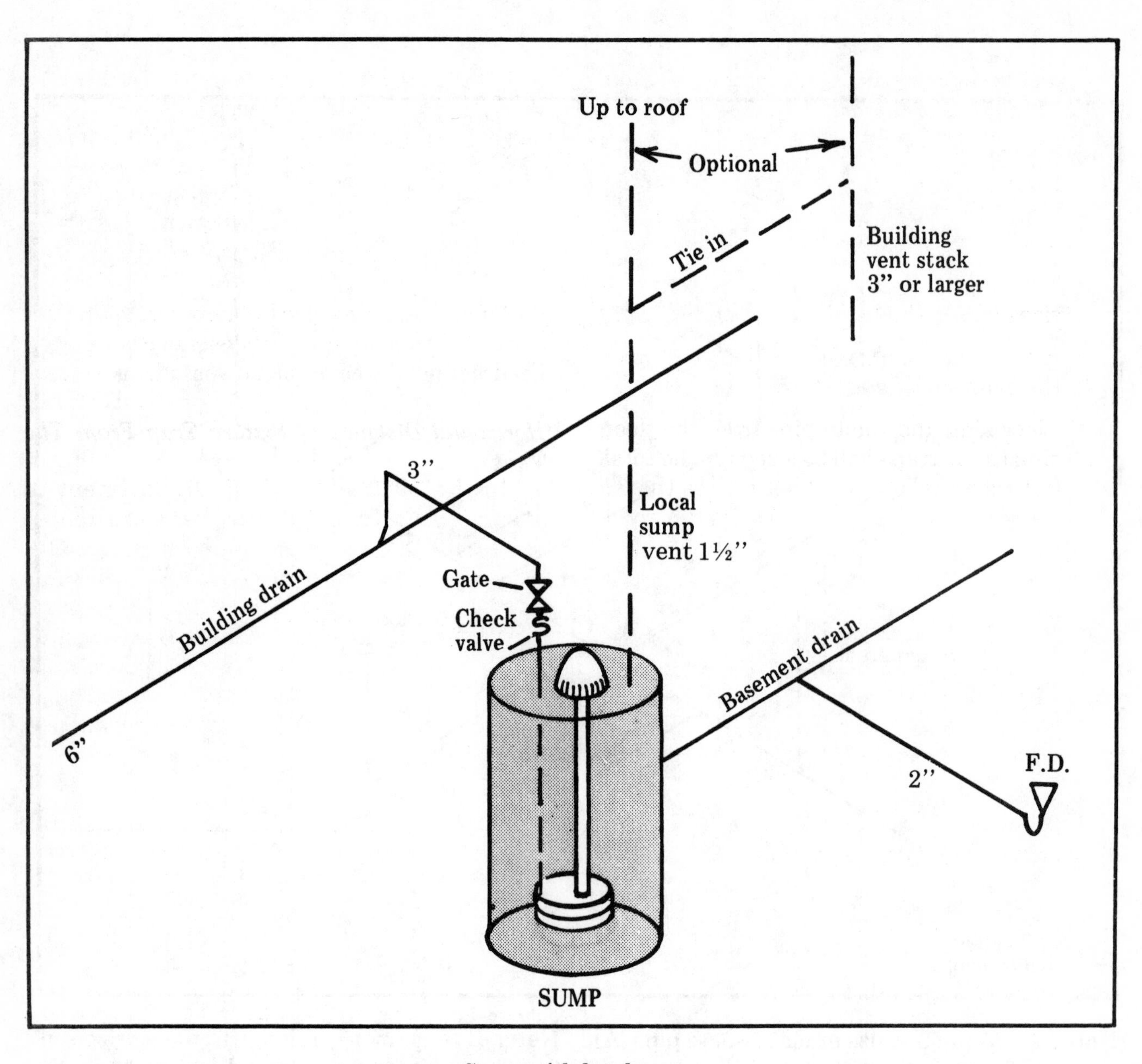

Sump with local vent
Figure 2-21

3 inch combination waste and vent pipe, it is necessary to use the next larger size as shown in Table 2-18, 4 inches. The length of the pipe always takes precedence over the number of fixture units involved.

Horizontal Combination Waste and Vent System

Certain plumbing fixtures (usually found in restaurants or similar establishments) are permitted to use this special system for receiving waste and for venting. This one-pipe method can be used to advantage where special plumbing fixtures and regular plumbing fixtures are not adjacent to walls or partitions.

This system is limited to sinks, dishwashers, floor sinks, indirect waste receptors, floor drains or similar applications. The trap of the fixture is not individually vented in this type installation. See Figure 2-20.

Because of the special role of the horizontal combination waste and vent system, the plumbing code has placed a number of restrictions on its use, as follows:

- Appurtenances delivering large quantities of water or sewage should not discharge into this type system, so that adequate venting will be maintained.

- The horizontal waste pipe and each fixture trap shall be at least two pipe sizes larger than required by code for regular systems.
- The vertical waste pipe shall be two pipe sizes larger than the fixture outlet.
- A cleanout shall be installed in the top of the connecting waste tee.
- Floor sink and waste pipe from the floor sink to the trap shall be sized for the total fixture units, but shall not be less than 2 inches when installed underground.
- A vent shall be provided at the upstream end of each branch and a vent shall be located downstream from all fixtures in the system.
- The system shall be designed to assure that the vertical distance from fixture or drain outlet to trap (weir) does not exceed 24 inches.
- In large installations where long runs cannot be avoided, relief vents shall be installed at intervals of not more than 100 feet.
- The size of vents shall be in accordance with code requirements, but shall extend the same size as the waste pipe to a point 6 inches above the flood level rim of the highest fixture before reducing in size.

Sump Vent

In a building with a sub-building drain in the basement which conveys body waste to a sump or retaining tank, the tank must have a local vent. See Figure 2-21. A sump or retaining tank provided with a gas and air tight metal cover would not be adequate. The local vent permits the air within the tank to escape as sewage enters. When the sewage is ejected the vent permits air to re-enter the tank. This prevents the sub-building drainage system from becoming air-locked and useless.

The following venting procedures are accepted by the plumbing code.

- The minimum size vent for a sump receiving body waste must be no less than 1½ inch in diameter.
- The sump vent may extend independently up and through the building roof.
- The sump vent may be connected to existing vent system of the same size (1½ inch) or to a vent pipe of a larger size.

Sumps receiving clear water waste from floor drains, air conditioning condensate drains, etc., need not have a cover or be vented.

Horizontal Distance of Fixture Trap From The Vent

In the 1920's the U. S. Department of Commerce made a substantial effort to establish standards for many plumbing requirements. Those standards are now reflected in most codes. Unfortunately there is still a great variation in distance and fixture branch (drain) size requirements from one model code to another. For this particular section of the code, you must refer to the table adopted by your local authority.

Size of fixture drain in inches	Maximum distance trap to vent opening in feet
1¼	5
1½	5
2	5
Floor connected fixture with integral trap	5
Floor drains and interceptors	15

Horizontal distance of fixture trap from vent opening
Table 2-22

Compare Tables 2-22 and 2-23. Table 2-22 is taken from one of the model plumbing codes and Table 2-23 is taken from the Standard Plumbing Code.

Table 2-22 is comparatively simple. It has been adopted by local authorities in many states and has proved to be economical and satisfactory in its use. The maximum distance from trap to vent is 5 feet, regardless of the trap size. There is an exception for floor drains and interceptors. Slope per foot of fixture branch (drain) is not considered.

The maximum distance from the fixture trap to the vent in Table 2-23 depends on three factors that are subject to interpretation. Thus, the installing plumber will be very wise to check the following points with his local plumbing inspector: 1) the size of the fixture drain, 2) the size of the fixture trap, 3) the slope (fall) per foot of the fixture branch (drain, in some codes).

Size of fixture drain	Size of trap	Fall per foot	Distance from trap
1¼"	1¼"	¼"	3' 6"
1½"	1¼"	¼"	5'
1½"	1½"	¼"	5'
2"	1½"	¼"	8'
2"	2"	¼"	6'

Horizontal distance of fixture trap from vent
Table 2-23

As mentioned earlier in this chapter, where a conflict seems to exist between two or more codes, it is always best to follow the tables provided in the code adopted by your local authority.

There are some principles for horizontal runs between the fixture trap and the vent that are accepted by all codes.

- **The closer the trap to the vent on a minimum slope, the better.**
- **Every fixture trap must be protected against siphonage and back pressure and must be provided with a vent piping system which permits the admission or emission of a free flow of air when under normal intended use.**
- **The developed length of a fixture branch (drain) is taken from the crown weir of a fixture trap to the vent pipe. The measurement must include offsets and turns and be within the prescribed limits as set forth in the code. See Figure 2-24.**
- **In some instances, because of the fixture location within a bathroom, kitchen or utility room, the fixture branch may have to exceed the limits. See Table 2-22.** When this occurs, you must install a relief vent as illustrated in Figure 2-25.

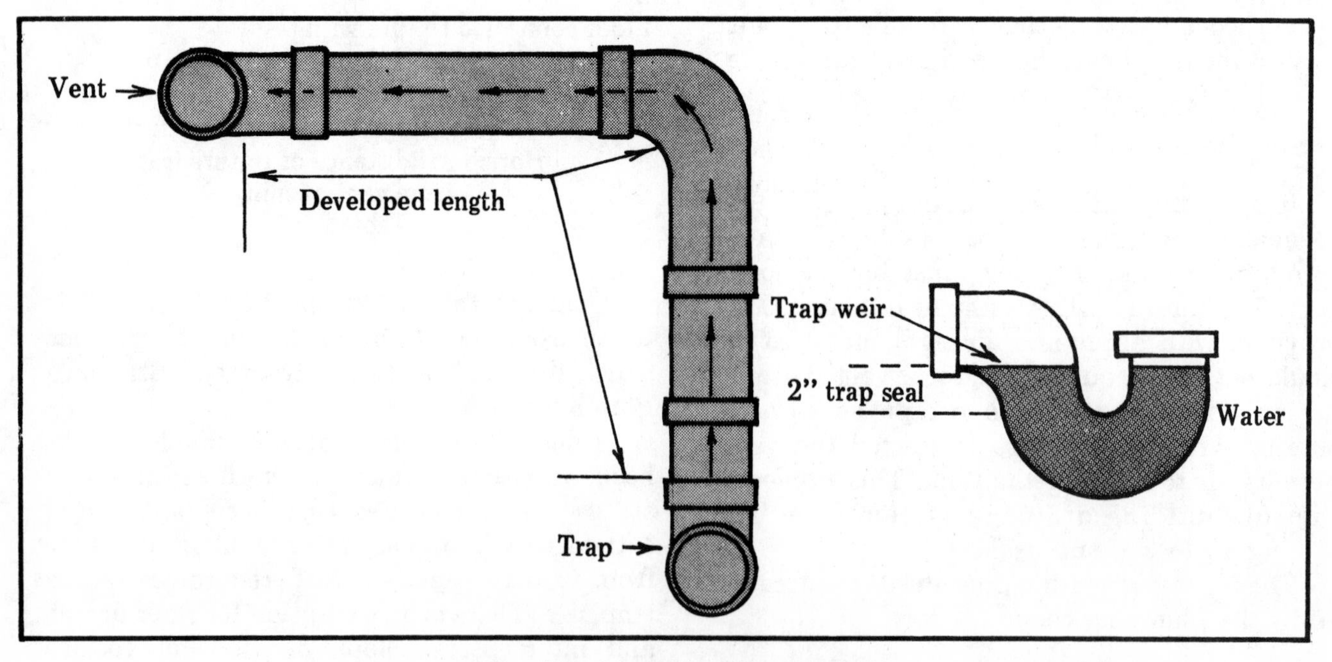

Developed length of fixture drain
Figure 2-24

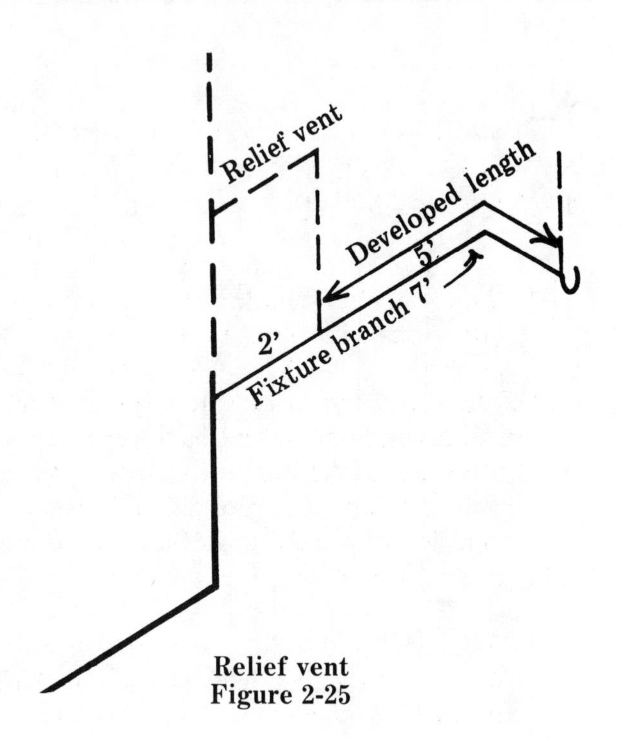

Relief vent
Figure 2-25

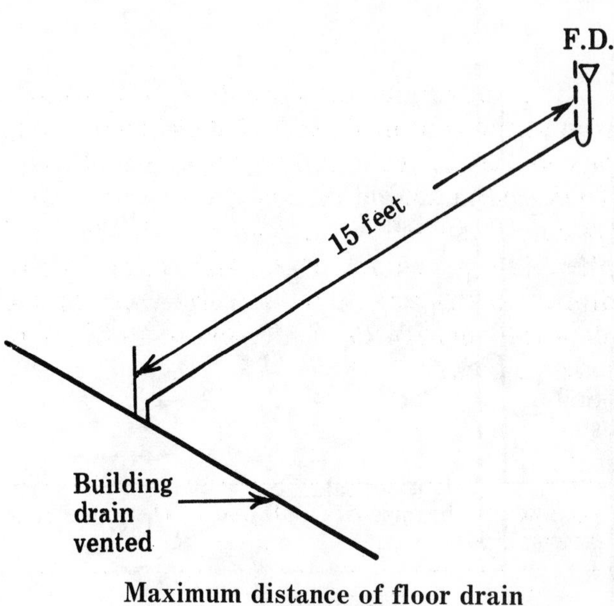

Maximum distance of floor drain
Figure 2-26

- Traps of floor drains may be located a maximum of 15 feet (less by some codes) from an individual vent or a vented drainage line. See Figure 2-26.

Any coverage of drainage, waste, and vent systems would not be complete without including the isometric drawings on the following pages. These piping diagrams are carefully selected to include various plumbing designs and definitions ranging from the simplest to the most difficult drainage, waste and vent installations, and their relation to one another.

It is well worth your time to try to understand these various definitions, as well as how, why, and where they are used in the plumbing system. For example, the definition of a *yoke vent* as given in the code does not adequately describe its use and purpose. However, the code definition, examined together with the yoke vents shown in Figures 2-34 and 2-35, make its use considerably more understandable.

These piping diagrams will be of utmost importance to the professional as a reference guide for code interpretation, for installation purposes, and for preparation for plumbing examinations.

The pipes are not sized in the following isometric drawings. These illustrations are for the purpose of clarifying the location and intended use of these code-related definitions. For practice, you should develop various plumbing designs from the following isometrics. You could then size the waste, drainage, and vent pipes according to Tables 2-11, 2-12 and 2-16. The last step in this practice would be to have your isometric drawings checked for accuracy by some qualified individual like your employer, job foreman, or the plumbing inspector who visits the job.

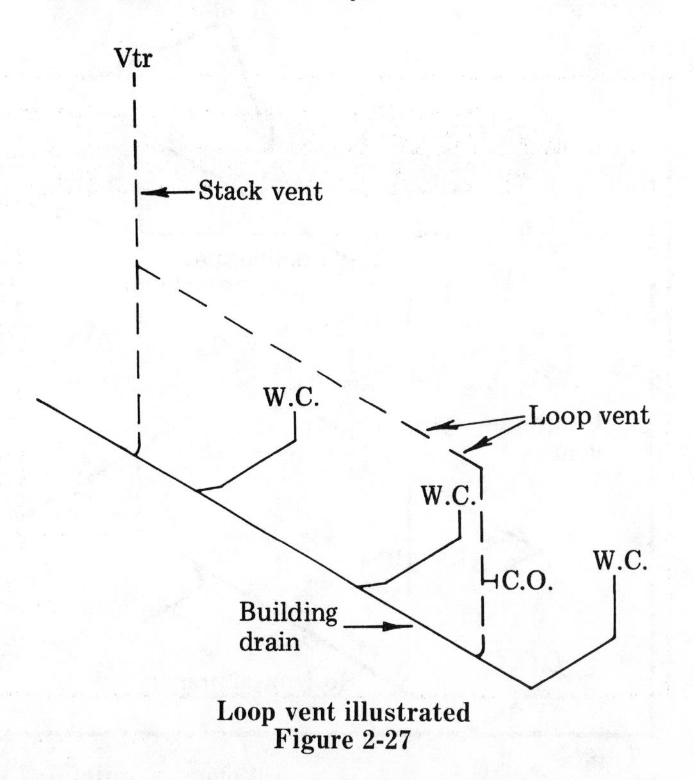

Loop vent illustrated
Figure 2-27

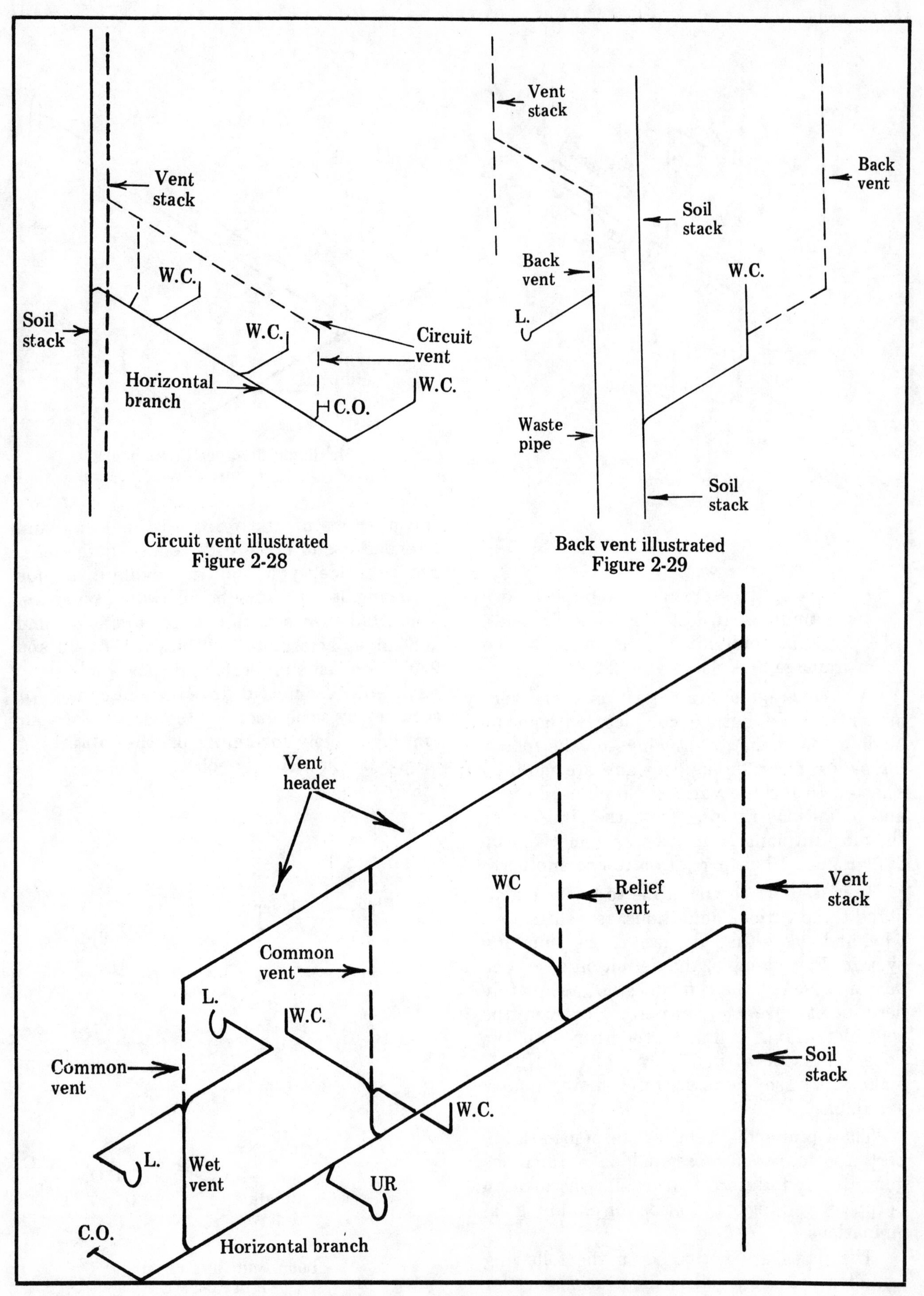

Circuit vent illustrated
Figure 2-28

Back vent illustrated
Figure 2-29

Common, relief and vent header illustrated
Figure 2-30

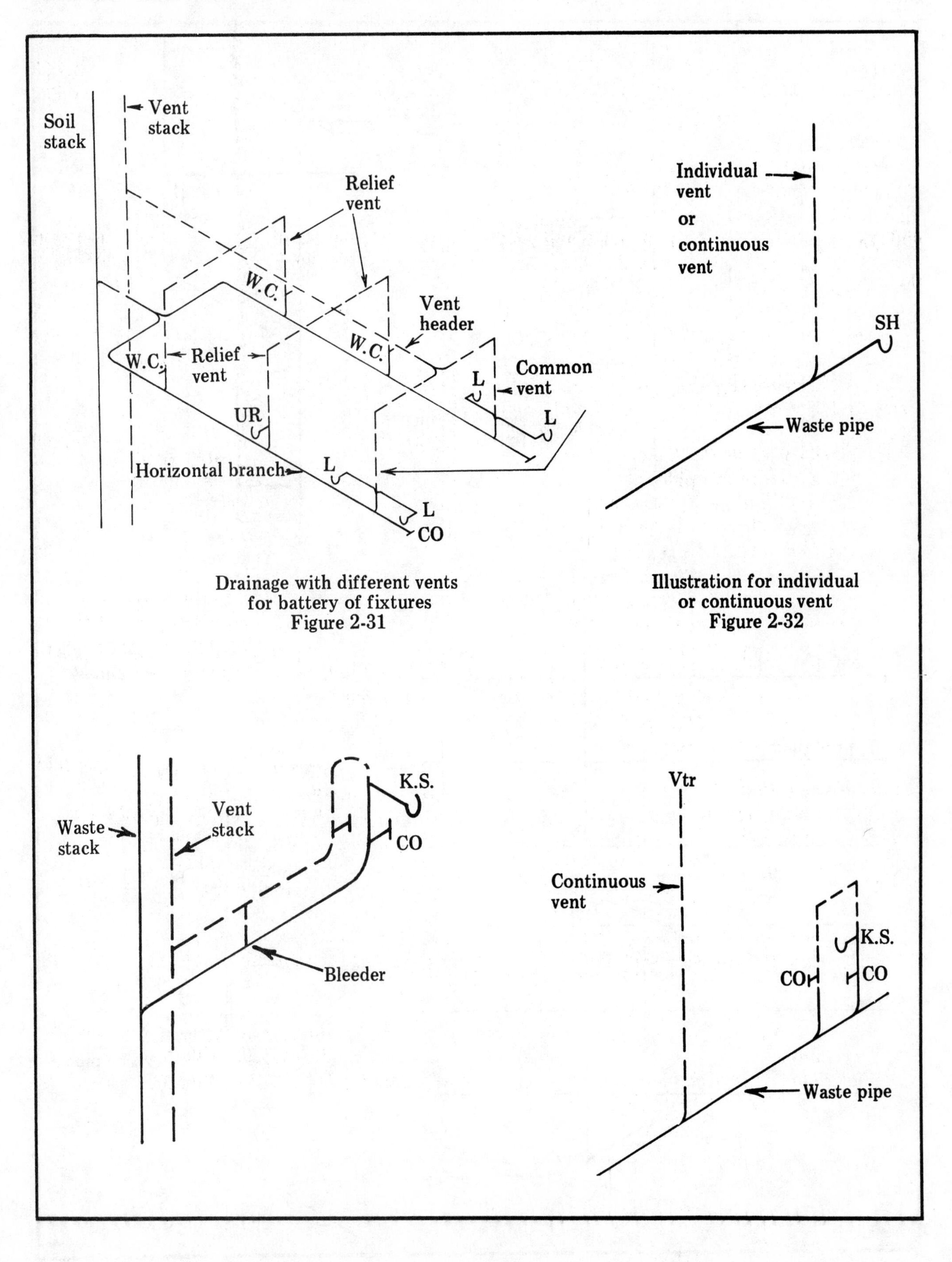

Drainage with different vents for battery of fixtures
Figure 2-31

Illustration for individual or continuous vent
Figure 2-32

Two types of looped vent systems illustrated for isolated plumbing fixtures
Figure 2-33

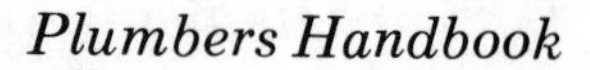

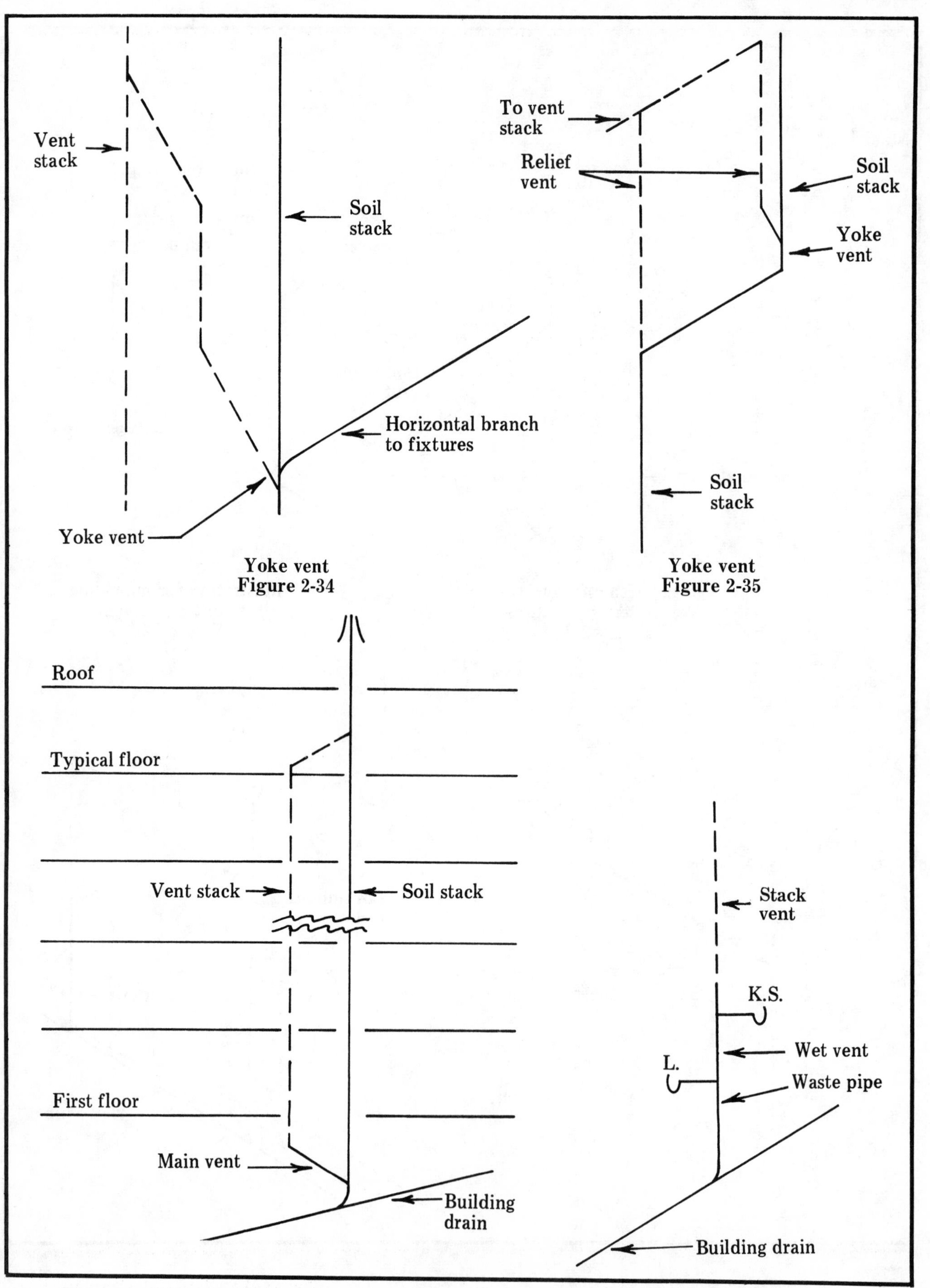

Yoke vent
Figure 2-34

Yoke vent
Figure 2-35

Main vent to connect at base of soil or waste stack
Figure 2-36

Figure 2-37

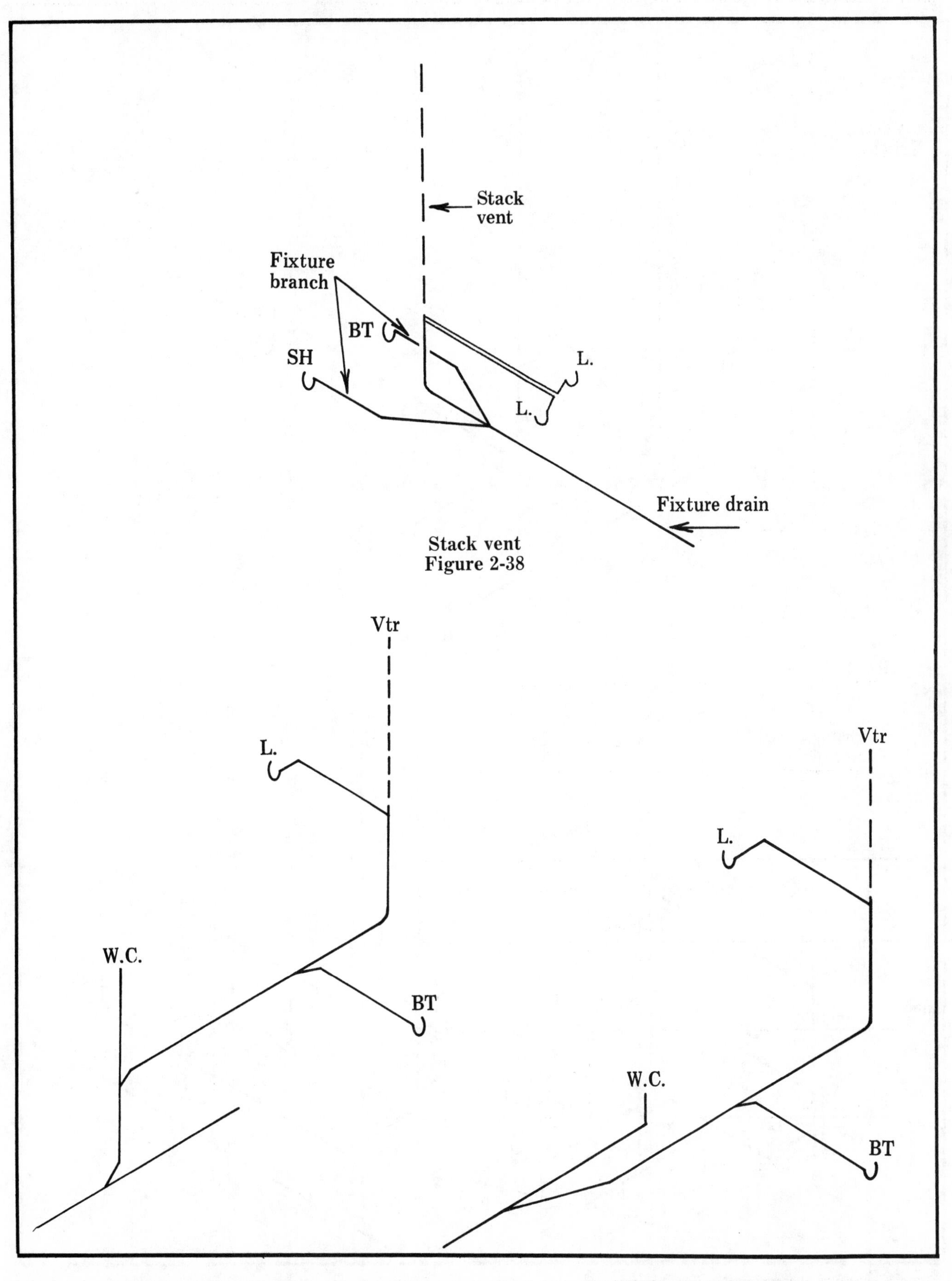

Stack vent
Figure 2-38

Wet vent system stacked
Figure 2-39

Wet vent system flat
Figure 2-40

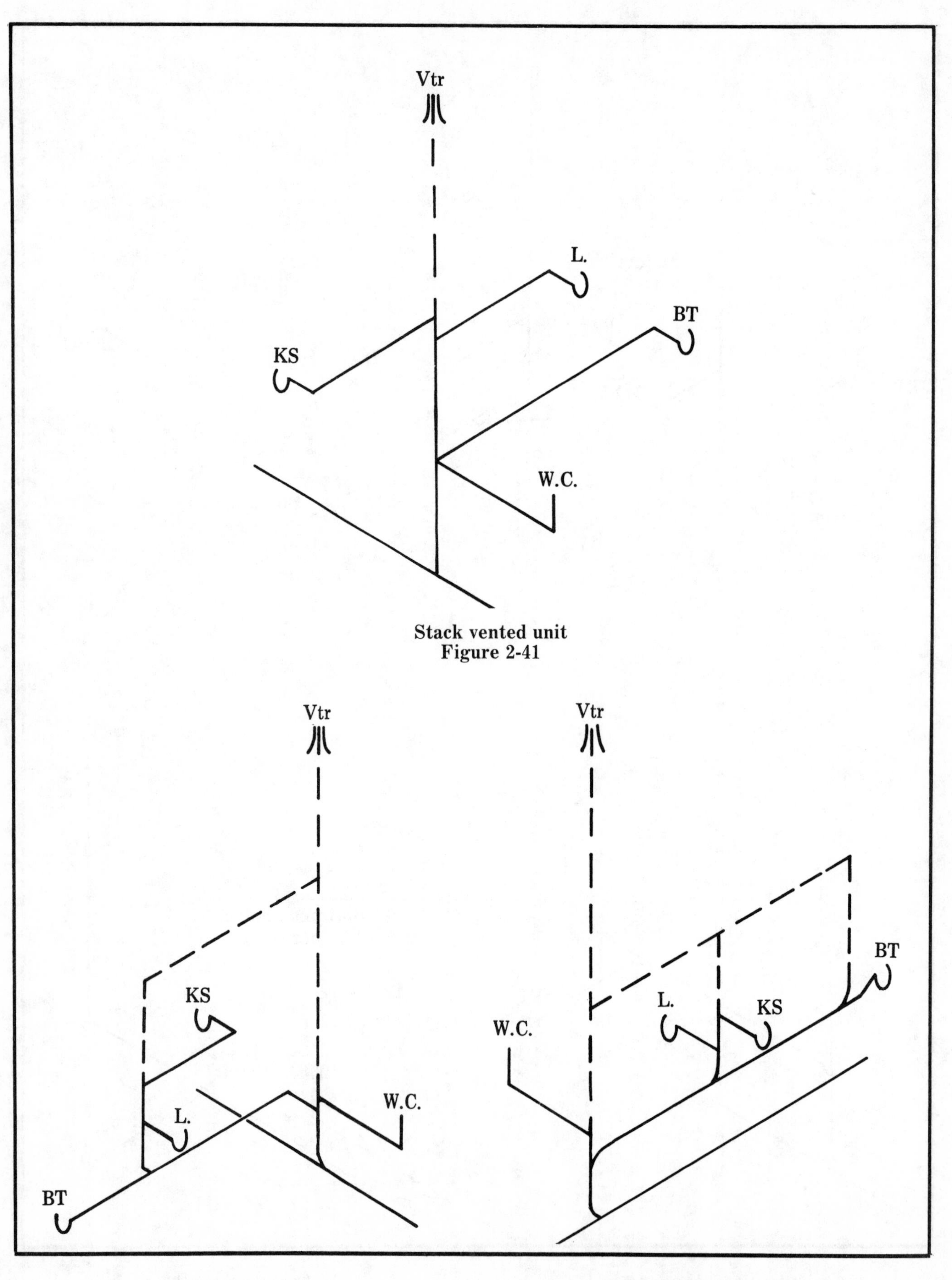

Stack vented unit
Figure 2-41

Wet vented unit
Figure 2-42

Individually vented unit
Figure 2-43

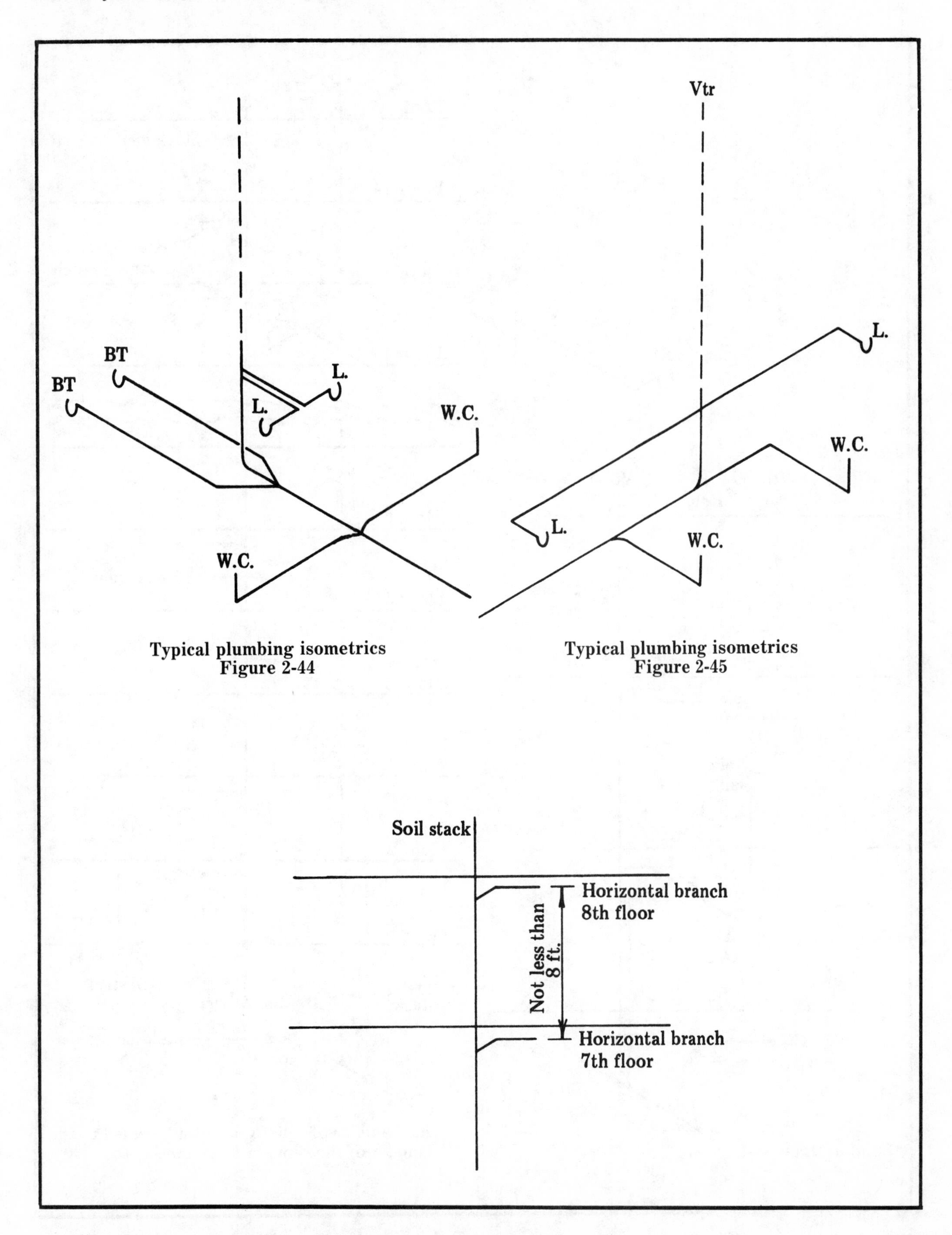

Typical plumbing isometrics
Figure 2-44

Typical plumbing isometrics
Figure 2-45

Horizontal branch must not be less than eight feet center to center
Figure 2-46

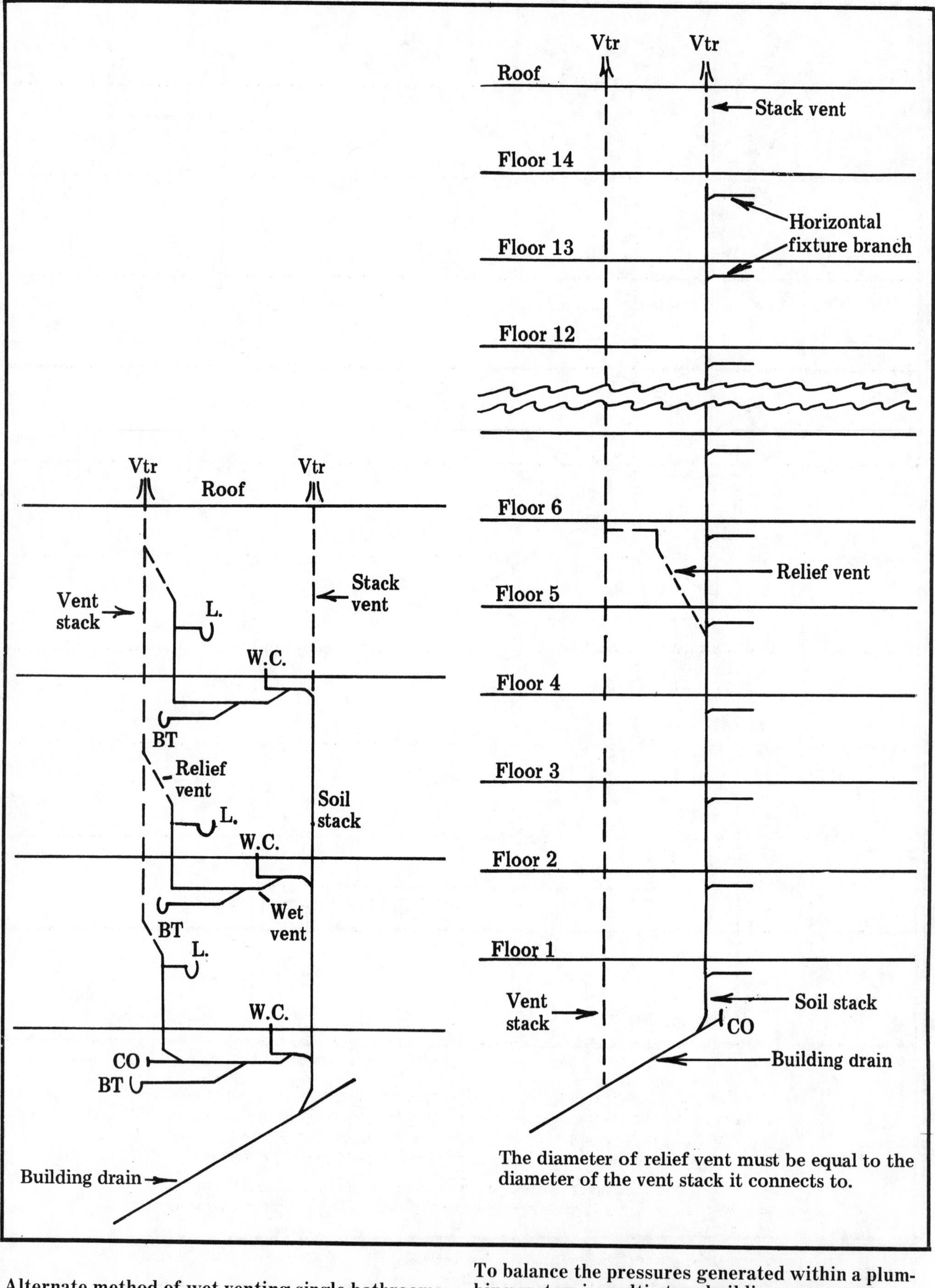

Alternate method of wet venting single bathrooms in multi-story building
Figure 2-47

To balance the pressures generated within a plumbing system in multi-story buildings, you must provide a relief vent at various intervals
Figure 2-48

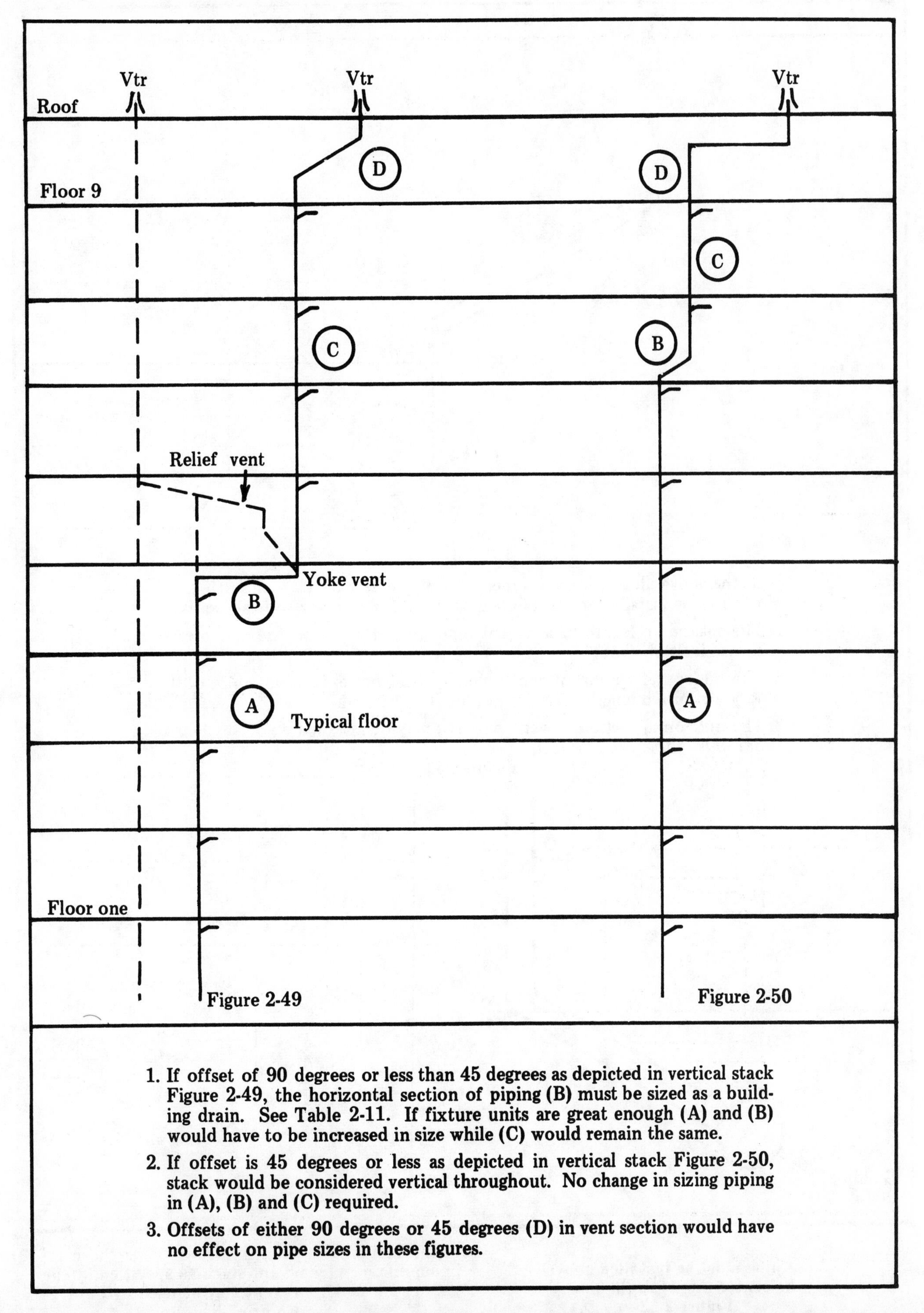

Figure 2-49

Figure 2-50

1. If offset of 90 degrees or less than 45 degrees as depicted in vertical stack Figure 2-49, the horizontal section of piping (B) must be sized as a building drain. See Table 2-11. If fixture units are great enough (A) and (B) would have to be increased in size while (C) would remain the same.
2. If offset is 45 degrees or less as depicted in vertical stack Figure 2-50, stack would be considered vertical throughout. No change in sizing piping in (A), (B) and (C) required.
3. Offsets of either 90 degrees or 45 degrees (D) in vent section would have no effect on pipe sizes in these figures.

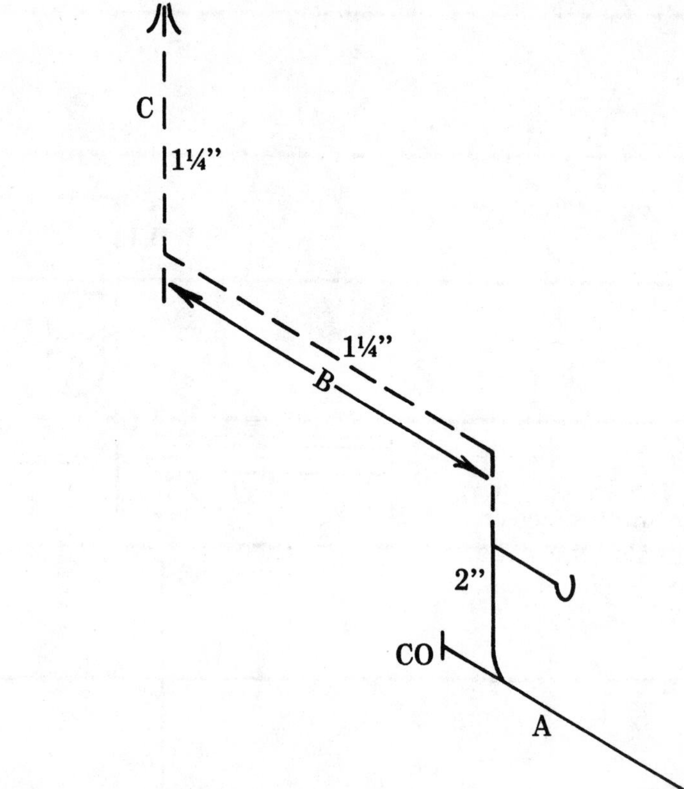

In the above illustration a 1¼ inch vent pipe is serving a 2 inch waste pipe. The following installation procedure must be adhered to for code acceptance.

1. The maximum length of any vent pipe is determined by the diameter as shown in Table 2-16.
2. The horizontal section of vent pipe (B) must not exceed one-third of the total permitted length of any vent pipe (C) installed.
3. The horizontal section of vent pipe (B) may be installed up to any length *but must not exceed 20 feet.*

Figure 2-51

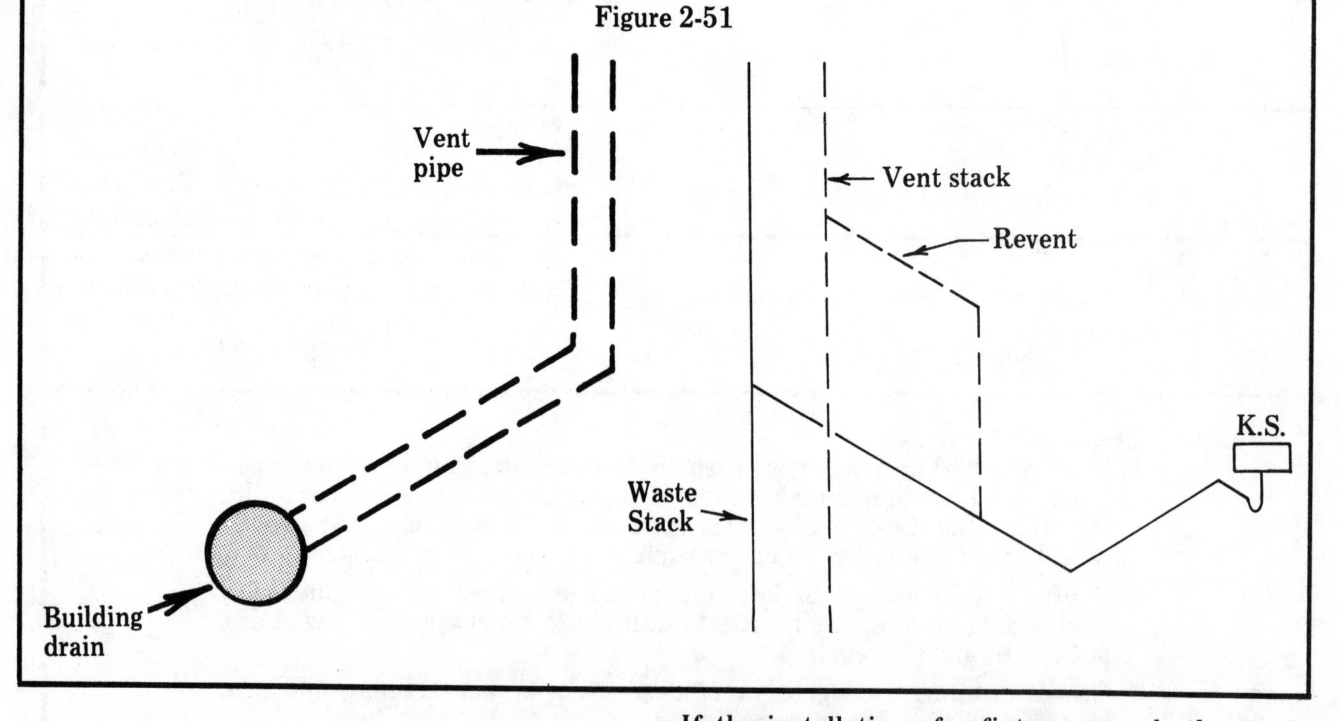

Vent pipe must be installed above the center line of drain pipe
Figure 2-52

If the installation of a fixture exceeds the maximum distance from vent stack as specified by the code, a revent may be used
Figure 2-53

Fixture Traps

Plumbing fixtures connected directly to the sanitary drainage system must be equipped with a water seal trap. Each fixture must be separately trapped, except those having integral traps fashioned within the fixture body such as a water closet.

The trap is designed and constructed to provide a liquid seal to prevent drainage system odors, gases and even vermin from entering the building at fixture locations. It must provide this liquid seal protection without materially affecting the flow of sewage or other waste liquids through it.

Because of the trap's unique importance in protecting the health of individuals, the code has placed many restrictions and limitations on its use. These are grouped and listed below.

- Fixture traps must be self-cleaning with the exception of interceptor traps.
- No trap outlet can be larger than the fixture branch to which it is connected. In other words, a 1½ inch trap cannot be connected to a 1¼ inch branch with a reducer.
- No trap can be used which depends on the action of movable parts to retain its seal.
- Traps prohibited by most model codes are bell traps, crown-vented traps, pot traps, ¾ S traps and traps with slip-joint nuts and washers on the discharge side of the trap above the water seal.
- Each fixture trap must have a water seal of not less than two inches nor more than four inches. An exception is made for interceptor traps that require deeper seals.
- Fixture traps installed below the floor or otherwise concealed cannot have trap cleanouts.
- All traps must be installed at a level in relation to their water seals. This is necessary to prevent negative action, self-siphonage, from taking place.
- Each plumbing fixture must be separately trapped by a water seal trap with the following exceptions. Water closets or similar fixtures having integral traps cannot be separately trapped. Two or three compartment sinks, laundry trays or other similar fixtures may be connected to a single trap with a continuous waste (providing the compartments are adjacent to one another). No fixture can be double trapped.
- **Materials for concealed fixture traps must be cast-iron, cast brass or lead. In a plastic system, plastic traps can be used.**
- **Trap materials exposed or otherwise accessible (with the exception of fixtures with integral traps) must be of cast-iron, cast brass, lead, 20 or 17 gauge chrome brass, or copper. In a plastic system, plastic traps must be used.**

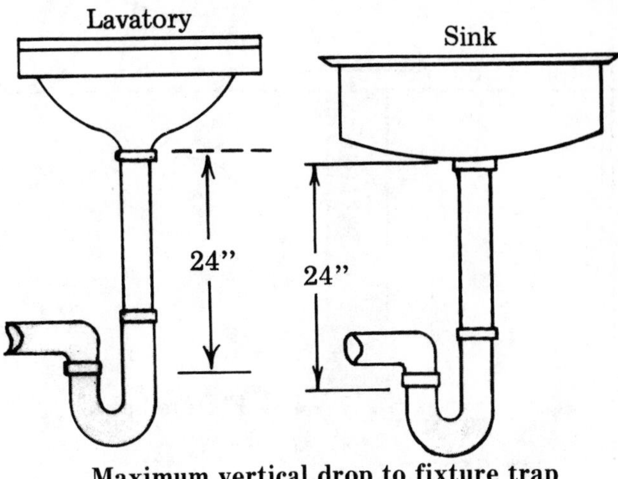

Maximum vertical drop to fixture trap
Figure 2-54

- There is a maximum vertical drop from a fixture waste outlet to the trap water seal. This further prevents self-siphonage of the fixture trap water seal. The shorter the distance between these two points, the more efficient the fixture trap will be. For sinks, lavatories, showers, bathtubs and all similar fixtures, the vertical drop (tail piece) cannot exceed 24 inches, as illustrated in Figure 2-54. The vertical drop of the pipe serving floor connected fixtures with integral traps, water closets and similar fixtures cannot exceed 24 inches, as illustrated in Figure 2-55.
- Floor drains are included in the fixture family and are therefore governed by the 24 inch tail piece limit as shown in Figure 2-54. Because many times there is difficulty in installing the floor drain within the required 24 inch limit, most codes have made an exception to the length limit. See Figure 2-56. You must follow the requirements carefully to stay within the exception *providing the following is adhered to.*

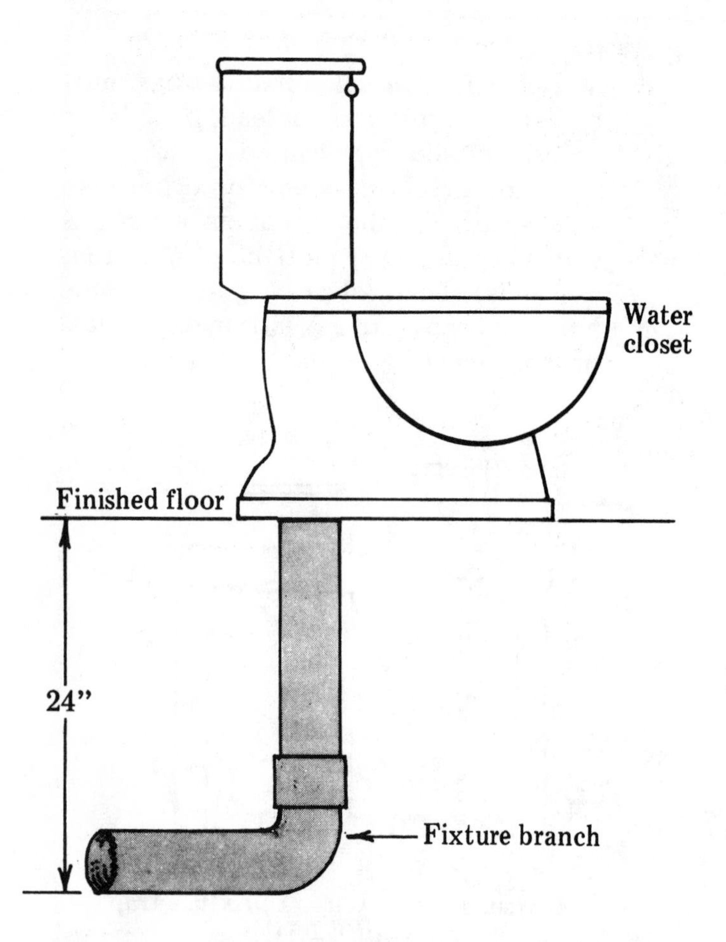

Maximum vertical drop for floor mounted fixtures
Figure 2-55

Figure 2-56 shows how to install a floor drain in a room where the horizontal building drain is 4 feet below the finished floor. The vertical rise A from a horizontal drainage line serving a floor drain cannot exceed 6 feet. The horizontal waste line B cannot be less than 3 feet from the vertical rise A. (Less than 3 feet would form an S-trap that is prohibited by the code.)

Building Traps

Before sanitary systems were properly designed with a venting system to protect the fixture traps, it was common for the water seal in these traps to be lost by the action of back pressure or siphonage or both. Rats were able to travel freely from one building to another. Decomposing sewage in the sewage collection system generated gas and offensive odors which were released into buildings at fixture locations. Health department officials of that day recognized this condition as a serious health menace to people living in fast-growing cities and towns. They required that a building trap, as in Figure 2-64, be placed on each building drainage line. These proved to be generally effective.

Building traps at that time provided a secondary safeguard to keep rats, vermin, sewer gases and odors out of a building. The individual fixture traps provided the primary safeguard. The building trap was deemed a necessity until the modernization of collection, drainage and venting systems. Today most model codes do not require (and actually prohibit) the installation of a building trap in a building drainage line. Refer to your local code for verification of its requirements.

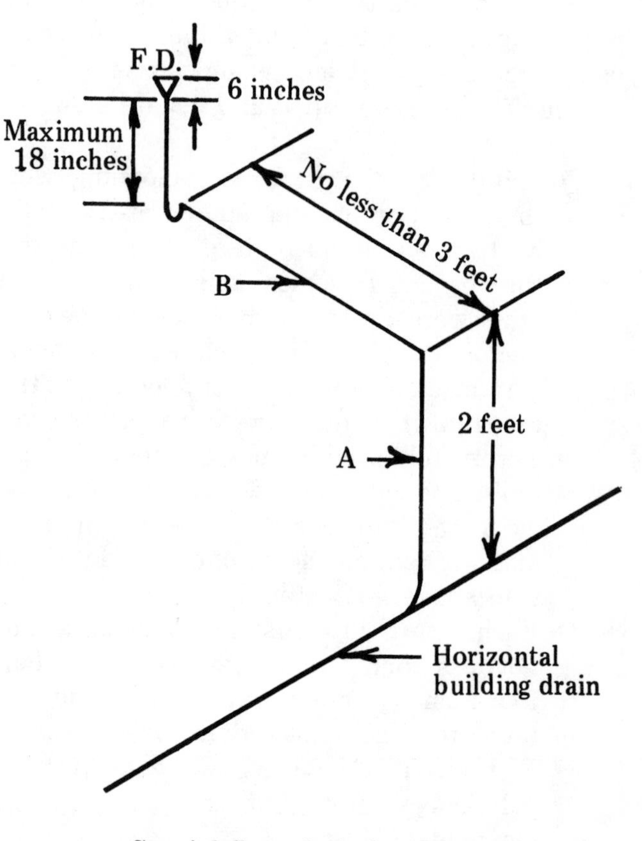

Special floor drain installation
Figure 2-56

Cleanouts

Years ago before cleanouts were required on drainage piping, the plumber had to cut a hole in blocked drainage pipe to insert a cleaning cable to remove the obstruction. The hole was then patched with a cement mixture or a similarly impervious material. These patch jobs often deteriorated and allowed raw sewage to seep out on or into the ground. This caused a health hazard for the building occupants and their neighbors.

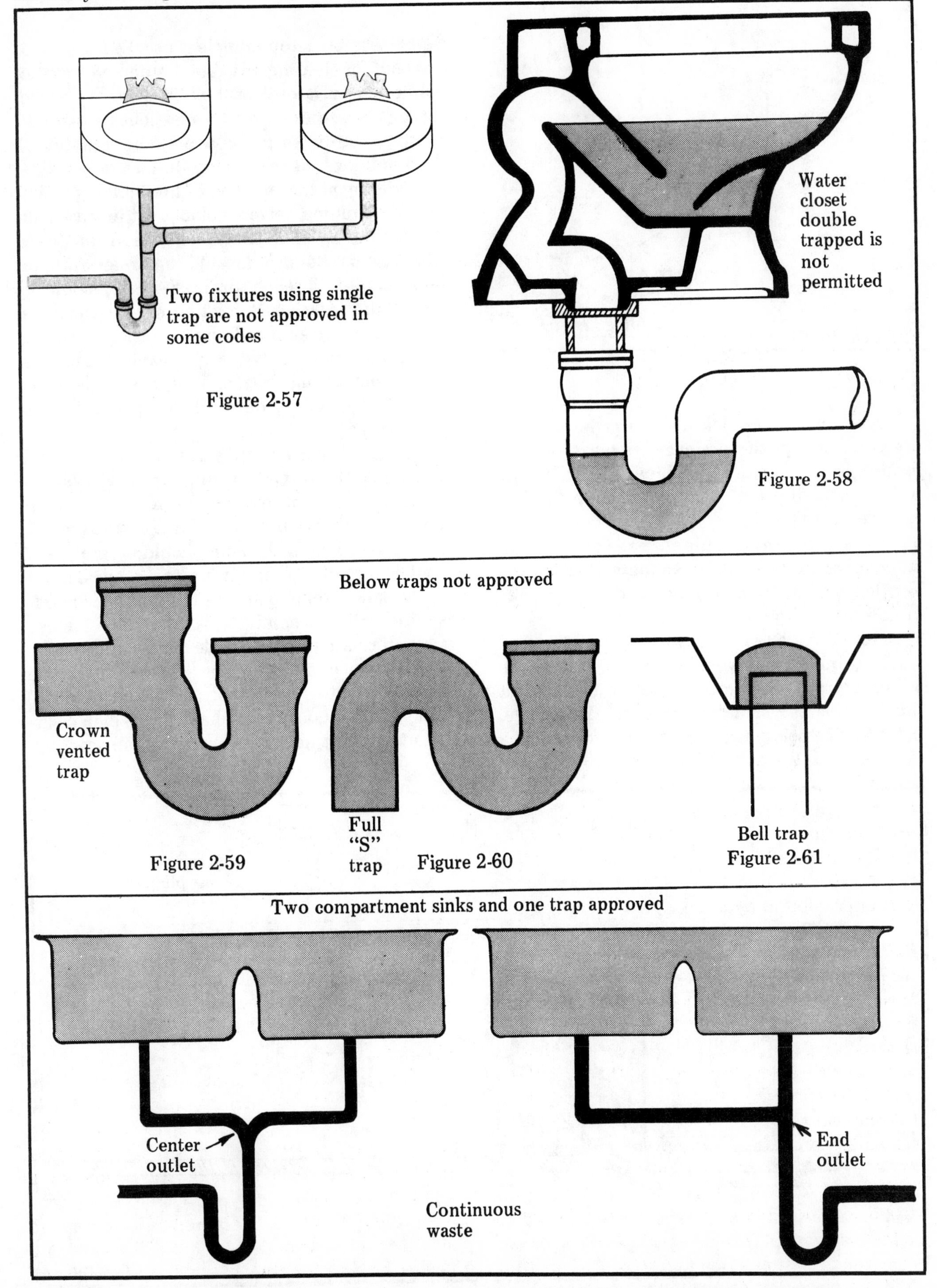

Figure 2-57

Figure 2-58

Figure 2-59

Figure 2-60

Figure 2-61

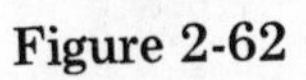
Figure 2-62

Figure 2-63

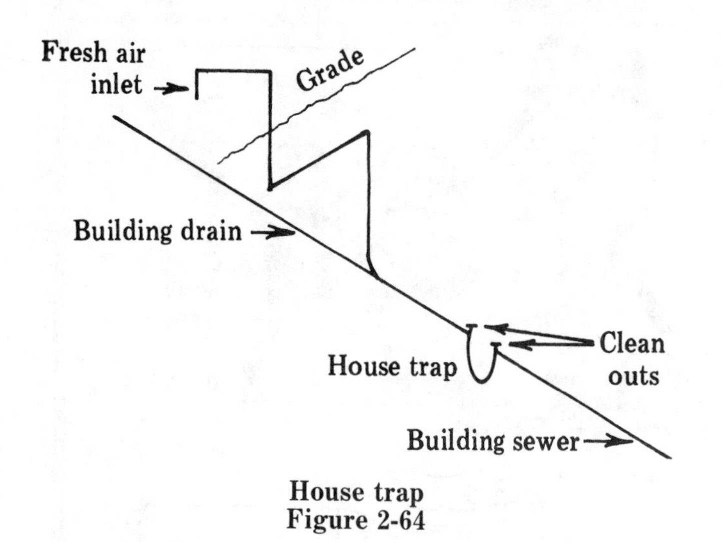

House trap
Figure 2-64

Over the years, cleanouts have become an essential part of the drainage system. Today's model codes specify the location, distance between cleanouts, size, and many other requirements.

As all drainage pipes are subject to stoppages, the accessible cleanouts will save the serviceman valuable time and the owner unnecessary expense.

Cleanouts or cleanout tees are required where a building sewer connects to the public sewer lateral at the property line. A cleanout at this location serves a dual purpose. This is the point for inserting a test plug for performing a water test on a building sewer. The tee is a cleanout for clearing any future stoppages in the public sewer lateral or building sewer. Some authorities require that this cleanout be extended up to the finish grade while others do not.

A full size cleanout *may* be located outside the building at the junction of the building drain and the building sewer, usually within five feet of the building line. This is not a requirement if other cleanouts are located upstream and a sewer cable will reach this area. If a cleanout is used at this location, it must permit upstream as well as downstream rodding. In other words, a fitting known as a two-way cleanout must be used. This fitting may or may not have to be brought to grade, depending on local code requirements.

Accessible cleanouts are required on all horizontal drainage piping and separation distances can not exceed 75 feet. For each change of direction in a building drain greater than 45 degrees, a cleanout should be provided.

The size of a cleanout, up to 4 inches, must be the same nominal size as the pipe to which it is joined. Pipe over 4 inches in diameter can use a 4 inch cleanout fitting. Cleanouts must be located to permit rodding of all portions of the sanitary waste system with a 75 foot cleaning cable. All cleanouts must have an 18 inch clearance to permit access for rodding purposes.

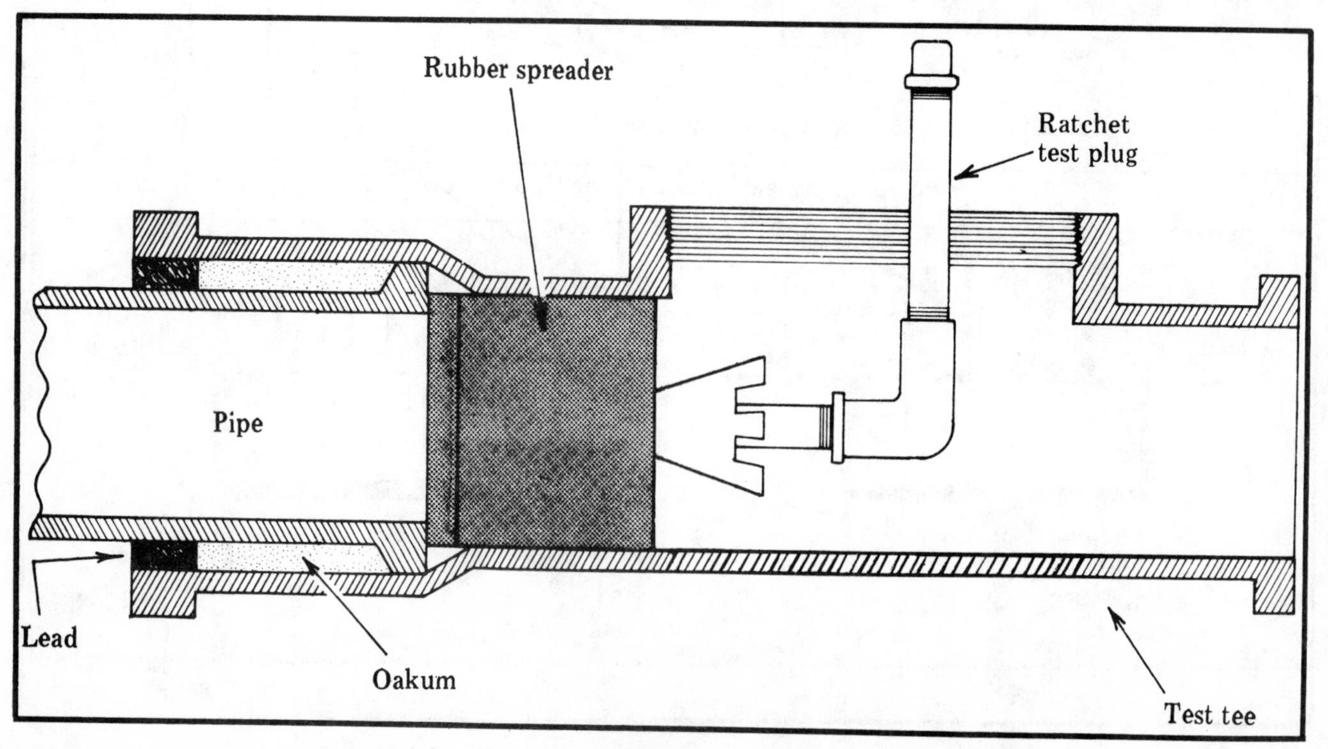

Test tee used for testing sewer
Figure 2-65

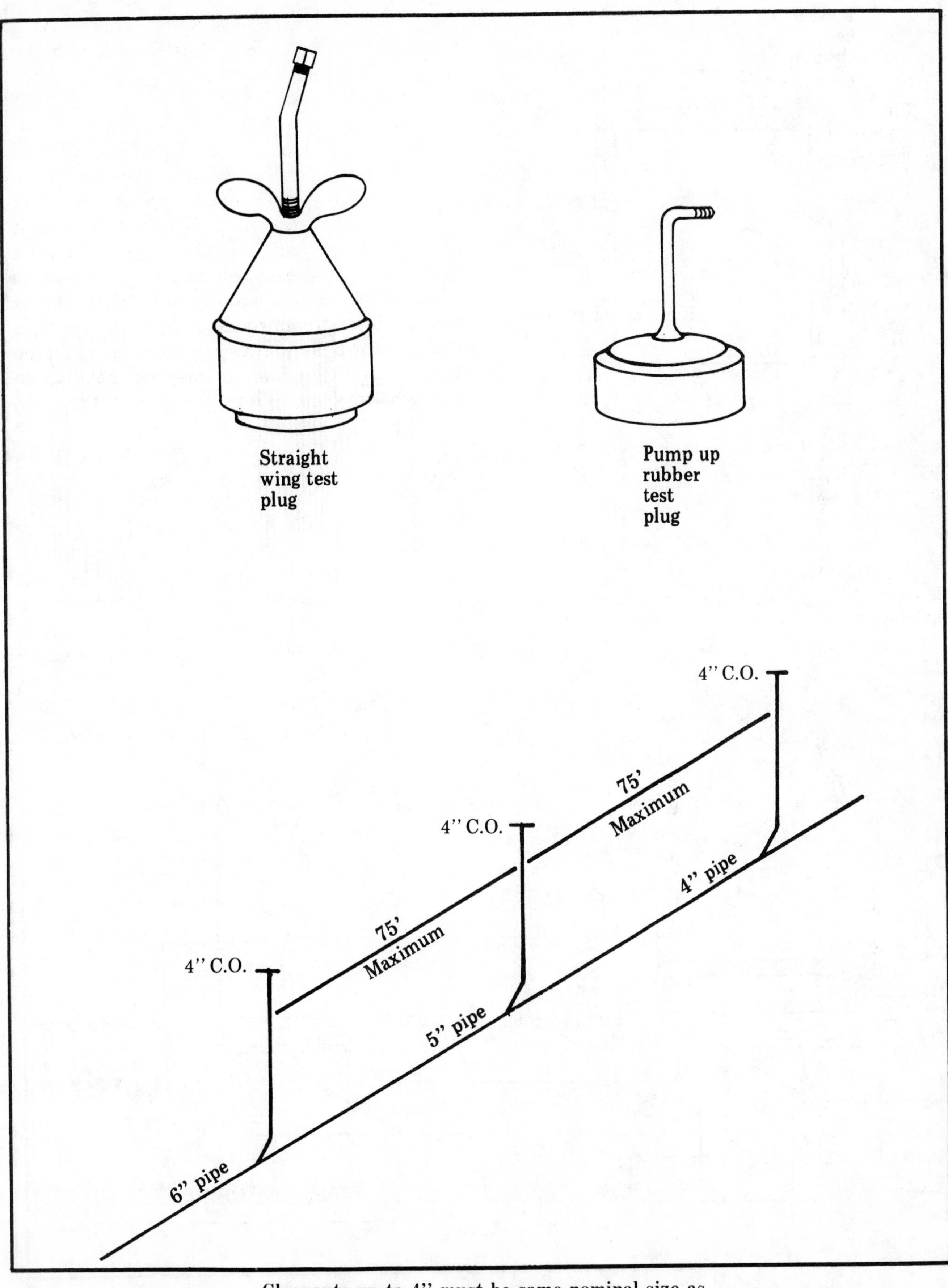

Cleanouts up to 4" must be same nominal size as the pipe
Figure 2-66

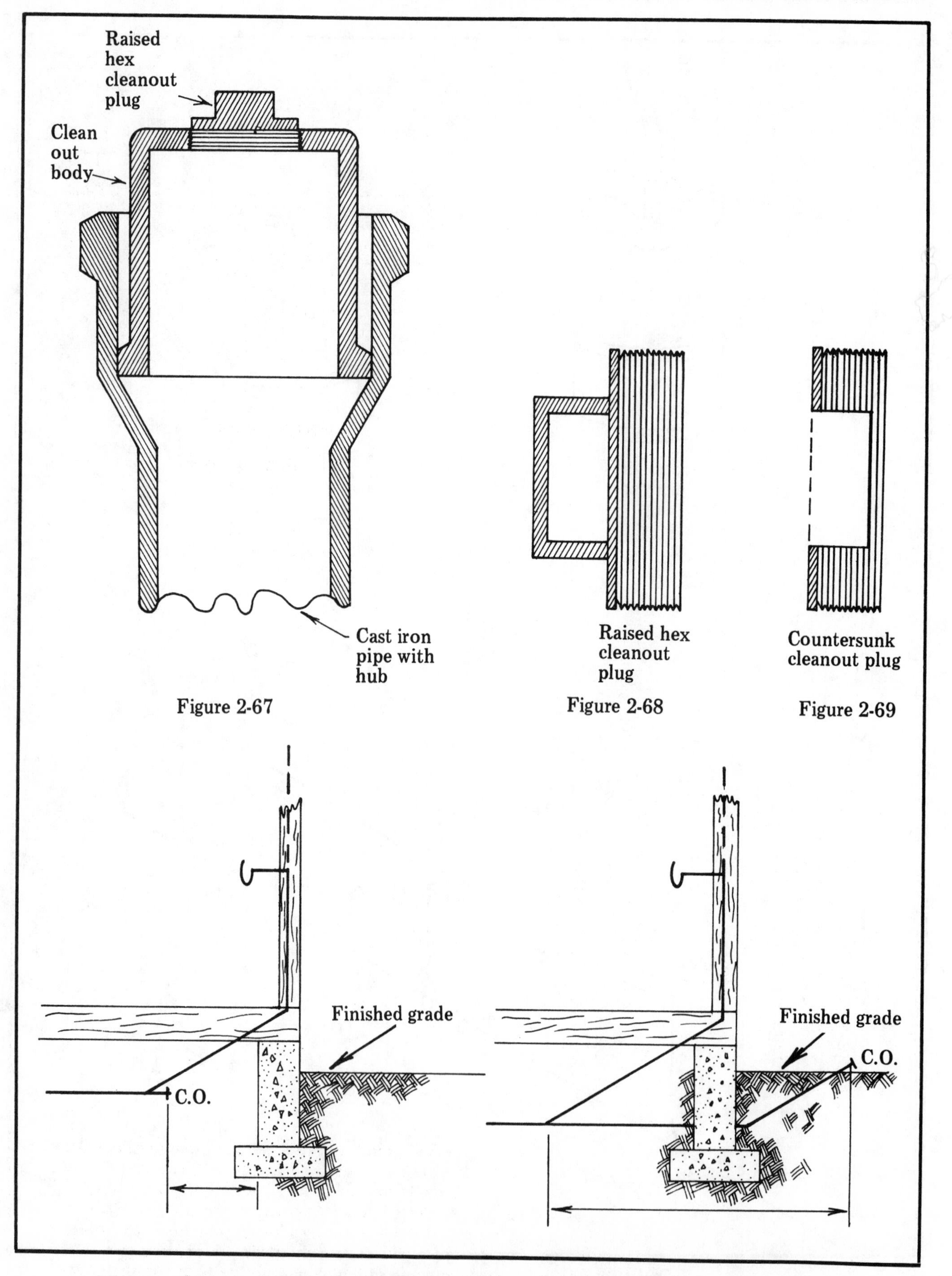

Figure 2-67

Figure 2-68

Figure 2-69

Minimum 18 inch clearance from wall
Figure 2-70

Dead end created with cleanout extension
Figure 2-71

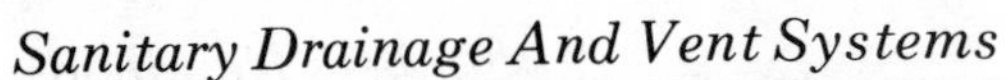

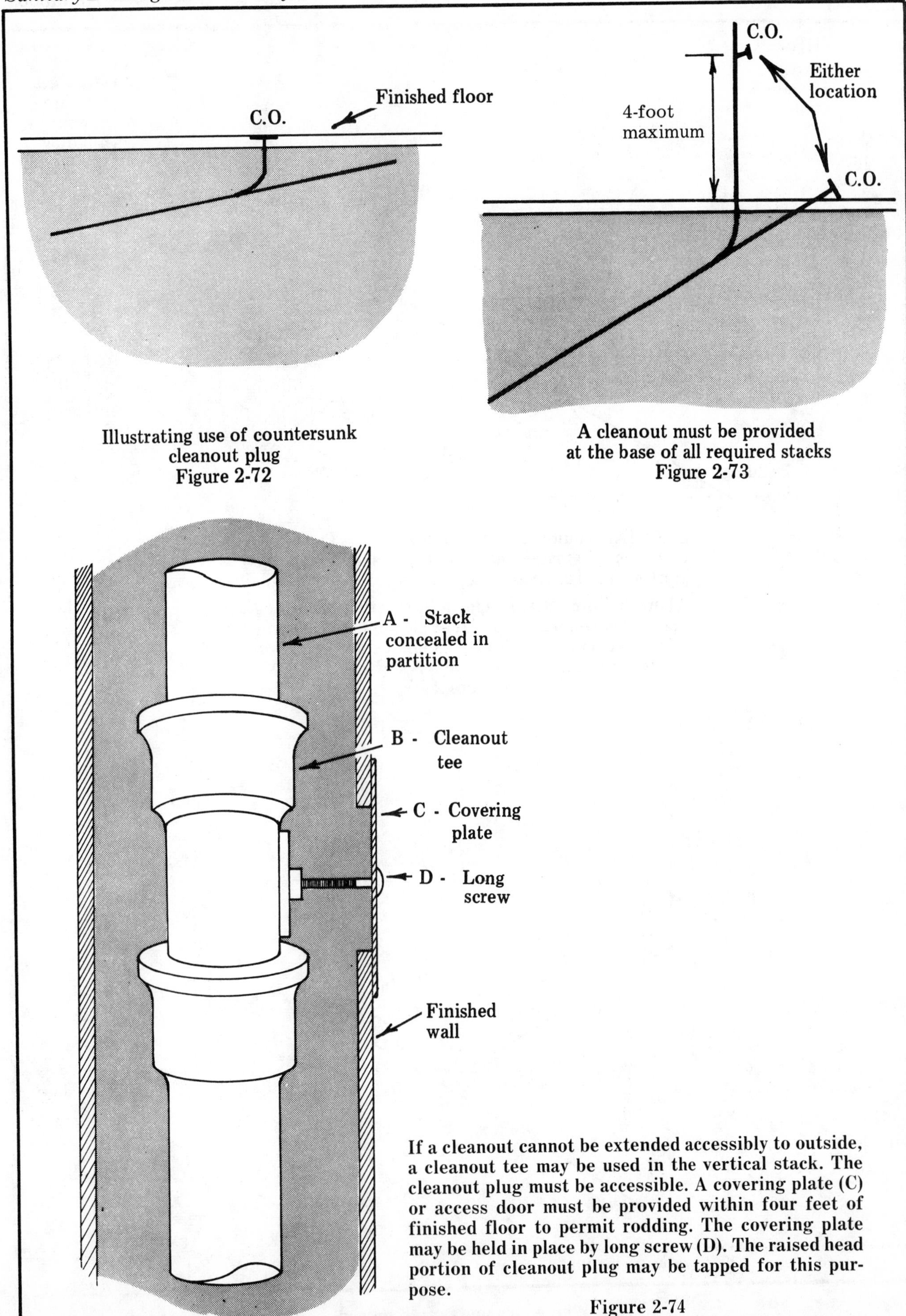

Illustrating use of countersunk cleanout plug
Figure 2-72

A cleanout must be provided at the base of all required stacks
Figure 2-73

If a cleanout cannot be extended accessibly to outside, a cleanout tee may be used in the vertical stack. The cleanout plug must be accessible. A covering plate (C) or access door must be provided within four feet of finished floor to permit rodding. The covering plate may be held in place by long screw (D). The raised head portion of cleanout plug may be tapped for this purpose.

Figure 2-74

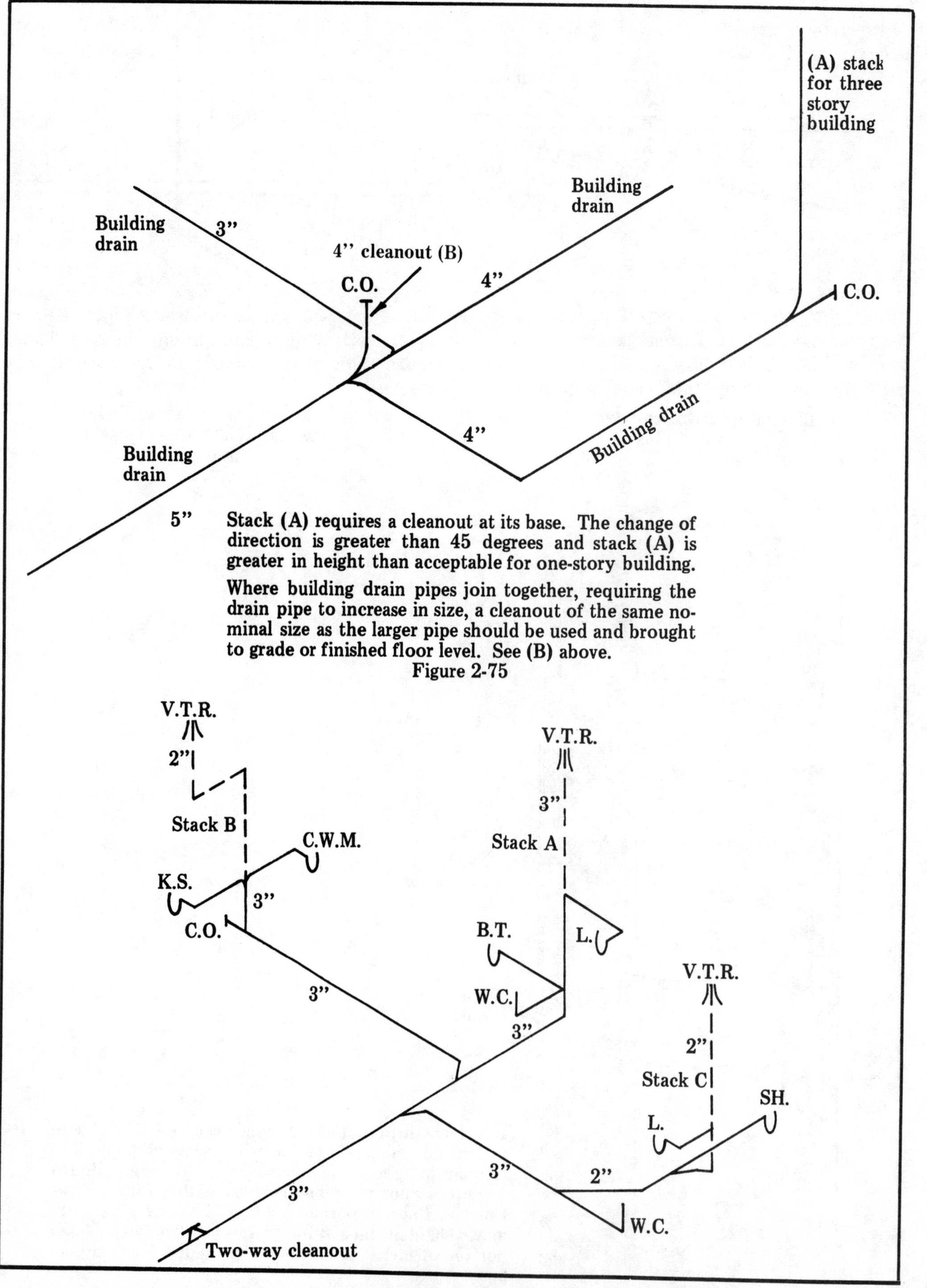

Stack (A) requires a cleanout at its base. The change of direction is greater than 45 degrees and stack (A) is greater in height than acceptable for one-story building.

Where building drain pipes join together, requiring the drain pipe to increase in size, a cleanout of the same nominal size as the larger pipe should be used and brought to grade or finished floor level. See (B) above.

Figure 2-75

Where cleanouts are required
Figure 2-76

Cleanouts may be located in a basement if accessible or flush with the finished floor. Cleanouts in walkways, hallways, and rooms must have countersunk plugs to prevent tripping or accidental injury.

A cleanout can be in the base of an exposed or concealed vertical stack not exceeding four feet above the finished floor. If concealed, the cleanout must have a removable cover plate or access door for easy access.

As discussed earlier, cleanouts in one story buildings can be omitted if certain code requirements are met. This section of the code is controversial. The following illustration may help you understand this code requirement. Refer back to the section that explains the basic principles of venting for a list of the requirements that must be met before the cleanout can be omitted.

In Figure 2-76 a two-way cleanout is located in the building drain and will permit rodding upstream to the base of Stack A. Thus, installing a cleanout in the base of Stack A is not necessary.

Stack B will require a cleanout at its base. It is properly sized, but not vertical throughout as required by the code. Stack C has the same diameter (2 inches) as the waste pipe it serves and is vertical up to and through the roof. Thus, installation of a cleanout in its base is not necessary.

Interceptors, Special Traps, And Drain Piping

Many substances are considered objectionable waste or harmful to the building drainage system, the public sewer, or the sewage-treatment plant. These substances include grease, flammable waste, sand, plaster, lint, hair and ground glass. Materials such as these must be intercepted or separated from the liquid waste before entering the collection system. An *interceptor or separator* is used for this purpose. These are also referred to as *traps.* Figure 3-1 shows the type of installation that would require a grease interceptor.

Interceptor traps are required in commercial buildings primarily to accumulate and recover objectionable substances from liquid waste, and to prevent their introduction into the drainage system. Each interceptor trap must be located to provide ready accessibility for maintenance and for removing the accumulated intercepted matter.

Many sizes and types of interceptors are available to perform various functions. Detailed drawings should be submitted to local authorities for advance approval before the interceptors are installed.

When interceptors are required in drain lines, waste *not* requiring separation must *not* be discharged through an interceptor.

Nearly all interceptors required by modern codes are for commercial establishments. The three most often used interceptors are the grease interceptor (Figure 3-2), the oil interceptor (Figure 3-3), and the lint interceptor (Figures 3-4 and 3-5).

Grease Interceptor

Commercial buildings such as restaurants, hotel kitchens and bars, factory cafeterias and restaurants. clubs, processing plants and the like must have a grease interceptor or trap to accumulate grease that might otherwise clog the waste lines. When required, a separate grease waste line is installed to serve pot, scullery, food scrap and vegetable preparation sinks. Floor drains that may receive kitchen waste spillage and floor drains that may receive waste from individual fixtures, appliances or other apparatus must go through an approved grease interceptor. However, waste from a commercial-type food grinder *must not pass through a grease interceptor.* It must pass directly into the main sanitary drainage system.

Grease interceptors are not generally required in single family residences, private living quarters or apartment buildings.

As a rule, grease interceptors for commercial establishments which prepare and serve food are sized according to the seating capacity. Sizing should also provide for a retention period of not less than 2.4 hours for the cooling and separation of grease.

Most codes do not provide established siz-

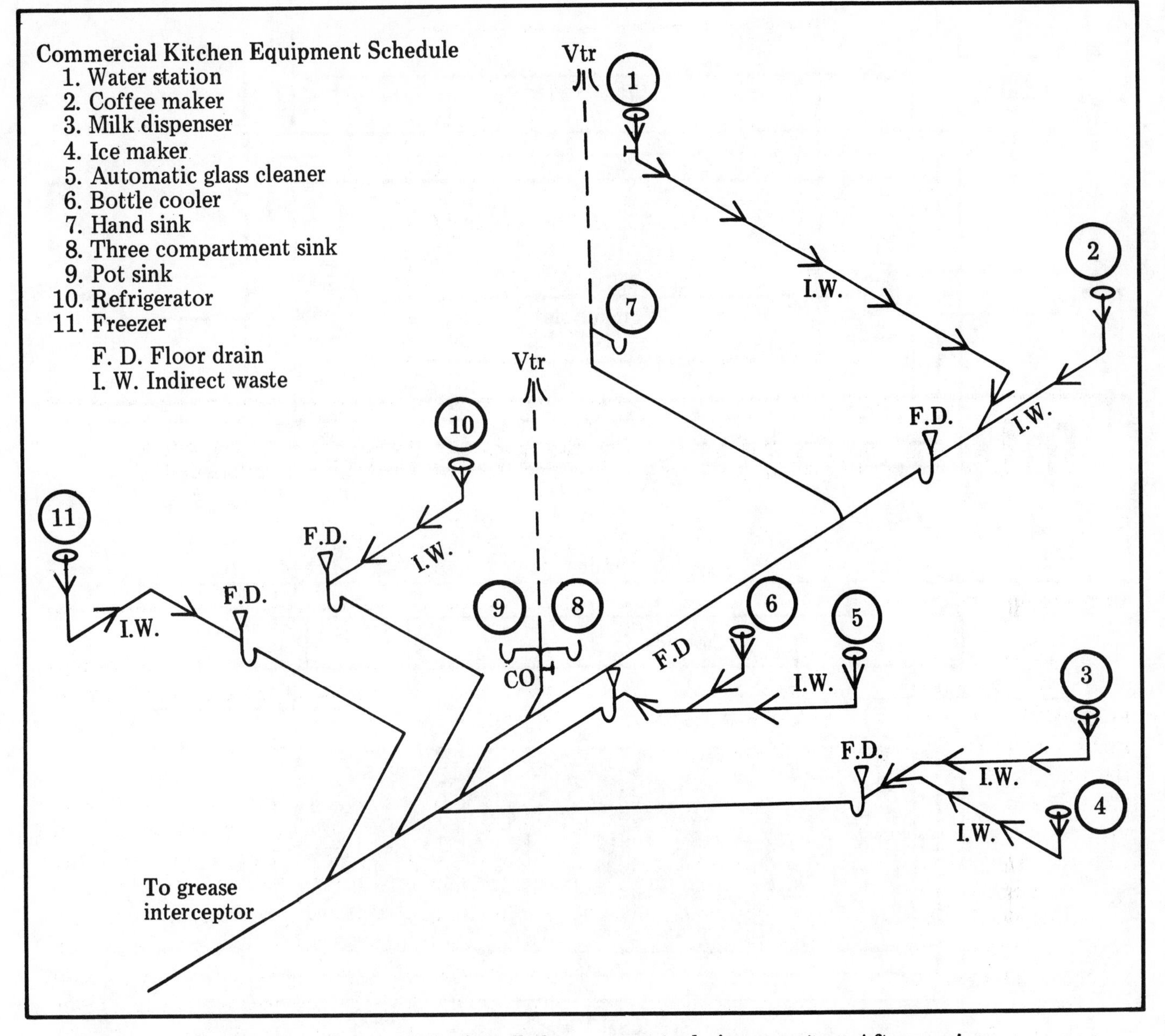

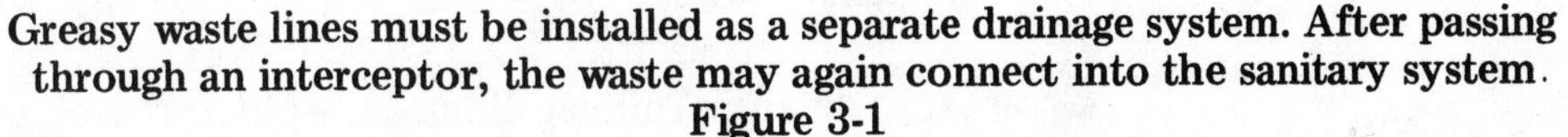

Greasy waste lines must be installed as a separate drainage system. After passing through an interceptor, the waste may again connect into the sanitary system.
Figure 3-1

ing methods, but recommend design criteria or leave it to the discretion of the Health Department or the local plumbing official. Small eating establishments having fixed seats or stools are usually required to install a grease interceptor of not less than 750 gallon liquid capacity.

The following figures for restaurants having seating capacities up to 200 persons generally meet the design criteria of most codes.

As a possible guide for restaurants with larger seating capacities than those listed above, multiply the number of seats by 15. This should provide you with acceptable capacities for grease interceptors for most restaurants.

Liquid capacities of grease interceptors for eat and/or drink establishments (in gallons)

Up to 50 persons	51 to 100 persons	101 to 150 persons	151 to 200 persons
750	1,500	2,250	3,000

Check with the local Health Department, plumbing official or local code for verification of sizes in your particular area.

A grease interceptor must usually be located outside the building it serves, because

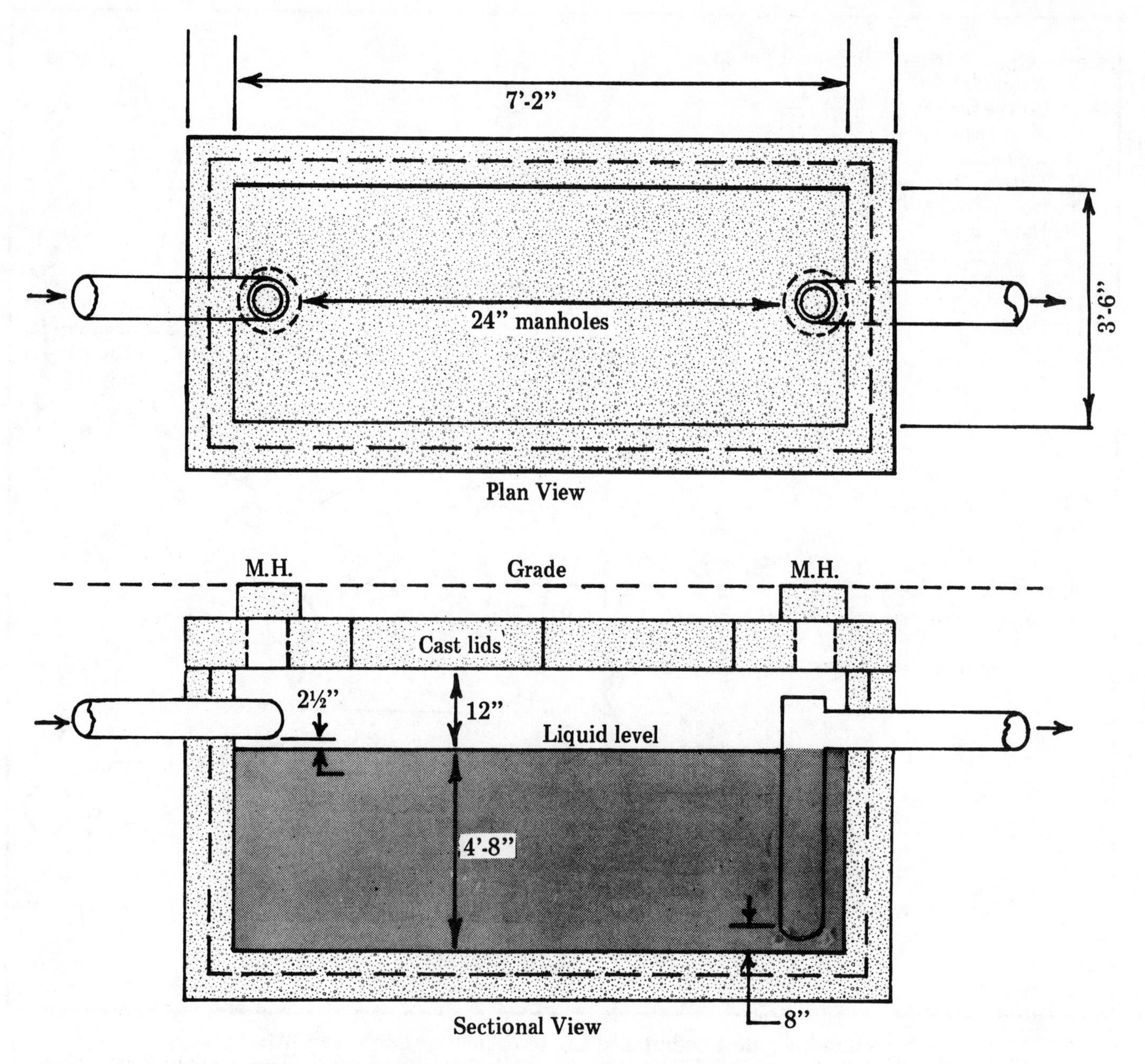

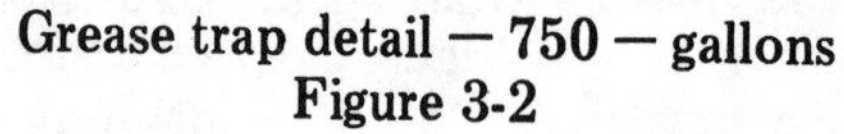

Grease trap detail — 750 — gallons
Figure 3-2

the capacity is necessarily large.

Some codes permit floor-mounted grease interceptors for commercial establishments generating small amounts of grease to be located within the building, close to the fixtures they serve. The capacity of this type interceptor is rather small, generally 14 to 50 pounds. Check local code for permitted use.

The materials from which grease interceptors may be constructed are governed by your local code. The most common material accepted is concrete. In some areas steel or fiberglass may also be approved. The structural design criteria for grease interceptors are fairly uniform across the country. Figure 3-2 shows the principal requirements:

- The inlet invert must discharge a minimum of 2½ inches above the liquid level line.
- The outlet tee must extend to within 8 inches of the bottom of the tank.
- For maintenance, a 24 inch diameter cleanout manhole (one over the inlet and one over the outlet tee) must be provided and brought to grade.
- Grease interceptors must be designed and installed so that they will not become air bound. (Grease interceptors should not require a local vent, as enough air can

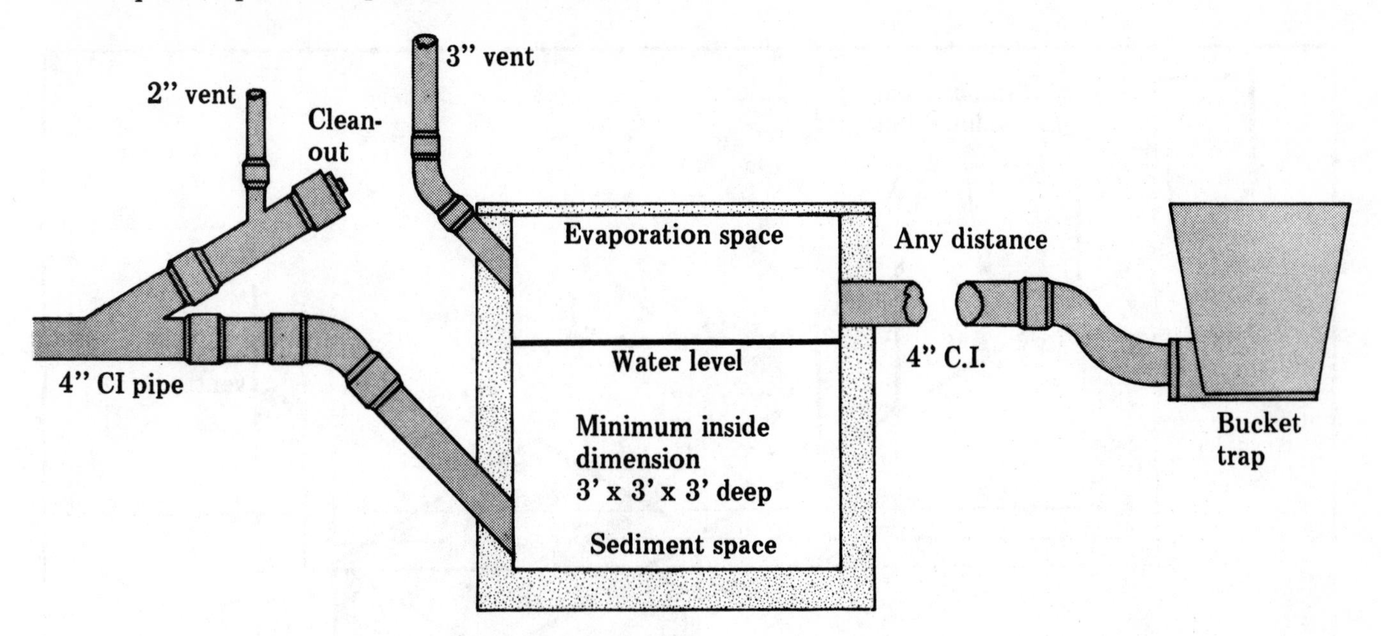

The drain from interceptors used for glass, sand, automobile wash floors or for liquid waste which does not convey steam, noxious odors or gases need not be vented provided that:

(A) It is discharged into a street catch basin.
(B) It is discharged into a vented building sewer or building drain and the interceptor branch does not exceed 15 feet developed length.

A 3" local interceptor basin vent must be provided and extended separately through a roof for all gasoline and oil interceptors or for interceptors receiving material which would emit offensive odors.

A local vent may be located on the drain from the bucket trap.

It is recommended that the interceptor be located outside building.

Interceptor for gasoline, oil, sand, auto repair, etc.
Figure 3-3

enter and leave through the vents of the fixtures they serve.)

- Other codes may require interceptors to have at least two compartments.
- Waste discharge from interceptors may be connected to existing building sewer or into a separate disposal system, if first approved by the plumbing official.

Gasoline, Oil and Sand Interceptors (Figure 3-3)

Interceptors must be provided to prevent the introduction of gasoline, grease, oil or sand into the drainage system in the following places:

- Any place where motor vehicles are repaired and floor drainage is provided.
- Any place where motor vehicles are washed. (Private car wash facilities are excluded.)
- Public storage garages where floor drainage is provided
- Any place where oil, gasoline or other volatile liquids can be discharged into the drainage system.
- Plants where parts are washed to remove oil or greasy substances.

Bucket type floor drains should be used where this type of interceptor is required and there should be a minimum 4 inch diameter outlet. The bucket is removable for cleaning and is made of the same material as the floor drain. The bottom portion of the bucket is solid to retain sand. Drainage holes near the top of the bucket let liquid waste pass out of the bucket and into the pipe or pipes leading to the interceptor.

Oil interceptors usually must have a minimum capacity of 18 cubic feet per 20 gallons of design flow per minute. Figure 3-3 shows how

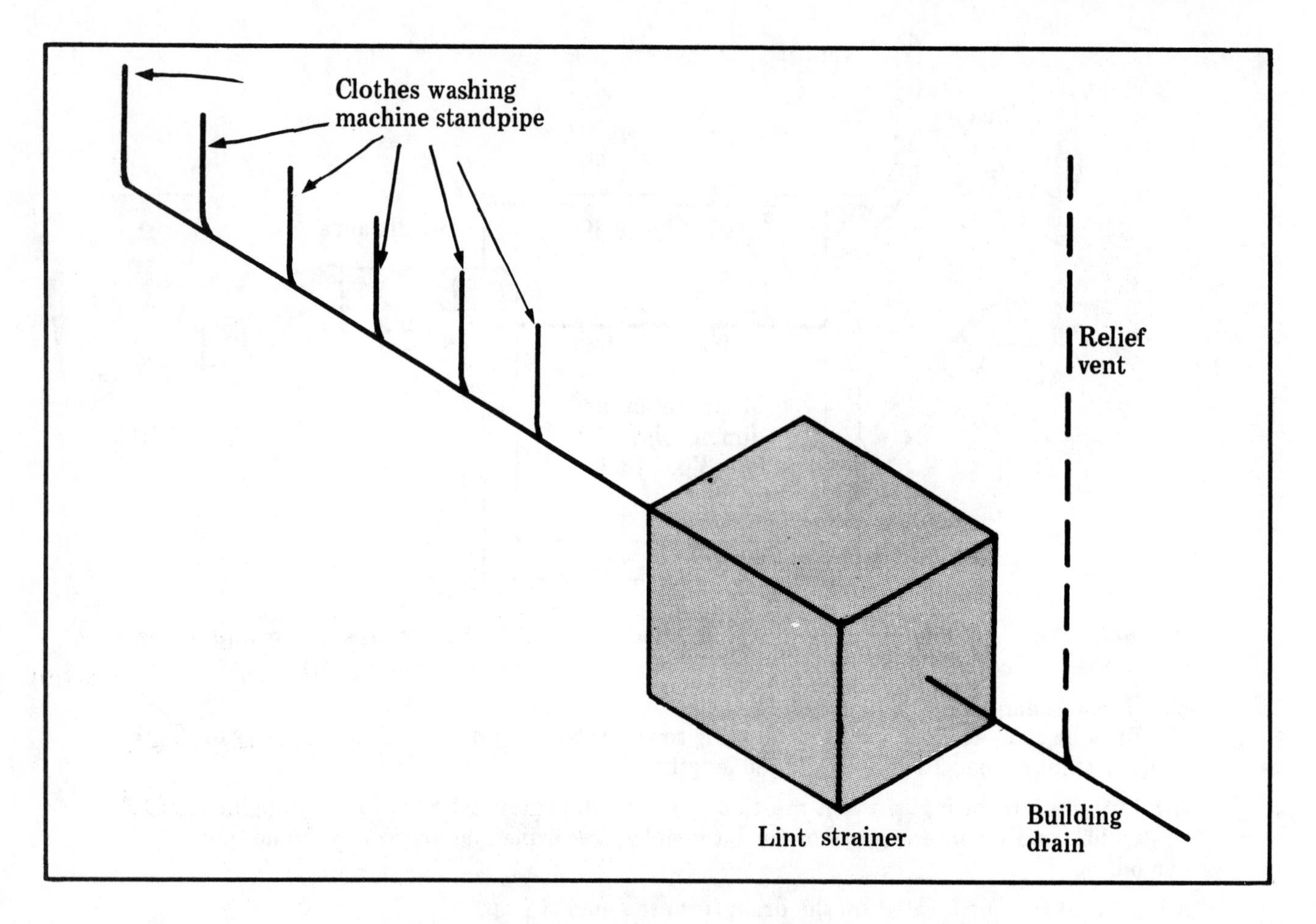

Install lint interceptors on the drainage pipes from laundries
Figure 3-4

the interceptor works and describes the conditions under which the venting may be omitted or required.

Slaughter House Interceptors

Where an establishment slaughters, prepares or processes meat, the waste from the floors must pass through a specially designed floor drain before entering the grease interceptor. These bucket type floor drains are equipped with metal screen baskets that keep solids more than ½ inch across out of grease interceptor traps. Fats and smaller solids are collected in the interceptor trap itself.

Laundry Interceptors

Commercial laundries discharge solids such as lint, string, and buttons with the liquid waste. Solids such as this should not enter the drainage system. Therefore lint interceptors must be installed on the drainage pipes from laundries.

The lint interceptor should have a *non-removable* ½ inch mesh screen metal basket or a similar device to collect the solids. The screen should be constructed so that it can be cleaned easily. See Figure 3-5.

The plumbing code considers the horizontal drainage pipes serving commercial clothes washing machines to be indirect waste pipes. This is a unique method of piping but is economical and practical in this application. The indirect waste system does not have to be trapped or vented as do most other plumbing fixtures. The washing machine standpipes are open-ended 3 or 4 inch diameter pipes extending to about 26 inches above the finished floor. These drain pipes receive the discharge from the washers through flexible hose. A three inch standpipe can accommodate two machines; a four inch standpipe will serve four machines.

Horizontal drain pipes collect waste from the standpipes and convey the waste to a lint interceptor which is generally one pipe size

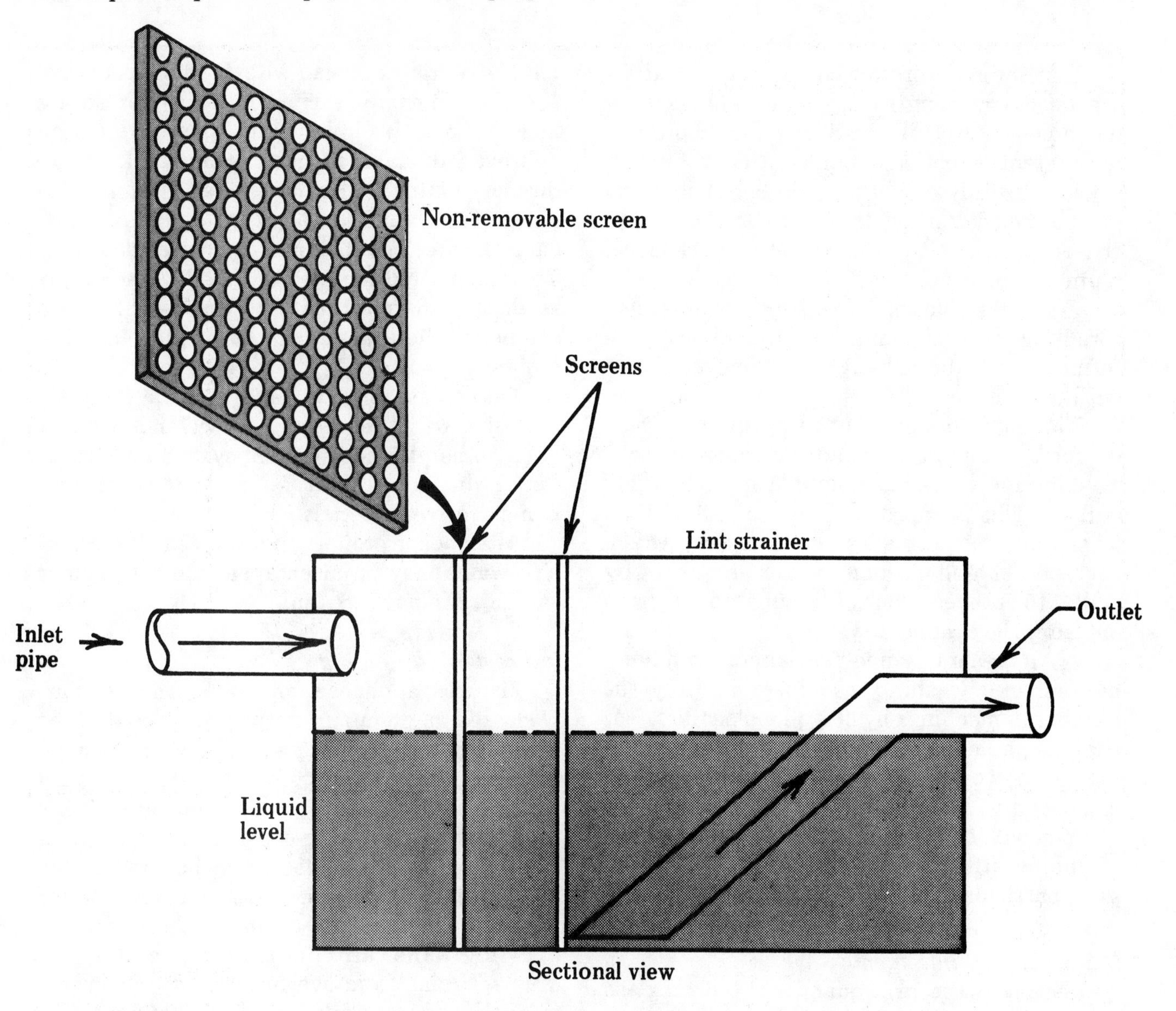

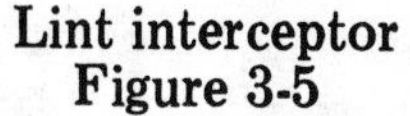
Lint interceptor
Figure 3-5

above the liquid level. This drainage system is not trapped because standpipes permit free circulation of air. See Figure 3-4.

The outlet or discharge pipe is extended down on a 45 degree angle to approximately two inches from the bottom of the lint interceptor. This serves two purposes: first, light objects that have passed through the screen and are floating in the top portion of the interceptor water will not sink into the outlet pipe and thus into the building's sanitary drainage system. Second, the extension of the outlet pipe toward the bottom of the lint interceptor creates a liquid seal. This prevents sewer gases from entering the building through the clothes washing machine standpipes. The screen or screens retain foreign objects that can not pass through them. These objects should be removed regularly when the system is in service.

The outlet pipe from the lint interceptor is connected to the regular sanitary drainage system serving other fixtures within the building. A vent must be installed as close as possible to the lint interceptor on the horizontal discharge (outlet) pipes. The vent pipe serves the drainage pipes between the lint interceptor and the main building sewer or drainage system, supplying and removing air as needed. This keeps the lint interceptor from becoming air locked and ensures a free flow of waste water.

Lint interceptors for commercial laundries are sized according to the number of washing machines in the laundry. Sizing should provide for a retention period of not less than 2.0 hours. Codes generally do not provide established sizing methods. Rather, they recommend design criteria, or leave it to the discretion of the local plumbing official.

For a possible guide in sizing lint interceptors for commercial laundries, the following information will be helpful and is in accordance with most codes.

The minimum size lint interceptor approved for a commercial laundry having up to 7 washing machines has a liquid capacity of 200 gallons. This is approximately a 3' x 3' x 3' interceptor. This provides the necessary retention period and storage factor, required by code, to prevent harmful substances from entering the drainage system.

As a general guide for laundries having more than 7 washing machines, multiply the number of machines by 30 gallons per day. For example, a laundry facility having 12 washing machines would require a 360 gallon liquid capacity lint interceptor.

You will, of course, need to check with local plumbing officials for verifications of sizes in your particular area.

Other Trap Requirements

Because of the large quantity of broken glass or other solids generated in bottling plants, they are required to discharge their process wastes through an interceptor trap. These interceptors are designed to separate broken glass and other solids from liquid wastes.

Where hair in large quantities may be introduced into the drainage system, interceptors similar to ones used for commercial laundries or swimming pools are required. Animal hospitals and dog grooming establishments, for example, may be required to install hair strainers (interceptors) on waste lines from bathtubs or other receptacles where animals are bathed.

In dental and orthopedic sinks where plaster, wax or other objectionable substances will be discharged into a drainage system, an interceptor trap must be installed in the waste line.

Dilution or Neutralizing Tanks

Sanitary drainage systems are not constructed of materials that can withstand the corrosive action from many chemicals, acids, or strong alkalis. Corrosive liquids must pass through an approved dilution or neutralizing tank before discharge into the regular sanitary system. The neutralizing tank is usually constructed of glass, earthenware, or other non-corrosive materials. The tank should have a controlled water supply so that the harmful material can be diluted to the point where it will not harm the plumbing system.

The pipes, fittings, and vent system for corrosive waste must be installed independent of any other drainage or vent system and has to be of duriron, plastic or other approved non-corrosive materials.

After being properly neutralized, the corrosive waste may be discharged into the regular sanitary drainage system.

Indirect Waste Piping

Fixtures, appliances and devices not regularly classed as plumbing fixtures may be drained by indirect means if they have drips or drainage outlets. These types of installations would include refrigerators, ice boxes, bar sinks, cooling or refrigerating coils, laundry washers, extractors, steam tables, egg boilers, coffee urns, stills, sterilizers, water stations, water lifts, expansion tanks, cooling jackets, drip or overflow pans, air conditioning condensate drains, drains from overflows, and relief vents from the water supply system. See Chapter 4, Figures 4-13 and 4-14 for an explanation of indirect waste systems.

The indirect drainage method is intended to prevent sewage from backing up into these special fixtures and contaminating their contents in case of a stoppage in the sanitary drainage system. For example, overflow and relief pipes on the water supply system and relief pipes on expansion tanks, sprinkler systems and cooling jackets must always be indirectly connected to the sanitary drainage system. This avoids the possibility of a cross connection which would contaminate the potable water supply system. A positive separation by indirect means between the waste outlet and the drainage inlet of hospital equipment and food storage and preparation establishments is also required to avoid contamination.

This method of piping is unique and very practical for fixtures with low discharge rates.

Nominal pipe size (inches)	Maximum roof area (square feet)				
	Building storm sewers and drains				
	1/8" per ft. slope	1/4" per ft. slope	1/2" per ft. slope	Gutters	Leaders
1½	127	190	222	--	222
2	270	380	460	--	460
2½	413	610	700	--	700
3	747	1,080	1,270	635	1,270
4	1,560	2,210	3,080	1,540	3,080
5	2,810	4,000	5,620	2,810	5,620
6	4,450	6,290	8,880	4,440	8,880
8	9,460	13,760	18,950	9,975	18,950

Size of storm water drain, leaders, and gutters
Table 3-6

Some codes do not require specified lengths for indirect waste piping, while others impose a maximum length of 15 feet. Check local code for specific requirements.

Storm Drainage System

The storm drainage system receives and conveys rain water to a point where discharge can legally occur. Storm drainage includes roof drains, area drains, catch basins, gutters, leaders, building storm drains, building storm sewers and ground surface storm sewers.

Rain water not properly collected and disposed of can become a nuisance and a health hazard. Storm water from paved parking lots and large residential and commercial buildings can collect into stagnant pools of water. This can produce offensive odors and is an ideal breeding ground for mosquitoes.

Collection and disposal of storm water is now considered one of the primary functions of the plumbing drainage systems. Many older cities located near large lakes, rivers or the ocean for many years used a combined sewer system to convey both storm water and sewage. The pipes of these combined systems were generally installed a considerable distance into large bodies of water through *outfall* lines. In most cases the waste products were released only partially treated or totally untreated. Builders usually must have special permission from local authorities for new construction which will discharge into these dual systems. These plants must meet the standard of 90% treated effluent by the early 1980's under regulations established by the Environmental Protection Agency. Separate sewage and storm water collection systems are now generally required by local codes.

Sizing Storm Water Drainage Systems

The sizing of all storm water drainage pipes and the method for disposal are usually determined by mechanical engineers. The calculation is based on the square footage of impervious areas (roofs, parking lots, etc.) being served. A second important determining factor is the maximum anticipated rainfall rate in inches in any one hour.

Although professional engineers are usually responsible for sizing storm drainage and disposal systems, you should know what principles are involved for several reasons. First, all phases of this work must be done by plumbers. Second, you will have to interpret the storm drainage tables and understand the regulations to pass the journeyman and master plumber's examination.

Differences in code requirements in this area are due to the great variation in maximum anticipated rainfall in any one hour. Rainfall varies from as little as two inches in some areas to more than eight inches in other areas. It follows, then, that the less rainfall, the smaller the pipe required; the greater the rainfall, the larger the pipe required.

Table 3-6 is intended for sizing building drains and leaders for roof drainage in a region that anticipates a maximum of five inches of rainfall per hour. You will note some differences in pipe sizes in the table in your particular code because your code reflects the rainfall expectation in your area.

There may be times when you will need to calculate storm water drains, leaders, or gutter size. The following formula will provide you with a sizing method for calculating storm water drainage systems *(when rainfall is a "given")* for any geographic location, whether above or below maximum rate of 4 inches per hour:

Square feet of area to be drained ÷ 4 x maximum rainfall per hour

For example, assume a roof area of 7,850 square feet in Birmingham, Alabama, a city listed in the table at the end of this chapter. The roof will have one rain leader and we will use Table 3-6 for the sizing. Birmingham, as you will note, has a maximum rainfall rate of 7 inches per hour. Following our suggested formula, this example would be worked as follows:

$$\frac{7850 \text{ ft}^2}{4} \times 7\text{'' rainfall/hr.} = 13.738 \text{ ft}^2 \text{ water to be drained}$$

Using Column 6 of Table 3-6, the rain leader would have to be 8 inches.

Generally speaking, tables in code books for sizing storm water drains, leaders, and gutters are already calculated for the particular geographic area involved. Upon establishing the square feet to be drained, as illustrated in Figure 3-7, the rain leaders, building drain and building storm sewer can be sized from Table 3-6. (Use the table in the local code for actual situations in your geographic location.)

You can also use Table 3-6 for sizing various storm water pipes.

The first column of Table 3-6 shows the pipe size of the horizontal building storm drain or the vertical leader pipe to the roof. The second, third, and fourth columns give the maximum square feet of roof area permitted, according to slope, for the listed pipe sizes in the first column. The sixth column gives the maximum square feet of roof area permitted for the leader size listed in column one. The fifth column is for gutters and downspouts which are installed by sheet metal workers and are inspected by the building inspector.

As an example, consider a roof with 900 square feet. This would be a building 20 feet wide by 45 feet long. Let's assume one leader is required and 1/8 inch slope per foot is the maximum pitch permitted. We now have the basic facts necessary to size the building storm drain and the leader pipe. By checking column two you can see that 900 exceeds 747, the maximum square footage of roof area allowed on a 3 inch horizontal storm drain. Yet it does not exceed the 1,560 square feet allowed on a 4 inch horizontal storm drain. Thus, the horizontal storm drain for this building would be 4 inches.

For sizing the leader we use column six. Again, we can see that 900 exceeds 700, the maximum square feet of roof area allowed on a 2½ inch leader. Nine hundred square feet does not exceed the 1,270 square feet permitted on a 3 inch vertical leader pipe. Therefore, the vertical leader pipe would be 3 inches.

Figure 3-7 shows a vertical wall that sheds rain water on a main building roof. The vertical wall section must be considered in sizing the rain leaders and the horizontal storm water pipes. Many building codes require that ⅓ of the total area of the vertical wall be added to the area of the main roof. Other codes require that more or less of the vertical wall section be included.

Refer to the illustration in the upper portion of Figure 3-7. Rain water must be disposed of from a roof area of 5,300 square feet. Since the roof is equally divided and has two roof drains and two leaders, each leader must be sized to carry the rain water from an area of 2,650 square feet (half of 5,300 square feet). In Table 3-6, sixth column, you see that the next highest square foot area is 3,080. The first column shows that a 4 inch leader would be required to serve each roof drain.

The horizontal storm water pipes to serve these leaders at 1/8 inch slope per foot are sized from column two. Since the end section will convey rain water from a single leader of 2,650 square feet, the figure 2,810 square feet (column two) and the corresponding 5 inch diameter in column one must be used. Therefore this section of horizontal pipe would be 5 inches.

The second section of horizontal storm water pipes will convey water from the complete roof

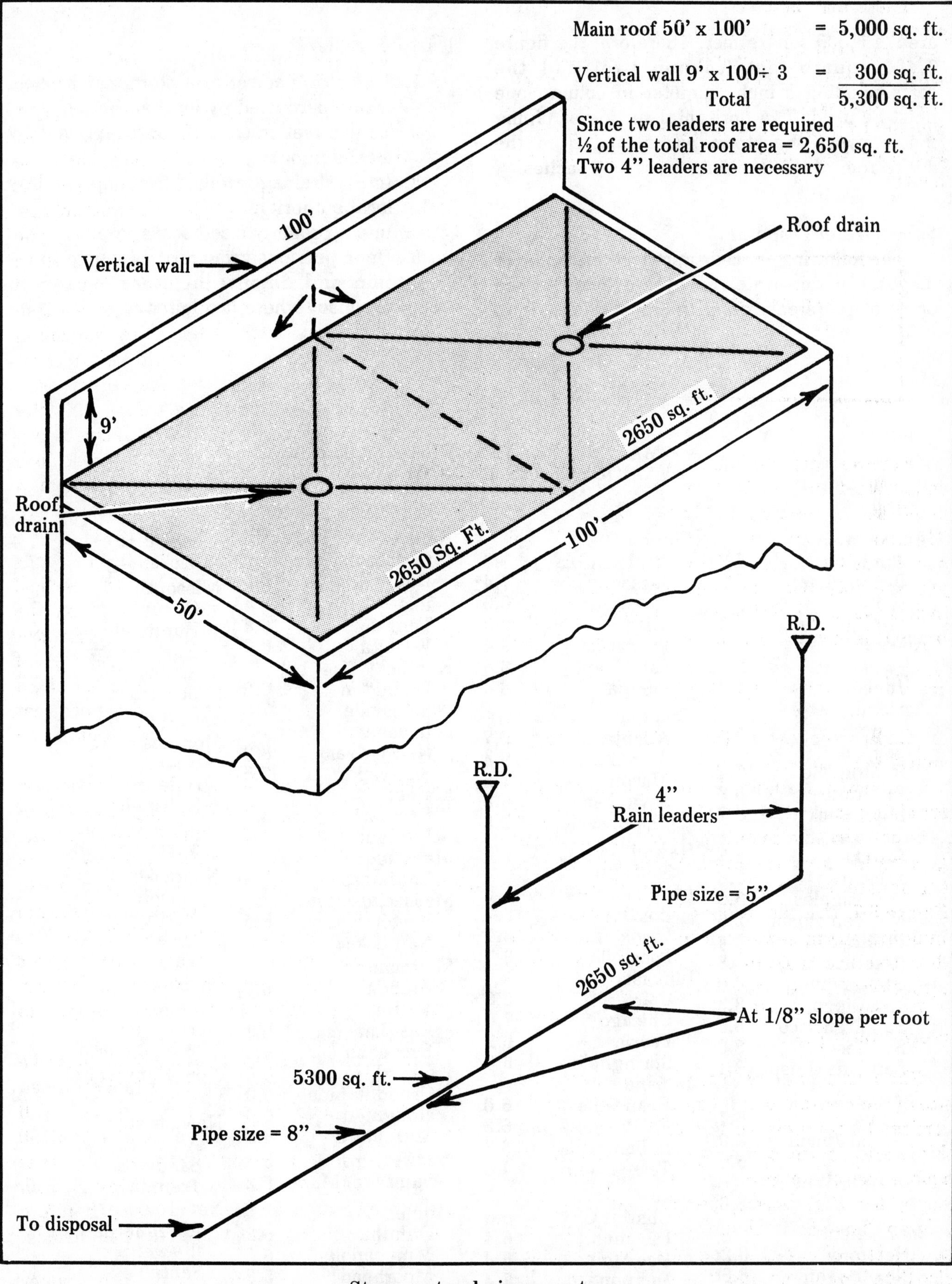

Storm water drainage system
Figure 3-7

area of 5,300 square feet. Therefore, the figure 9,460 square feet (column two) and the corresponding 8 inch diameter in column one must be used. The horizontal section of piping that conveys the storm water load from the entire roof of this building must be 8 inches.

Storm Water Disposal

The following are code approved methods for disposal of rain water. They are listed in the order of preference from the most desirable to least desirable.

1. To a storm sewer or a storm catch basin where permitted by local authority.
2. To a street gutter, if first approved by local authority.
3. Into a drainage well, if first approved by local authority.
4. Into a properly sized soakage pit.
5. Upon pervious ground, if adequate to retain and dispose of storm water on premises where it originates.

MAXIMUM RATES OF RAINFALL FOR VARIOUS CITIES, IN INCHES PER HOUR

Alabama:	
Anniston	7.2
Birmingham	7.0
Mobile	8.4
Montgomery	7.0
Alaska:	
Fairbanks	3.7
Juneau	1.7
Arizona:	
Phoenix	4.3
Arkansas:	
Bentonville	7.4
Ft. Smith	6.2
Little Rock	6.7
California:	
Eureka	2.7
Fresno	3.6
Los Angeles	3.6
Mt. Tamalpais	2.5
Pt. Reyes	2.4
Red Bluff	3.8
Sacramento	3.0
San Diego	3.3
San Francisco	3.0
San Jose	2.0
San Luis Obispo	3.1
Colorado:	
Denver	5.7
Grand Junction	3.0
Pueblo	5.0
Wagon Wheel Gap	3.6
Connecticut:	
Hartford	6.2
New Haven	6.6
District of Columbia:	
Washington	7.2
Florida:	
Apalachicola	7.3
Jacksonville	7.4
Key West	6.6
Miami	7.5
Pensacola	9.4
Sand Key	6.6
Tampa	8.4
Georgia:	
Atlanta	7.7
Augusta	8.4
Macon	7.2
Savannah	6.8
Thomasville	7.3
Hawaii:	
Honolulu	5.2
Idaho:	
Boise	2.7
Lewiston	3.1
Pocatello	3.7
Illinois:	
Cairo	6.6
Chicago	7.0
Peoria	6.2
Springfield	6.6
Indiana:	
Evansville	6.0
Ft. Wayne	6.3
Indianapolis	6.3
Terre Haute	7.5
Iowa:	
Charles City	6.5
Davenport	6.4
Des Moines	6.4
Dubuque	7.4
Keokuk	6.8
Sioux City	7.0
Kansas:	
Concordia	7.5
Dodge City	6.3
Iola	8.4
Topeka	6.8
Wichita	6.9
Kentucky:	
Lexington	6.0
Louisville	7.0
Louisiana:	
New Orleans	8.2
Shreveport	7.5
Maine:	
Eastport	4.7
Portland	4.7
Maryland:	
Baltimore	7.8
Massachusetts:	
Boston	5.5
Nantucket	4.8
Michigan:	
Alpena	6.1
Detroit	6.4
East Lansing	6.1
Escanaba	5.4
Grand Haven	5.0
Grand Rapids	6.0
Houghton	5.0
Marquette	6.0
Port Huron	5.3
Sault Ste. Marie	5.2
Minnesota:	
Duluth	6.2
Minneapolis	6.6
Moorhead	5.8
St. Paul	6.3
Mississippi:	
Meridian	7.4
Vicksburg	7.5
Missouri:	
Columbia	7.0
Hannibal	6.5
Kansas City	6.9
St. Joseph	6.5
St. Louis	6.5
Springfield	7.0
Montana:	
Havre	4.3
Helena	3.8
Kalispell	3.3
Miles City	7.0
Missoula	2.7
Nebraska:	
Lincoln	6.6
North Platte	6.0
Omaha	7.0
Valentine	6.3
Nevada:	
Reno	3.2
Tonopah	3.0
Winnemucca	2.7
New Hampshire:	
Concord	6.2
New Jersey:	
Atlantic City	6.2
Sandy Hook	7.0
Trenton	6.4
New Mexico:	
Albuquerque	3.7
Roswell	5.4
Santa Fe	4.4
New York:	
Albany	6.0

Binghamton	5.0
Buffalo	5.5
Canton	5.6
Ithaca	6.0
New York	6.6
Oswego	5.9
Rochester	5.4
Syracuse	6.3
North Carolina:	
Asheville	6.7
Charlotte	7.0
Greensboro	6.6
Hatteras	6.8
Raleigh	7.5
Wilmington	7.0
North Dakota:	
Bismarck	6.7
Devils Lake	6.8
Williston	6.5
Ohio:	
Cincinnati	6.5
Cleveland	6.9
Columbus	6.1
Dayton	6.0
Sandusky	6.2
Toledo	6.0
Oklahoma:	
Oklahoma City	6.7
Oregon:	
Baker	3.3
Portland	3.0
Roseburg	3.6
Pennsylvania:	
Erie	6.5
Harrisburg	7.0
Philadelphia	6.5
Pittsburgh	6.4
Reading	6.5
Scranton	6.1
Puerto Rico:	
San Juan	5.7
Rhode Island:	
Block Island	5.3
Providence	4.8
South Carolina:	
Charleston	7.0
Columbia	6.6
Greenville	6.6
South Dakota:	
Huron	6.2
Pierre	6.5
Rapid City	5.5
Yankton	5.8
Tennessee:	
Chattanooga	7.2
Knoxville	6.2
Memphis	6.8
Nashville	7.2
Texas:	
Abilene	7.2
Amarillo	6.8
Austin	7.4
Brownsville	7.5
Corpus Christi	6.6
Dallas	7.2
Del Rio	7.6
El Paso	4.2
Fort Worth	6.6
Galveston	8.2
Houston	8.0
Palestine	6.6
Port Arthur	7.5
San Antonio	7.5
Taylor	7.7
Utah:	
Modena	3.8
Salt Lake City	3.4
Vermont:	
Burlington	5.4
Northfield	6.2
Virginia:	
Cape Henry	7.4
Lynchburg	6.0
Norfolk	6.8
Richmond	7.2
Wytheville	5.7
Washington:	
North Head	2.8
Port Angeles	2.2
Seattle	2.2
Spokane	3.1
Tacoma	2.8
Tatoosh Island	3.2
Walla Walla	2.7
Yakima	2.6
West Virginia:	
Elkins	6.2
Parkersburg	6.7
Wisconsin:	
Green Bay	5.1
LaCrosse	6.5
LaCrosse	6.5
Madison	6.0
Milwaukee	6.2
Wyoming:	
Cheyenne	5.6
Lander	3.7
Sheridan	5.2
Yellowstone Park	2.5

Rates given are intensities for a 5 minute duration and a 10 year return period, from Technical Paper Number 25, Rainfall Intensity-Duration-Frequency Curves, U. S. Dept. of Commerce, Weather Bureau.
Taken directly from *National Standard Plumbing Code,* 1973. Published by National Association of Plumbing-Heating-Cooling Contractors. Co-sponsored by National Association of Plumbing-Heating-Cooling Contractors, 1016 20th Street, N.W., Washington, D.C. 20036 and American Society of Plumbing Engineers, 16161 Ventura Blvd., Suite 105, Encino, California 91316.

Drain, Vent, And Waste Systems

The plumbing code is written to protect the public health, welfare, and safety by the proper design, installation and maintenance of plumbing systems. While details of plumbing installations vary, the basic principles of sanitation and safety remain the same. This chapter illustrates installation requirements of most drain, vent and waste systems. The principles outlined here will not cover every situation but should make the intent of the code clear and understandable. These are the basics that every plumber must know. If you see one of these principles violated, you can bet that the plumbing inspector will not pass the work that has been done.

Plumbing fixtures are the end of the potable water supply system and the beginning of the sewage system. Therefore the drainage systems should be designed to prevent fouling or depositing of solids along the walls of drainage pipes. The drainage system must be properly vented to provide a free circulation of air. This prevents the possibility of siphonage or forcing of trap seals. However, stoppages occur in the best designed system. Therefore, adequate cleanouts must be provided so that all portions of the drainage system are accessible by cleaning equipment. Changes in direction in drainage should be made with 45 degree wyes, long or short sweep quarter bends, sixth, eighth, or sixteenth bends, or by a combination of these or other approved fittings. Single and double sanitary tees, quarter bends, and one-fifth bends may be used in vertical sections of drainage lines only where the direction of flow is from the *horizontal* to the *vertical.*

Neither a tee branch nor a fitting having a hub in the direction opposite to flow should be used as a drainage fitting. No running threads, bands or saddles should be used in the drainage system. No drainage or vent piping should be drilled or tapped.

Plumbing drainage pipes must be installed in open trenches and must remain open until the piping has been inspected, tested, and accepted by the plumbing inspector.

All the piping of a drainage system should be water tested and submitted to a test of not less than a five-foot head. The plumbing inspector may require the removal of cleanout plugs or caps to ascertain whether the pressure has reached all parts of the system.

Permitted Usage

All materials for drainage, vent, and special waste systems have been listed in Table 4-1. Note that certain materials cannot be used in some situations and that the use of other materials is limited or restricted. Actual conditions determine the final choice: building type and location, the type of fill material, traffic expected over piping and the waste conveyed.

Location Or Use	Clay pipe	Concrete pipe	Bituminous fiber pipe	Asbestos cement pipe	Cast iron soil pipe XH	Cast iron soil pipe SV	Cast iron hubless soil pipe	Cast iron threaded pipe	Wrought iron pipe	Galvanized pipe	Red brass or copper pipe	DWV copper	Type L copper	Type K copper	Type M copper	Stainless steel tubing	ABS plastic pipe	PVC plastic pipe	Lead pipe	PE plastic pipe
Building house sewer	X		X		X	X	X										X	X		
Underground drain, waste and vent in building					X	X	X				X		X	X	X		X	X	X	
Above-ground drain, waste and vent in building					X	X	X	X	X	X	X	X	X	X	X	X	X	X	X	
Storm and subsoil drain	X	X	X	X	X	X	X		X	X	X		X	X			X	X		X
Rainwater leaders					X	X	X	X	X	X	X	X	X	X	X	X	X	X	X	
Septic tank tight lines	X	X	X	X	X	X	X										X	X		
Chemical and acid drain, waste and vent																	A	A		A

Remarks:
A - Each job must be approved by plumbing official.
X - Approved for location or use as shown.

Schedule of drainage, waste, and vent materials
Table 4-1

Standards and specifications for plumbing materials are subject to change. Check with your local authority about code changes that occur from time to time. Always buy the updated code supplement sheets and insert them in your code book.

Materials For Building Sewers

Cast-iron The building sewer may be of cast-iron tar-coated soil pipe of no less weight than that which is used underground within the building. In other words, if extra-heavy soil pipe is required underground within the building and you want to install the building sewer in cast-iron pipe, it too must be extra-heavy. The cast-iron pipe joints may be lead and oakum, compression or no-hub joint type.

Vitrified clay pipe The building sewer may be of vitrified clay pipe and fittings with compression type joints.

Plastic pipe The building sewer may be plastic pipe and fittings with compression or cemented joints.

Bituminized fiber pipe The building sewer may be bituminized fiber pipe and fittings with tapered friction joints.

Underground Drainage Piping Within a Building

All underground piping for a drainage system within a building *not more than three stories* high must be either cast-iron soil pipe, lead pipe, brass pipe, copper pipe (type K, L or M), or plastic pipe. All underground piping for a drainage system within a building *exceeding three stories* in height must be centrifugally spun service weight or extra-heavy cast-iron pipe, lead pipe, brass pipe, copper pipe (type K, L or M), or plastic pipe. See Table 4-1.

There are special requirements where buildings are constructed in areas originally below high tide or where fill with hydrogen sulphide gas is used (known as deleterious fill). In this case all underground piping and fittings for the drainage and vent system within a building must be centrifugally spun service weight cast-iron, extra heavy cast-iron or plastic pipe. This type of material must be continued to the point of disposal at a public sewer or septic tank.

Fittings must conform to the material and type of piping used in the drainage system. Fittings on screwed pipe, plastic pipe or copper pipe should be of the recessed drainage type.

Subsoil drains, when required under a cellar or basement floor or around the outer walls of a building, must be at least four inches in diameter. Only open-jointed or horizontally split pipe, perforated clay pipe, perforated bituminized fiber pipe, or asbestos cement pipe can be used.

Above Ground Drainage and Vent Pipes Within a Building

Piping for a drainage system must be of centrifugally-spun service weight cast iron, galvanized, wrought iron, galvanized steel, lead, brass, copper pipe (type DWV, K, L, or M) or plastic pipe.

Piping for vents must be cast-iron, galvanized steel, lead, brass, copper pipe, or plastic pipe. All fittings should conform to the materials and type of piping used in the drainage and vent system. When galvanized, copper, or plastic fittings are used within a drainage system above ground, these fittings should also be of the recessed drainage type.

It is important that you be aware of the four major restrictions placed upon the drainage, waste and vent piping:

(1) Plastic pipe or fittings can not support the weight of any plumbing fixture.
(2) Different types of plastic materials (ABS or PVC) cannot be mixed or combined in any plumbing system.
(3) To extend, relocate, or add to any existing soil, waste or vent pipe, material of like grade and quality must be used.
(4) Materials for drainage or vent systems receiving chemicals, acid, or other corrosive waste must be ABS, PVC, PE plastics or other materials that are approved in your code.

Indirect Waste Piping

Indirect waste piping, when installed below or above a floor, must be of cast-iron, galvanized steel, lead, brass, copper, or plastic. Fittings must be the recessed drainage type. When used for acids and chemicals, the pipe must be of a material that will be unaffected by such waste. In any event the waste must be properly neutralized or diluted before being discharged into any building drainage system.

Storm Drainage

Building storm sewers must be of cast-iron, vitrified clay, bituminized fiber, asbestos cement, concrete or plastic pipe. The fittings must be of a similar grade and quality.

Storm drains and leaders within a building must be of cast-iron, lead, copper, brass, galvanized wrought iron, galvanized steel or plastic pipe with fittings of a similar grade and quality.

Standards For Plumbing Materials

All plumbing materials used in the construction, installation, alteration, or repair of any plumbing or drainage system should comply with the standards in Table 4-2.

Installation Methods

In recent years new materials have been developed, approved, and accepted by local authorities for use in drainage, waste, and vent systems. The new concepts in pipe and fitting materials have the advantages of performance, versatility, low-cost installation, and product availability. Because of the fragility of the substances from which some of these pipes and fittings are manufactured, certain installation methods must be carefully adhered to.

All piping should be securely supported within trenches and above ground within buildings to prevent sagging, misalignment and breaking. This will be discussed in more detail later.

Building Sewer

Cast-iron pipe There are three types of cast-iron pipe and fittings: (1) The lead and oakum joint, (2) The compression joint (rubber gasket), (3) The no-hub joint (stainless steel shield and retaining clamps) Two grades commonly used today are centrifugally spun service weight and extra-heavy cast-iron.

Certain characteristics of cast-iron soil pipe make it superior as a building material for sewers. Its strength, durability and resistance to trench loads are generally considered superior.

When installing cast-iron soil pipe, be sure to keep the pipe barrel in firm contact with solid ground. This means that you must excavate for the hub (bell). The weight is thus evenly distributed along the full length of the pipe. (See Figure 4-3.)

Because of cast-iron's high resistance to trench loads the depth is not too important. However, don't backfill with large boulders, rocks, cinder-fill or other materials which would

Description	ANSI	ASTM	Federal Spec	Other Standards
Ferrous Pipe, Fittings and Valves				
Cast iron drainage fittings, threaded	B16.12-1977	None	WW-P-491b-1967	None
Cast iron screwed fittings (threaded)	B16.4-1977	None	WW-P-501d-1967	None
Cast iron pipe (threaded) D.W.V.	A40.5-1943	None	WW-P-356a-1967	None
Cast iron pipe, thickness design of	A21.1-1967(R1977)	None	WW-P-421c-1967	AWWA C101-67R77
Cast iron soil pipe and fittings hub and spigot	A112.5.1-1973	A74-75	WW-P-401e-1974	None
Cast iron water pipe (2")	A21.12-1971	A377-79	WW-P-360b-1968	AWWA C112-71
Cast iron water pipe (cast in metal molds)	A21.6-1975	None	WW-P-421c-1967	AWWA C106-75
Cast iron water pipe fittings	C110-77	A377-79	None	AWWA C110-77
Ductile-iron pipe	A21.51-1976	A377-79	WW-P-421c-1967	AWWA C151-76
Hubless cast iron sanitary system with no-hub system fittings	None	None	WW-P-401e-1974	CISPI 301-78
Malleable iron screwed fittings, 150 lbs. and 300 lbs.	B16.3-1977	None	WW-P-521f-1968	None
Nipples, pipe, threaded	None	None	WW-N-351b(1)-1970	None
Pipe fittings, ferrous (bushings, plugs, and locknuts) threaded, 125 and 150 lbs.	B16.14-1977	None	WW-P-471b-1970	None
Pipe threads (except dry seal)	B2.1-1968	None	None	None
Steel pipe, stainless	B125.16-1975	A312-79a	None	None
Steel, stainless, water-DWV tubes	None	A651-79	None	None
Steel pipe, welded or seamless (for coiling) black or galvanized	B125.1-1976	A53-78	WW-P-471b-1970 Int Amend 3-1971	None
Steel pipe, welded or seamless black or galvanized	B125.2-1976	A120-79	WW-P-404c(1)1962	None
Steel pipe (cement mortar lining and reinforced cement mortar coating)	A21.4-1974	None	SS-P-305a(1)1968	AWWA C104-71
Steel pipe, wrought, welded & seamless	B36.10-79	None	None	None
Steel pipe (coat tar enamel or cement mortar lining and coal tar enamel coated and wrapped)	None	None	WW-P-1432-1970	AWWA C203-78
Unions, pipe, steel or malleable iron	B16.39-1977	None	WW-U-531c-1965	None
Valves, ball	None	None	WW-V-35a-1965	None
Valves, cast iron, gate 125 & 250 lb.	None	None	WW-V-58b-1971	None
Valves, cast iron, swing check	None	None	None	AWWA C508-76
Wrought iron pipe, welded, black or galvanized	B36.2-1969	A72-68	None	None
Non-Ferrous Metallic Pipe, Fittings and Valves				
Brass tube, seamless	H36.1-1973	B135-74	WW-T-791a-1971	None
Brass, red, seamless, pipe, standard sizes	H27.1-1973	B43-79	WW-P-351a-1963	None
Bronze flanges and flanged fittings	B16.24-1979	None	None	None
Cast bronze fittings for flared copper tubes	B16.26-1975	None	None	None
Cast bronze solder-joint pressure fittings	B16.18-1978	None	WW-T-725-1967	None
Cast bronze solder-joint drainage fittings	B16.23-1976	None	None	None
Copper pipe, seamless, standard sizes	H26.1-1976	B42-78	WW-P-377d-1962	None
Copper pipe, threadless	H26.2-1976	B302-76	WW-P-377d-1962	None
Copper tube, drainage DWV	H23.6-1976	B306-78	None	None
Copper tube, seamless	H23.3-1976	B75-79	WW-T-797c-1963	None
Copper tube, water, seamless, types K, L, and M	H23.1-1976	B-88-78	WW-T-799d-1971	None
Copper and copper alloy pipe and tube, general requirements	H23.4-1976	B251-76	None	None
Copper and copper alloy tube (welded)	None	B543-78a	None	None
Copper alloy water tube (welded)	None	B586-75	None	None
Copper tube, welded	None	B447-77	None	None
Copper tube, seamless and welded, distribution, type "D"	None	B641-78	None	None
Copper and copper alloy tube (welded)	None	B507-76	None	None
Lead pipe, bonds and traps	None	None	WW-P-325a-1967	None
Pipe fittings, brass or bronze, 125 and 250 lbs. cast or wrought	B16.15-1978	None	WW-P-460b-1967	None
Solder-joint fittings, pressure, copper alloy	B16.22-1973	None	WW-T-00725-1967	None
Solder-joint fittings, drainage, copper alloy	B16.29-1973	None	None	None
Unions, brass or bronze, 250 lbs.	None	None	WW-U-516a-1967	None
Valve, ball, copper alloy, iron or steel	None	None	WW-V-35a-1965	None
Valves, gate, bronze, threaded and flanged	None	None	WW-V-54c-1966 Int Amend 1-1970	None

Materials for plumbing installations
Table 4-2

Description	ANSI	ASTM	Federal Specs	Other Standards
Valves, angle, check and globe, bronze, 125 lb. screwed, flanged or solder	None	None	WW-V-51d-1967	None
Non-Metallic Pipe and Fittings				
Asbestos cement non-pressure sewer pipe	A165.3-1973	C428-78	SS-P-331d-1972	None
Asbestos cement non-pressure small diameter sewer pipe (4", 5", 6")	A165.4-1974	C644-78	None	None
Asbestos cement perforated underdrain pipe	A165.2-1973	C508-78a	SS-P-340b-1969	None
Asbestos cement pressure pipe	None	C296-78	SS-P-351c-1974	AWWA C400-77
Asbestos cement storm drain pipe	A165.7-1974	C663-78	None	None
Bituminized fiber drain and sewer pipe, homogeneous wall	A176.1-1971	D1861-77	SS-P-1540a-1969 Int Amend 1-1970	None
Bituminized fiber pipe, perforated, for septic tank disposal fields, homogeneous wall	A176.4-1971	D2312-77	SS-P-1540a-1969 Int Amend 1-1970	None
Clay drain tile, specification for	A6.1-1963(R1972)	C4-62(1975)	SS-P-1299a-1968	None
Clay drain tile, perforated	None	C498-75	SS-P-359b-1960	None
Clay pipe, standard and extra-strength	A106.8-1978	C700-78a	SS-P-361E	None
Clay pipe, perforated, standard and extra-strength	A106.8-1978	C700-78a	SS-P-361E	None
Concrete drain tile	None	C412-78	None	None
Concrete low head pressure pipe reinforced	None	C361-78	None	None
Concrete pipe perforated	None	C444-77	None	None
Concrete pipe (sewer, storm drain and culvert) non-reinforced	None	C14-78	SS-P-371e-1968	None
Concrete pipe, pressure, reinforced concrete, pre-tensioned reinforcement (steel cylinder type)	None	None	SS-P-381A(2)-1972	AWWA C30378
Concrete pipe (culvert, storm drain and sewer) reinforced	None	C76-79	SS-P-375d-1970	None
Acrylonitrile-butadiene-styrene (ABS) plastic pipe, schedule 40 and 80	B72.5-1971	D1527-77	None	None
Acrylonitrile-butadiene-styrene (ABS) plastic pipe, (SDR-PR and class T)	B72.3-1975	D2282-77	None	NSF 14
Socket-type acrylonitrile-butadiene-styrene (ABS) plastic pipe, fittings, schedule 40	K65.164-1971	D2468-76	None	NSF 14
Socket-type acrylonitrile-butadiene-styrene (ABS) plastic pipe, fittings, schedule 80	K65.163-1971	D2469-76	None	NSF 14
Threaded acrylonitrile-butadiene-styrene (ABS) plastic pipe fittings, schedule 80	K65.165-1971	D2465-73	None	NSF 14
Acrylonitrile-butadiene-styrene (ABS) plastic drain, waste, and vent pipe & fittings, schedule 40	B72.18-1971	D2661-78	L-P-332D-1973	NSF 14
Solvent cement for acrylonitrile-butadiene-styrene (ABS) plastic pipe and fittings	B72.23-1971	D2235-76A	None	NSF 14
Polyethylene (PE) plastic pipe schedule 40	B72.8-1971	D2104-74	None	NSF 14
Polyethylene (PE) plastic pipe schedule 40 and 80 based on outside diameter	B72.13-1971	D2447-74	None	NSF 14
Polyethylene (PE) plastic pipe, (SDR-PR)	B72.1-1975	D2239-74	L-P 315c-2-1975	NSF 14
Butt fusion polyethylene (PE) plastic pipe fittings, schedule 40	K65.160-1971	D3261-78	None	NSF 14
Butt fusion polyethylene (PE) plastic pipe fittings, schedule 80	K65.159-1971	D3261-78	None	NSF 14
Plastic insert fittings for polyethylene plastic pipe	None	D2609-74	L-F-001546-1968	NSF 14
Polybutylene (PB) plastic hot and cold water distribution systems	None	D3309-79	None	NSF 14
Type PSP PVC sewer pipe & fittings	None	D3033-78	None	NSF 14
Polyethylene (PE) plastic tubing	None	D2737-74	None	NSF 14
Chlorinated polyvinyl chloride (CPVC) plastic hot-water distribution systems	None	D2846-79	None	NSF 14
Chlorinated polyvinyl chloride (CPVC) solvent cement	None	P-493	None	None
Polyvinyl chloride (PVC) plastic pipe, schedule 40, 80 and 120	B72.7-1971	D1785-76	L-P1036A-1974	NSF 14
Socket-type polyvinyl chloride (PV) plastic pipe fittings, schedule 40	None	D2466-78	None	NSF 14

Materials for plumbing installations (continued)
Table 4-2

Description	ANSI	ASTM	Federal Specs	Other Standards
Socket-type polyvinyl chloride (PVC) plastic pipe fittings, schedule 80	None	D2467-76a	None	NSF 14
Solvent cement for polyvinyl chloride (PVC) plastic pipe and fittings	B72.16-1971	D2564-78a	None	NSF 14
Threaded polyvinyl chloride (PVC) plastic pipe fittings, schedule 80	K65-166-1971	D2464-76	None	NSF 14
Bell-end polyvinyl chloride (PVC) pipe	B72.20-1971	D2672-78	None	NSF 14
Polyvinyl chloride (PVC) plastic drain, waste, and vent pipe and fittings	K65.56-1971	D2665-78	L-P 320a-1966	NSF 14
Type PSM PVC sewer pipe and fittings	None	D3034-78	None	None
Styrene-rubber plastic drain pipe and fittings	None	D2852-77	None	None
Solvent cement for styrene rubber plastic pipe, and fittings	None	D3122-78	None	None
Poly (vinyl-chloride) PVC plastic pipe (SDR-PR)	None	D2241-78	None	NSF 14
Thermoplastic accessible and replacement plastic tube and tubular fittings	None	F409-77	None	NSF 14
Polybutylene (PB) plastic pipe and tubing for cold water service pipe (SDR)	None	D2662-78	None	NSF 14
Tubing	None	D2666-75	None	NSF 14
Pipe (SDR)-00)	None	D3000-73	None	NSF 14
Recommended practice for making solvent-cemented joints with polyvinyl chloride (PVC) plastic pipe and fittings	None	D2855-78	None	None
Pipe Jointing Materials and Gaskets, and Supports				
Caulking, lead wool and lead pig	None	None	QQ-C-40(2)1970	None
Compression joints for vitrified clay bell and spigot pipe	A106.6-1977	C425-77	None	None
Flexible elastromeric joints	None	D3139-77 D3212-73T	None	None
Fixture setting compound	None	None	TT-P-1536(1968) Revision of HHC 536a-1954	None
Hubless stainless steel couplings	None	None	None	CISPI 301-78
Non-metallic gaskets for pipe flanges	B16.21-1978	None	None	None
Neoprene rubber gaskets for hub and spigot cast iron soil-pipe & fittings	None	C564-76	None	CISPI HSN-76
Rubber gaskets for cast iron soil-pipe & fittings	None	C564-76	None	None
Rubber gasket joints for cast iron pressure pipe and fittings	A21.11-1979	None	None	AWWA C111-79
Rubber gaskets, molded or extruded, for concrete non-pressure sewer pipe	None	C443-79	HH-G-160b-1968	None
Rubber rings for asbestos cement pipe	18.7-1971	D1869-78	None	None
Rubber gaskets, sheet	17.2-1971	D1330-70	None	None
Pipe Joining Materials and Gaskets, and Supports				
Sealing compound, preformed plastic, for expansion joints and pipe joints	None	None	SS-S-210(1965)	None
Sealing compound, sewer, bituminous, two-component, mineral-filled, cold applied	None	None	SS-S-168(2)1962	None
Pipe hangers and supports	None	None	WW-H-171d-1970	None
Plumbing Appliances and Appurtenances				
Dishwashing machines commercial	None	None	000-431c(2)1970	UL 921-1978 ASSE 1006
Dishwashing machines, household	None	None	None	UL 749-1978
Drinking water coolers, self contained mechanically refrigerated	None	None	None	ARI-1010-73 UL 399-1978
Food waste disposal units, household	None	None	QQ-G-001513-1968	UL 430-1978 ASSE 1008
Home laundry equipment	None	None	None	UL 560-1978 ASSE 1007
Water heaters, automatic storage type	z21.10.1a-1978	None	None	None

Materials for plumbing installations (continued)
Table 4-2

Description	ANSI	ASTM	Federal Specs	Other Standards
Water heaters, circulating tank	z21.10.3a-1978	None	None	None
Water heater, electric storage tank household	None	None	W-H-196j-1973	UL 174-1977 ASSE 1002-73
Water heater, instantaneous	None	None	WW-H-191b-1970	None
Water heater, side arm type	z21.10.1-1975	None	None	None
Water heater oil fired storage type	None	None	None	UL 732-1975
Plumbing Fixtures and Appurtenances				
Accessories for plumbing fixtures	None	None	WW-P-541/0a-1974	None
Bathtubs	A112.19.1H-1979	None	WW-P-541/3a-1971	None
Bathtubs units, gel-coated, glass fiber reinforced polyester resin	s124.1-1974	None	WW-P-541/3a-1971	None
Drinking fountains	A112.11.1-1973	None	WW-P-541/6a-1971	None
Fittings, plumbing fixtures, finished and rough brass	A112.18.1H-1979	None	WW-P-541 ALL-1971	None
Floor drains	A112.21.1-1968	None	None	None
Hand-held showers, performance requirements	None	None	None	ASSE 1014-79
Individual control valves, anti-scald type	None	None	None	ASSE 1016-79
Lavatories	None	None	WW-P-541/4a-1971	None
Lavatory, cultured marble	None	None	None	IAPHO PS-18-74
Plumbing fixtures, general specification	None	None	WW-P-541/GEN-1971	None
Plumbing fixtures, enameled cast iron	A112.19.1-1973	None	WW-P-541/3A+ 5A-1971	None
Plumbing fixtures, stainless steel	A112.19.3-1976	None	WW-P-541/5A-1971	None
Plumbing fixtures, vitreous china	A112.19.2-1973	None	WW-P-541/1A, 2A, 4A, 6A - 1971	None
Plumbing fixtures, enameled steel	A112.19.4-1977	None	None	None
Shower baths and heads and water control valves	None	None	WW-P-541/7B-1974	None
Shower receptors, shower enclosures, and non-metallic bathtubs prefabricated	None	None	None	IAPHO PS-11-76
Shower receptors, and shower stall units, gel-coated, glass fiber reinforced polyester rosin	Z124.2-1967	None	None	None
Sinks, kitchen and service, and laundry tub	A112.19.1-1973	None	WW-P-541/5A-1971	None
Supports for off-the-floor plumbing fixtures for public use	A112.6.1M-1979	None	None	None
Thermostatic mixing valves, self-actuated primarily for domestic use	None	None	None	ASSE 1017-79
Urinals	A112.19.2-1973	None	WW-P-541/2A-1971	None
Water closets	A112.19.2-1973	None	WW-P-541/1A-1971	None
Water closets bowls, trim for	A112.19.5-1978	None	None	None
Water hydrants	None	None	None	ARSR 1010-78
Backflow Preventers	**A40.6-1943**			
Vacuum breakers, anti-siphon	A112.1.1-1971	None	None	ASSE 1001
Vacuum breakers, hose connection	A112.1.3-1976	None	None	ASSE 1011-70
Double check with atmospheric vent	None	None	None	ASSE 1012
Reduced pressure zone device	None	None	None	ASSE 1013
Double check valve assembly	None	None	None	ASSE 1015
Vacuum breakers, pressure type	A112.1.7-1976	None	None	ASSE 1020
Diverters for plumbing faucets w/hose spray, anti-siphon type, residential applications	None	None	None	ASSE 1025-78
Miscellaneous				
Air gap standards	A112.1.2-1942 R-1979	None	None	None
Arrestors, water hammer	A112.26.1-1969 R-1975	None	None	ASSE 1010
Asbestos cement pressure pipe, installation	None	None	None	AWWA-C603-78
Enamel, coat-tar, (protective coating)	None	None	None	AWWA-C203-78 AWWA-C210-78
Clamps, hose	None	None	WW-C440B(a)1969	None
Coating pipe, epoxy, fusion bond	None	None	None	AWWA-C213-79

Materials for plumbing installations (continued)
Table 4-2

Description	ANSI	ASTM	Federal Specs	Other Standards
Coating, pipe, thermoplastic resin or thermosetting, epoxy	None	None	L-C-530B-1970	None
Copper, sheet or strip for building construction	None	B370-77	None	None
Clay pipe, installation	A106.2-1977	C12-77	None	None
Clay pipe, testing	A106.5-1978	C301-78c	None	None
Drain, roof	A112.21.2-1971	None	None	None
Interceptors, grease	None	None	None	PDI G 101
Drain, floor	A112.21.1-1968 R-1974	None	None	None
Lead, sheet, grade A	None	None	QQ-L-201f(2)1970	None
Relief valves, automatic	Z21.22-1979	None	None	None
Reducing valves, water pressure for domestic water supply system	A112.26.2-1975	None	None	ASSE 1003
Solder, soft	None	None	QQ-S-571d-1963	None
Tape pipe coating, pressure sensitive polyethylene	None	None	L-T-0075(1)-1966	None
Thermo plastic pressure piping, underground installation	None	D2774-72	None	None
Trap seal primer valves	None	None	None	ASSE 1018-7
Valve, backwater	A112.14.1-1975	None	None	None
Valve, drain, water heater	None	None	None	ASSE 1005
Water closet, flush tank, ball cocks	None	None	None	ASSE 1002-7

Footnote* Standards on materials in this table do not imply that these materials may be used for a specific service. Materials permitted for a specific service shall be specified under the various sections of your particular Code. Revised January, 1984.

Materials for plumbing installations (continued)
Table 4-2

Excavation for piping hub projection
Figure 4-3

damage or corrode the pipe.

Vitrified clay pipe, plastic pipe, and bituminized fiber pipe Because of their fragility, these types of pipe require special installation methods when used for building sewers. Here are some precautionary measures. The bottom quarter of these pipes should be continuously and uniformly supported by the trench bottom. Support the pipe with 4 inches of fine, uniform material that will pass through a ¼ inch screen. Hub and coupling projections should be excavated as shown in Figure 4-3 so that no part of the pipe load is supported by the hub or coupling. The supported material should extend 4 inches on each side of the pipe.

Backfill should be firmly compacted with selected material which will pass through a ¼ inch screen. The backfill should extend from the trench bottom to a point six inches over the top of the pipe. The minimum pipe depth below ground level must be twelve inches. See Figure 4-4.

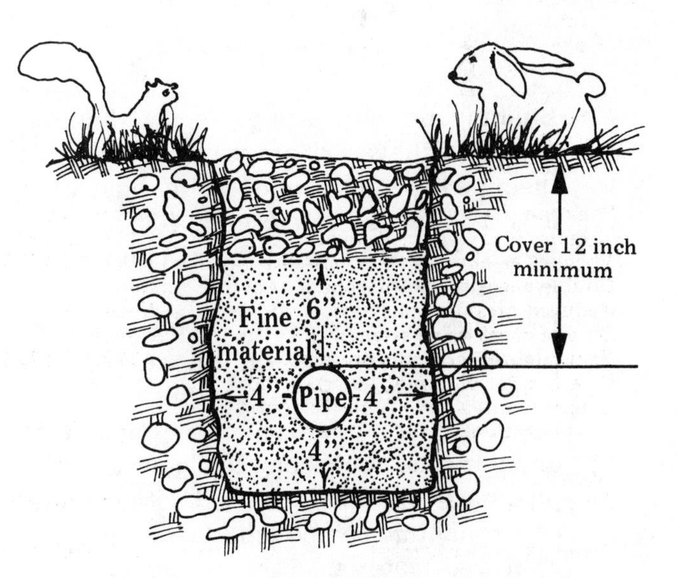

Protecting fragile piping materials
Figure 4-4

Drainage, Waste, and Vent Piping Within a Building

You have a greater variety of piping materials to choose from for underground or above ground installations within a building. However, restrictions have been placed on the use of certain pipe and fitting materials. This is especially true in regard to the building height. These restrictions will be discussed as this section progresses.

- Underground or horizontal drainage, waste and vent piping must be adequately supported. Approved hangers or masonry supports may be needed to keep the pipe in alignment and to prevent sagging.

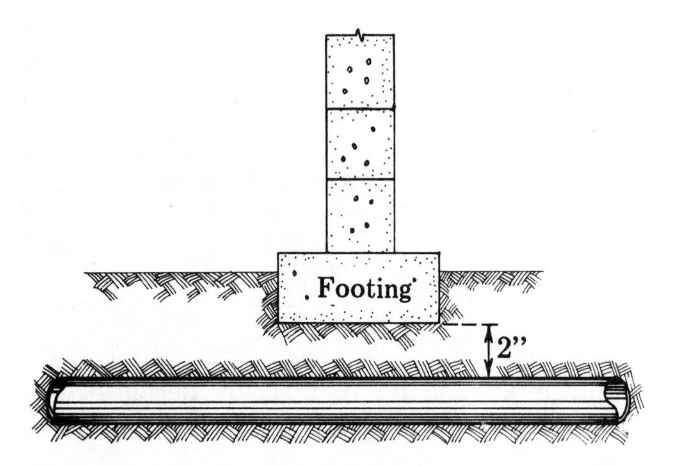

Clearance for building drains beneath foundations
Figure 4-5

- Building drains passing beneath foundations must have a clearance from the top of the pipe to the bottom of the footings of at least two inches. See Figure 4-5.
- Bases of stacks must be supported by masonry or concrete.
- Drainage pipe passing through cast-in-place concrete should be sleeved to provide ½ inch annular space around the entire circumference of the pipe. This annular space between the sleeve and the pipe must be tightly caulked with coal tar or asphaltum compound, lead, or other approved materials.
- Unless a special design is approved by the building official, no excavation for drainage piping can be placed within a 45 degree angle of pressure from the base of an existing structure to the sides of the trench. (See Figure 4-6.) In other words, a drainage pipe trench deeper than the base of the structure cannot be excavated

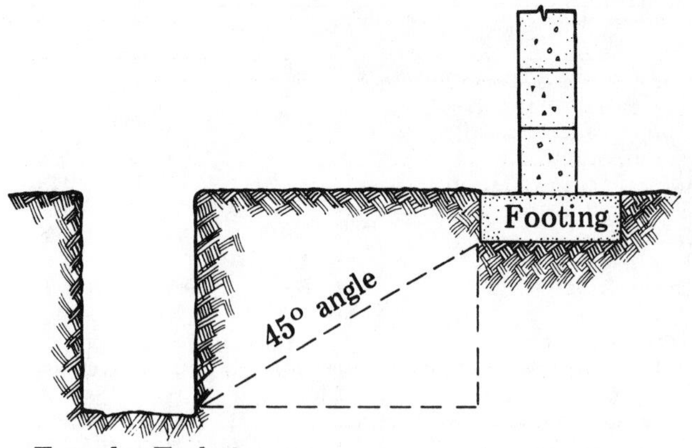

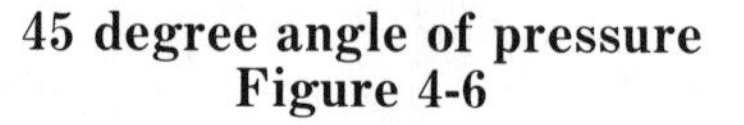
45 degree angle of pressure
Figure 4-6

parallel to the structure's foundation in such a way as to place the trench within the 45 degree angle illustrated.
- Drainage piping, when installed in contact with cinders or other corrosive materials, must be protected by sleeves, coating, wrapping or other approved methods.

There are certain limitations you should know:

(1) Cast-iron soil pipe is limited to buildings not over three stories high.
(2) Centrifugally spun service weight or extra-heavy cast-iron pipe should be used in buildings over three stories high.

Horizontal Drainage, Waste, and Vent Piping Supports

Cast-iron soil pipe with lead and oakum joints must be supported with hangers or other approved means at not more than 5 foot intervals. Pipe lengths over 5 feet may be supported at intervals of not more than 10 feet. See Figure 4-7. Supports must be placed within 18 inches of the hub or joint. Cast-iron soil pipe with hubless or compression gasket joints must have supports immediately adjacent to each joint or coupling if the developed length exceeds 4 feet.

Screwed pipe is stronger at the joint but still must be supported at approximately 12 foot intervals when used within a horizontal drainage and vent system. Horizontal copper pipe or tubing 1½ inches and smaller must be supported at approximately 6 foot intervals.

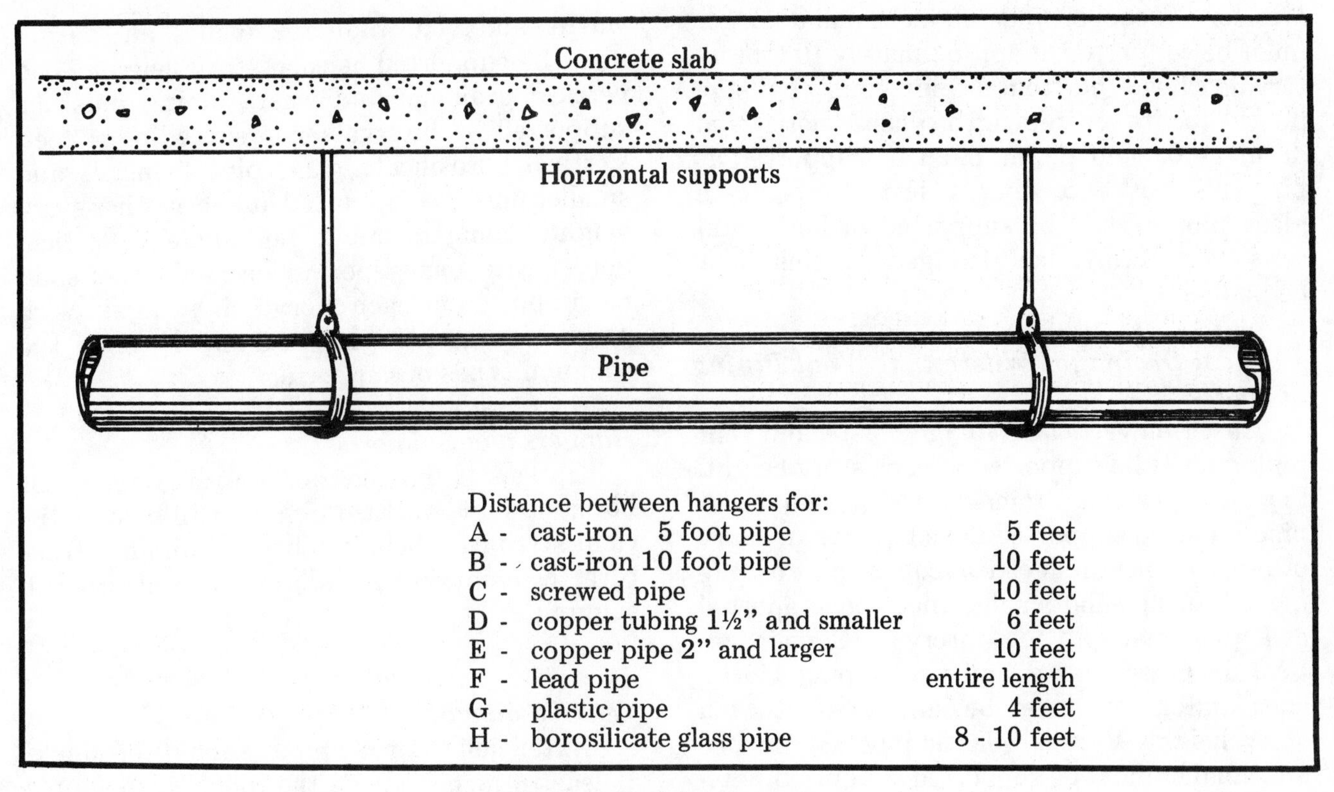

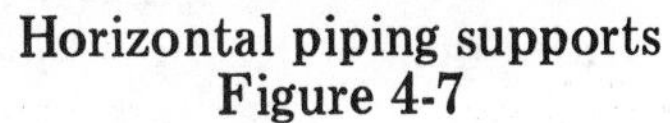
Horizontal piping supports
Figure 4-7

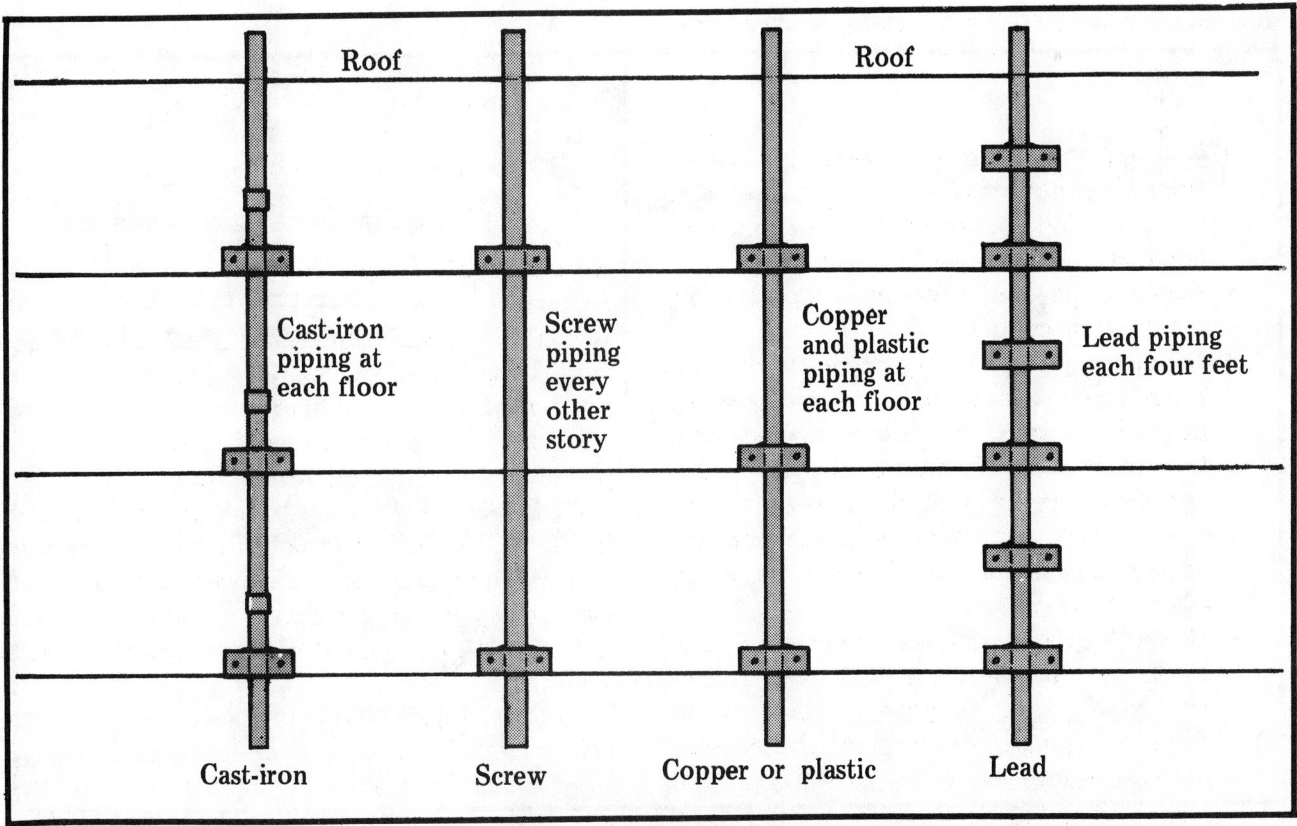

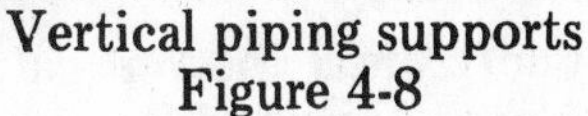
Vertical piping supports
Figure 4-8

Horizontal copper pipe 2 inches and larger must be supported at approximately 10 foot intervals. Lead horizontal pipe must be supported for its entire length. Plastic horizontal drainage or vent piping must be supported at intervals not exceeding 4 feet. Borosilicate glass piping shall be supported in horizontal runs every 8 to 10 feet. Hangers for this must be padded.

Vertical Drainage, Waste, and Vent Piping Supports

Cast-iron vertical drainage, waste, and vent piping must be supported at each story height. Screwed vertical drainage, waste, and vent pipe must be supported at not less than every other story height. Vertical copper pipe or tubing 1¼ inches and smaller must be supported at approximately 4 foot intervals. Copper vertical drainage, waste, and vent piping 1½ inches and larger must be supported at each story height. Vertical plastic pipe 1½ inches and smaller must be supported at approximately 4 foot intervals. Plastic vertical drainage, waste, and vent piping 2 inches and larger must be supported at each story height. Lead vertical drainage and waste piping must be supported at intervals not exceeding 4 feet. Vertical borosilicate glass pipe 2 inches and smaller must be supported at every other story height. Hangers must be padded. Vertical borosilicate glass pipe 3 inches and larger shall be supported at each story height, again with padded hangers. Figures 4-8 and 4-9 show the common types of supports.

Hangers and Supports

Hangers and anchors for support of horizontal and vertical piping shall be of sufficient strength when attached to building structure to maintain pipe alignment and prevent sagging.

Vent Terminals

Extensions of vent pipes should terminate at least 6 inches above the roof. This ensures that gases and odors in all parts of the drainage

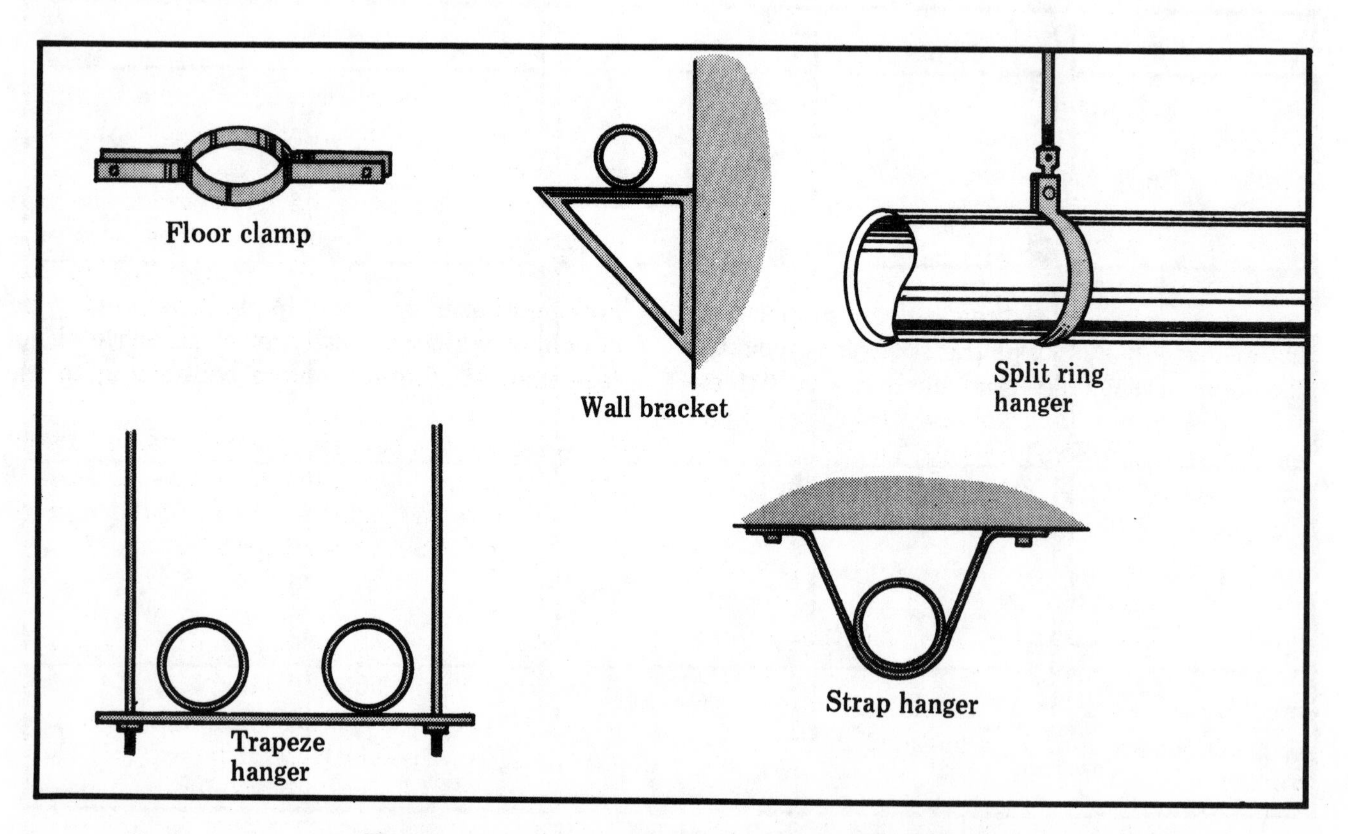

Five common horizontal and vertical pipe supports
Figure 4-9

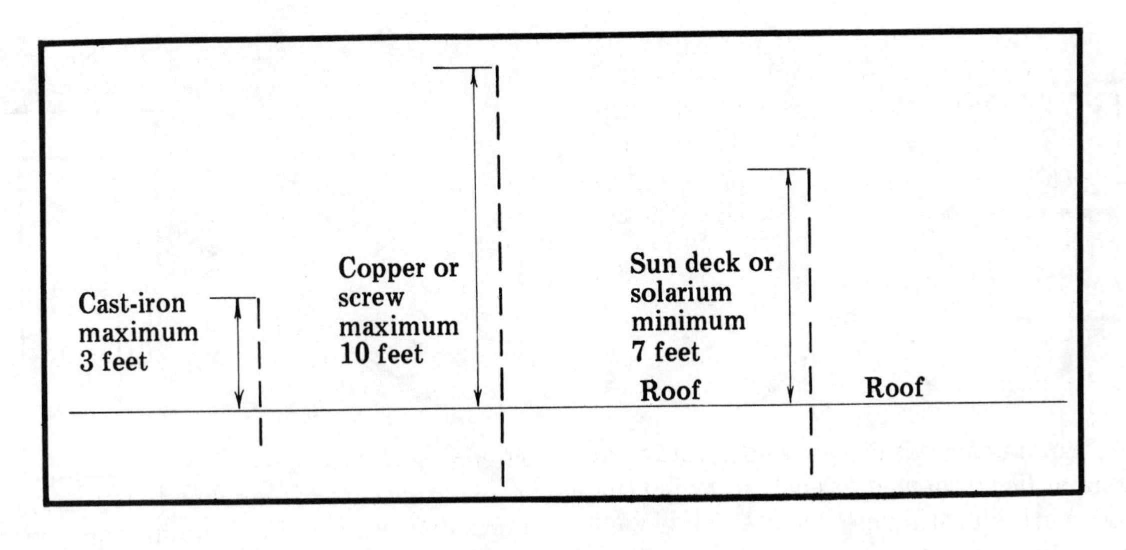

Vent extensions
Figure 4-10

system discharge well above the roof surface.

Horizontal vent piping should be installed and sloped to drain back to the soil or waste pipe by gravity. Improper grading or sags allow condensation to collect in low places, thus restricting air circulation and reducing the venting capacity. Vertical vent piping, when connected to a stack vent or vent stack, should be installed in an upward slope to prevent the entrapment of warm, moist air and the restriction of free air circulating within the system. Trapped moist air can accelerate corrosion of the pipe and greatly reduces its useful life.

Vent extensions, if made of cast-iron, should not exceed 3 feet. If a screw pipe or copper tube extension is used above the roof of a building, the vent extension should not exceed 10 feet. Where roofs are used for sun decks, solariums or similar functions, the vent should extend a minimum of 7 feet above the roof deck. See Figure 4-10.

The vent terminal of a sanitary system within 10 feet of any door, window or ventilating opening should extend at least 3 feet above the height of the opening. A vent pipe terminal must be at least 10 feet from any mechanical air intake opening for air conditioning units. See Figure 4-11.

Vent terminals should not be used to support flag poles, TV aerials, clotheslines or similar items and must not terminate under the overhang of the building.

Horizontal vent terminals extending through a wall shall not terminate within 10 feet of a lot line. Approved horizontal vent terminals shall turn upward and be effectively screened.

Where danger of frost closure exists, the vent extension through a roof must have a minimum diameter of 3 inches. When necessary to increase the size of the vent terminal, the change in diameter must be made at least one foot inside the building.

Vent pipes must not be connected to the side or bottom of soil or waste pipes. The vent pipe should connect above the center line of the horizontal soil or waste pipe. The vent pipe should then rise vertically, or at an angle of not less than 45 degrees before connecting to the branch vent.

A connection between a vent pipe and a vent stack or stack vent must be made at least 6 inches above the flood-level rim of the highest fixture served by the vent. See Figure 4-12.

Indirect Waste Piping and Special Waste

Indirect waste piping is sized and installed to accommodate the outlet drainage of the fixture or appliance. Indirect waste piping without its own trap need not be vented but should never connect directly to the sanitary drainage system.

Special indirect waste piping usually carries waste to the receiving fixture such as a floor sink or floor drain which does not receive

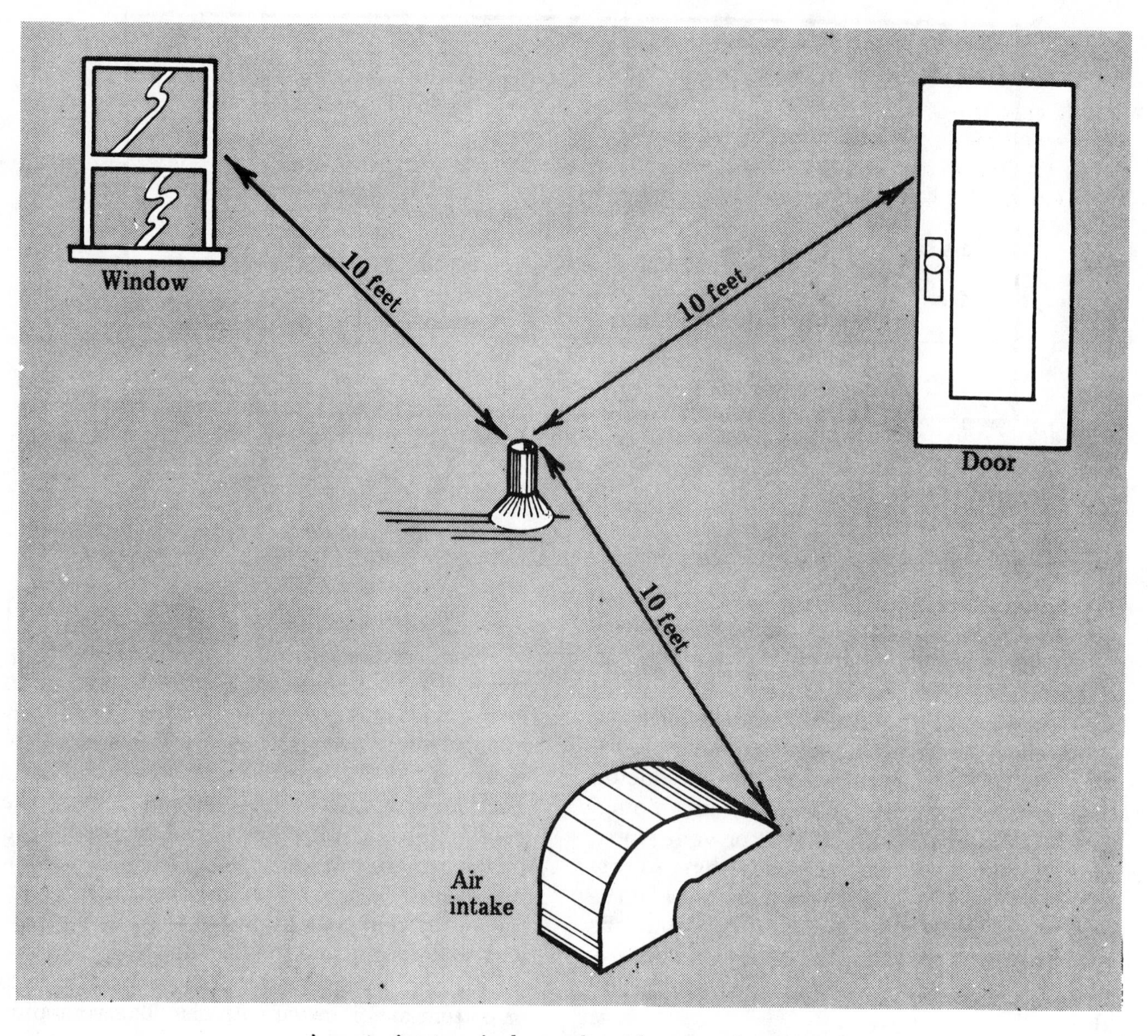

A vent pipe terminal must be at least ten feet from windows, doors or air intake openings for air conditioning units.

A vent terminal within ten feet of an opening must extend at least three feet above the opening.

Figure 4-11

material with highly offensive odors. Because of the relatively slow rate of movement of the liquid within the pipe, slime deposits can build up and eventually cause stoppages. It is important to install indirect drain piping so that it does not have sags or low spots. Gravity should cause it to drain dry. Accessible cleanouts should be provided for cleaning and flushing purposes. To further reduce the possibilities of slime forming in the indirect waste piping, locate the receiving fixture in a well ventilated area.

Indirect waste pipe should be installed below the floor where practical. If an indirect waste pipe is installed, it should connect through the receiving fixture above the water seal of the trap. See Figure 4-13.

When above-the-floor installation is required, the pipe should be installed at least 3 inches above the floor to allow room for floor cleaning. The outlet should then terminate with an air gap of one to two inches above the receiving fixture.

See Figure 4-14.

Indirect waste pipe installed above the floor should be a minimum of ¾ inch in diameter, but must never be smaller than the outlet drains of the fixtures or appliances it serves. When installed below the floor, indirect waste pipe

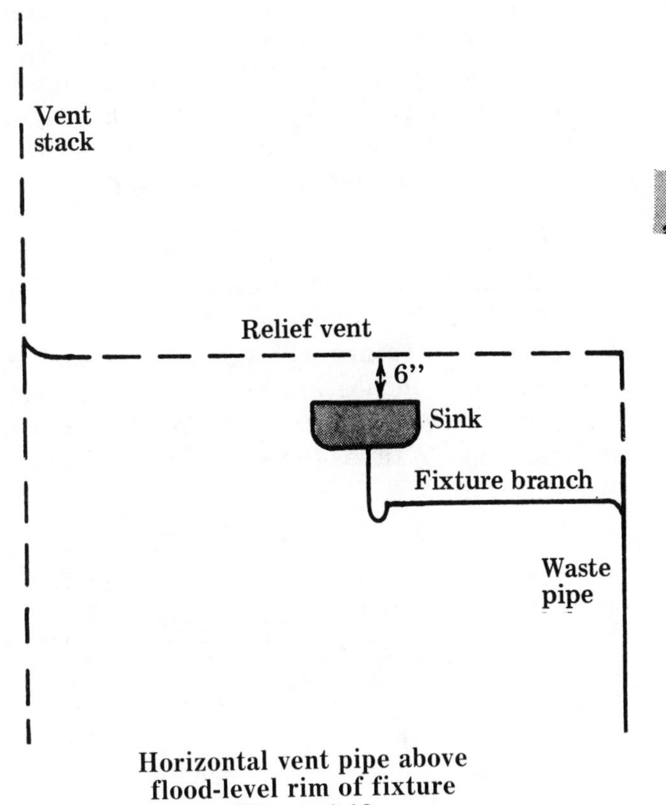

Horizontal vent pipe above flood-level rim of fixture
Figure 4-12

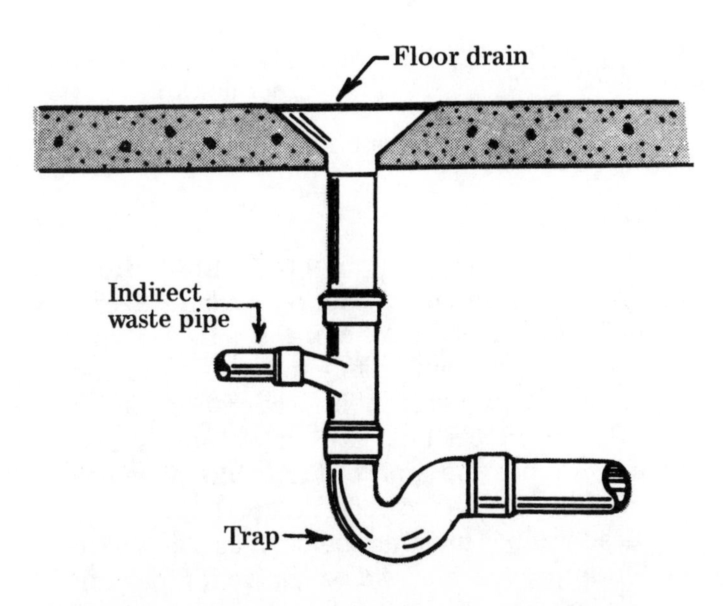

Indirect waste pipe connection beneath floor
Figure 4-13

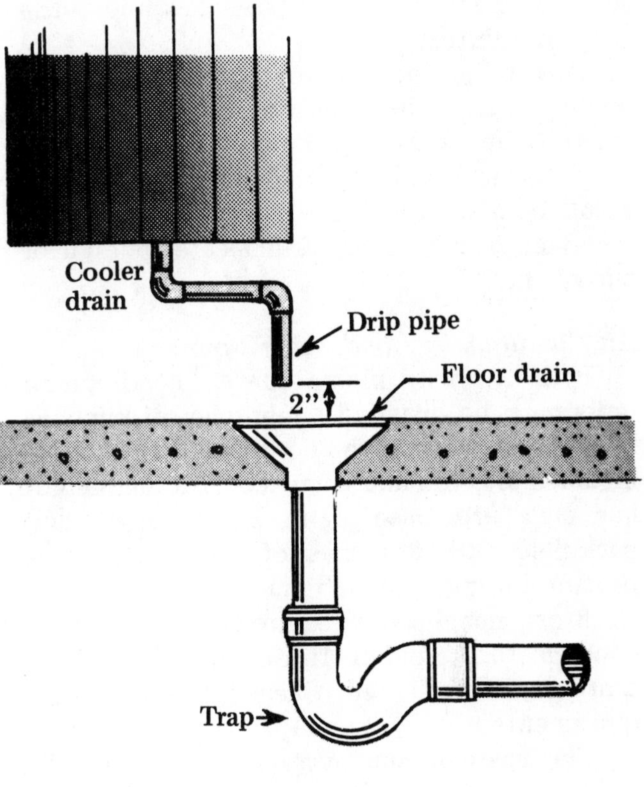

Air gap
Figure 4-14

must be at least 1¼ inch in diameter.

The size of indirect waste pipe serving appliances rated at up to one fixture unit should be at least 1¼ inch. For a greater number of fixture units, the indirect waste pipe must be sized in accord with Table 2-10, special fixtures.

Some codes specify that the maximum developed length of the indirect waste pipe must not exceed 15 feet.

Commercial dishwashing machines must be indirectly connected to the building greasy waste drain line.

The receiving fixture (floor drain or floor sink) serving indirect waste pipes shall not be installed in toilet rooms.

The drip pipe from a walk-in refrigerator or a cooler room refrigeration coil should terminate a minimum of two inches above receiving fixture installed outside of the room, as shown in Figure 4-14.

Floor drainage from a walk-in cooler, freezer or food storage room floor should be indirectly connected to the receiving fixture (floor drain) installed outside of room. It is required that the indirect waste pipe be equipped with a flap check valve and that the floor of the room be a minimum of two inches above the receiving fixture.

A drinking fountain may be indirectly connected to a floor drain for the purpose of resealing the trap when located outside a toilet room.

Floor drainage from the air handling equipment room must be indirectly connected to the receiving fixture (floor drain) located outside of the room.

Air Conditioning Condensate Drains

Indirect waste piping for air conditioning units must be installed a minimum of 2 inches below the bottom of the floor slab. It should be installed after fill and compaction is completed, laid on a firm base for its entire length and backfilled with 2 inches of sand. All risers passing through the slab must be sleeved.

Most model codes require that the vertical condensate drain line to an air conditioning unit or units be vented. (Check local code for requirements.)

The waste or condensate from an air conditioning unit is classified as a plumbing fixture only if connected to the plumbing drainage system.

Note: This is important when sizing waste piping. See special fixtures, Table 2-10.

Air conditioning condensate waste must not be used for the purpose of resealing the trap of the floor drain.

Air conditioning units not exceeding 5 ton capacity may discharge their waste on a pervious area such as bare soil. Air conditioning units over 5 tons but under 10 tons may discharge their waste into a buried 10 inch diameter by 24 inch long pipe filled with 3/4 inch rock. No cover is required.

The main condensate drain line from a 10 ton or larger unit must discharge into a drainage well, storm sewer, adequate size soakage pit, drainfield or the building drainage system. Note that air conditioning equipment or a condensate drain can not discharge its waste on the roof of a building, into rain leaders, or allow waste to flow from any surface to the curb gutter at grade level.

Building Storm Drains

Cast-iron pipe used as a storm drain in a building is installed using bedding and backfill the same as for building sewers. (See Figure 4-3.) Vitrified clay pipe, plastic pipe, bituminized fiber pipe, asbestos cement and concrete pipe are more fragile. They require the same special installation and backfill methods as building sewers of like materials. See Figure 4-4.

Horizontal storm drains and leaders and vertical leaders have the same installation methods, limitations, restrictions and horizontal and vertical support requirements as drainage, waste and vent piping within a building for the same type of pipe. See Figures 4-5 to 4-9.

Roof Drain Strainers

A strainer must be provided where roof surfaces drain into the inlet pipe of an inside leader. The strainer cover must extend a minimum of 4 inches above the roof surface. The strainer must have an available inlet area of not less than 2½ times the area of the leader to which the drain is connected. Roof drain strainers for use on sun decks, parking decks, and similar areas must be of flat surface type. These drains are more likely to be serviced and maintained because of their location near public use. The available inlet area must not be less than 2½ times the area of the leader to which the drain is connected. See Figure 4-15.

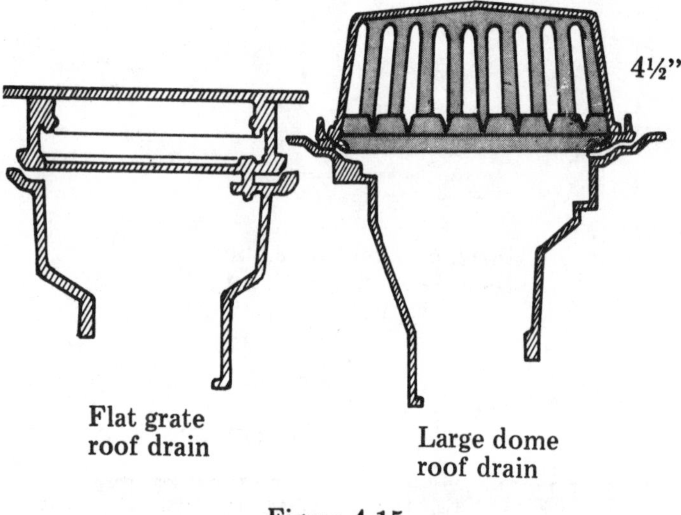

Figure 4-15

Leader pipes must not be used as soil, waste or vent pipes and soil, waste and vent pipes can not be used as leaders. Rainwater leaders and drains can not be reduced in cross-sectional area in the direction of flow.

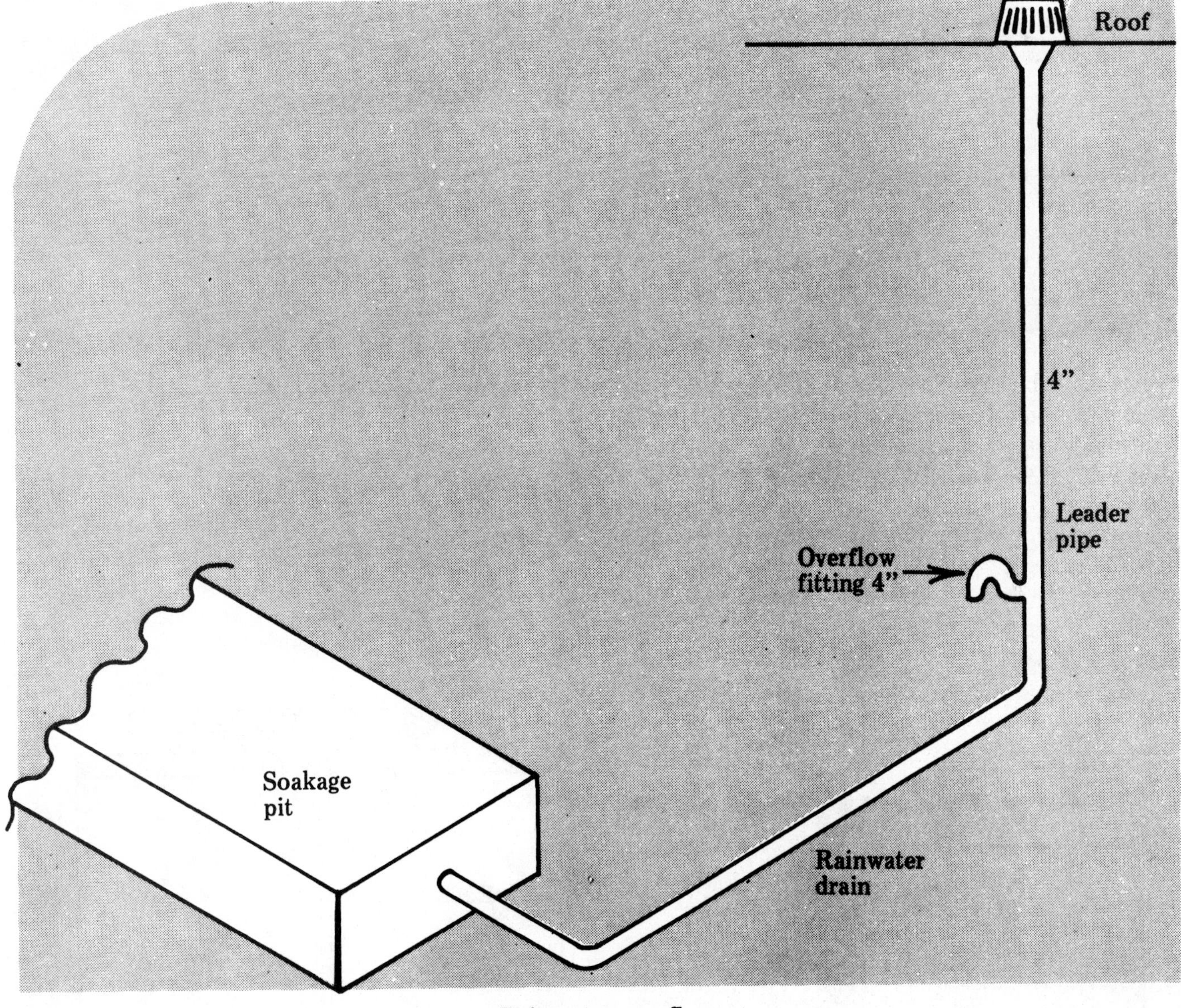

Rainwater overflow
Figure 4-16

Rainwater drains conveying runoff from leaders and discharging directly into soakage pits must have overflow fittings. Overflow fittings must also be provided at the base of rainwater leaders. The overflow fitting must be the same size as the drain up to 4 inches. See Figure 4-16.

Joints and Connections

Joints and connections in the drain, vent and waste system must be gas and watertight for the pressure required by test or by intended use. An exception to this requirement is made for those portions of a drainage system which use perforated or open-joint piping which is intended to collect and convey underground water to underground storm drains.

Types of Joints

Caulked joints Every lead caulked joint for cast-iron bell-and-spigot soil pipe should be firmly packed with oakum or hemp and filled with molten lead not less than 1 inch deep. The lead must not extend more than 1/8 inch below the rim of the hub. Lead must be run in one pouring and caulked tight. Paint, varnish, or other coatings are not permitted on the jointing material until after the joint has been tested and approved. See Figure 4-17.

Compression-type joints Neoprene rubber gaskets for use with bell-and-spigot cast-iron soil pipe and fittings must be made with an approved compression gasket that is compressed when the spigot is inserted into the hub of the pipe. See Figure 4-18.

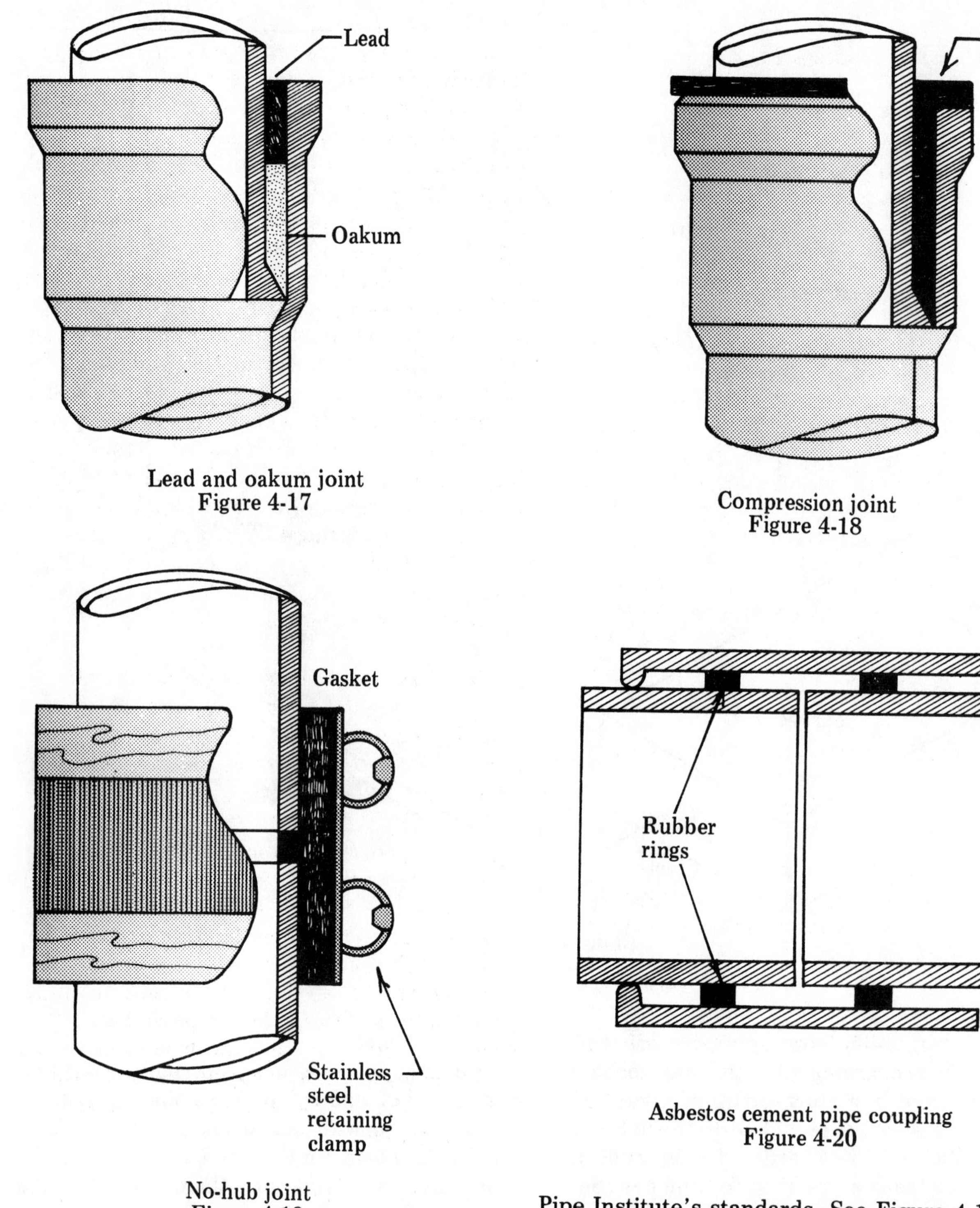

Lead and oakum joint
Figure 4-17

Compression joint
Figure 4-18

No-hub joint
Figure 4-19

Asbestos cement pipe coupling
Figure 4-20

Hubless joints Joints for hubless cast-iron soil pipe and fittings must be made with an approved elastomeric sealing sleeve and stainless steel clamp. The clamp assembly used in joining a hubless cast-iron sanitary system for soil, waste, vent and house or building sewer lines must comply with the mechanical and dimensional requirements of the Cast-Iron Soil Pipe Institute's standards. See Figure 4-19.

Plastic pipe joints Joints connecting plastic pipe and fittings must be solvent-welded or fusion-welded connections using only procedures recommended by the manufacturer or the Plastic Institute.

- Schedule 40 pipe can not be threaded on the job site.
- Only approved male or female threaded adapter fittings can be used in transitions.

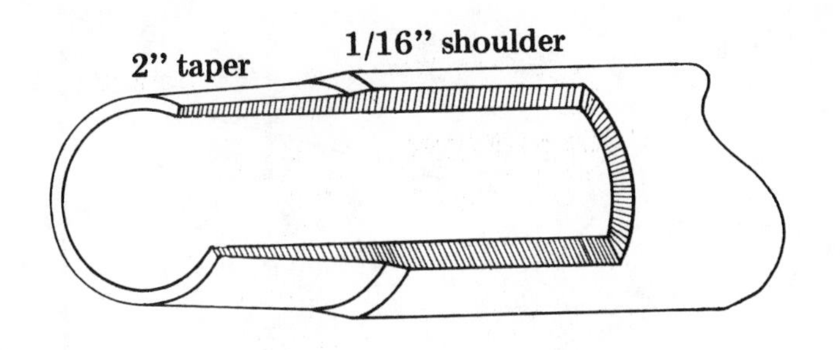

Bituminized-fiber pipe
Figure 4-21

Threaded joints must not be over-tightened.

- When connections between plastic pipe and other types of materials are required, you must use approved fittings. For example, within the same job installation you can not mix ABS and PVC plastic pipe or fittings with solvent-weld joints

Compression-type joints for non-metallic pipe Neoprene preformed elastomeric rings can be used to join vitrified clay pipe with bell-and-spigot connections. Hot poured bitumastic compound can also be used if it has a bond strength of not less than 100 psi in shear. Approximately 25 percent of the joint space at the base of the socket must be filled with jute or hemp. Each joint must be poured in one operation until it is filled. Joints must not be tested until one hour after pouring.

Asbestos cement sewer pipe Joints in asbestos cement pipe must be made with sleeve couplings of the same composition as the pipe and sealed with preformed rubber rings. See Figure 4-20. Joints between asbestos cement pipe and metal pipe must be made with an adapter coupling caulked with lead and oakum. Joints between asbestos cement pipe and clay pipe must be made with an adapter coupling using approved rubber rings or by a preformed bituminous ring approved by your local authority. Joints between asbestos cement pipe and plastic pipe should be made with an approved adapter coupling with an approved rubber ring.

Bituminous fiber pipe joints Joints in bituminized fiber pipe should be made with tapered-type couplings of the same material as the pipe. Joints between bituminized fiber pipe and metal pipe should be made with an adapter coupling caulked with lead and oakum. (See Figure 4-21.)

Cement mortar joints Cement joints should have a layer of jute or hemp rammed into the base of the joint space to prevent mortar from entering the interior of the pipe. Jute or hemp should be dipped in a slurry of portland cement in water prior to inserting into the bell. Only 25% of the joint space can be used for jute or hemp. The remaining space must be filled in one continuous operation with a thoroughly mixed mortar composed of one part cement and two parts sand. Use only enough water to make the mixture workable by hand. After half an hour of setting the joint should be rammed around the entire periphery with a blunt tool to force the partially stiffened mortar into the joint and to repair any cracks that may have formed during the initial setting period. The interior of the pipe should be swabbed to remove any material that fell into the pipe. Additional mortar of the same composition should then be troweled on to form a 45 degree taper with the barrel of the pipe.

Slip joints Slip joint connectors may be used on both sides of the trap and in the trap seal in a drainage system. Note: Some codes prohibit the use of slip joint connectors on outlet side of the trap.

Wiped joints A connection of lead pipe to brass or copper pipe or fittings (this includes ferrules, solder nipples, or traps) must be fully wiped. Wiped joints must have an exposed surface on each side of the joint of not less than ¾ inch, and at least as thick as the material being joined. Joints between lead pipe and cast-iron, steel or wrought iron pipe must be made with a caulking ferrule, soldering nipple or bushing. The minimum length of lead from wiped joint to fixture connection is 4 inches.

Borosilicate glass joints Glass to glass connections shall be made with a bolted compression type stainless steel coupling with contoured acid-resistant elastomer compression ring and a flourocarbon polymer inner seal ring.

Joints between glass pipe and other types of piping materials shall be made with approved adapters or according to manufacturer's recommendation.

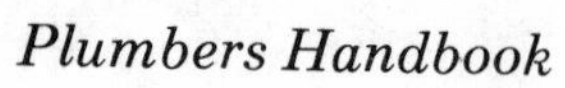

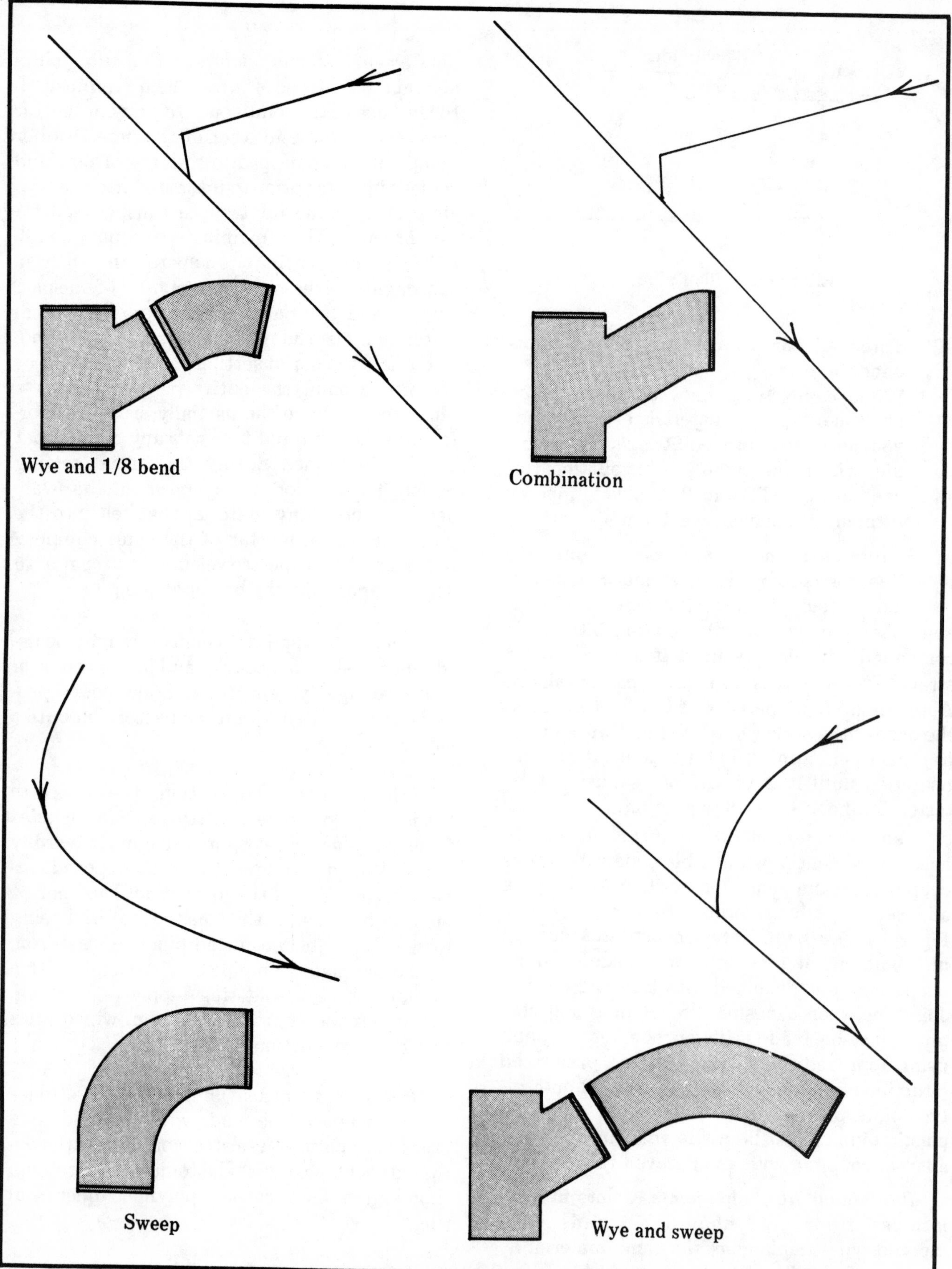

Horizontal to horizontal change of direction
Figure 4-22

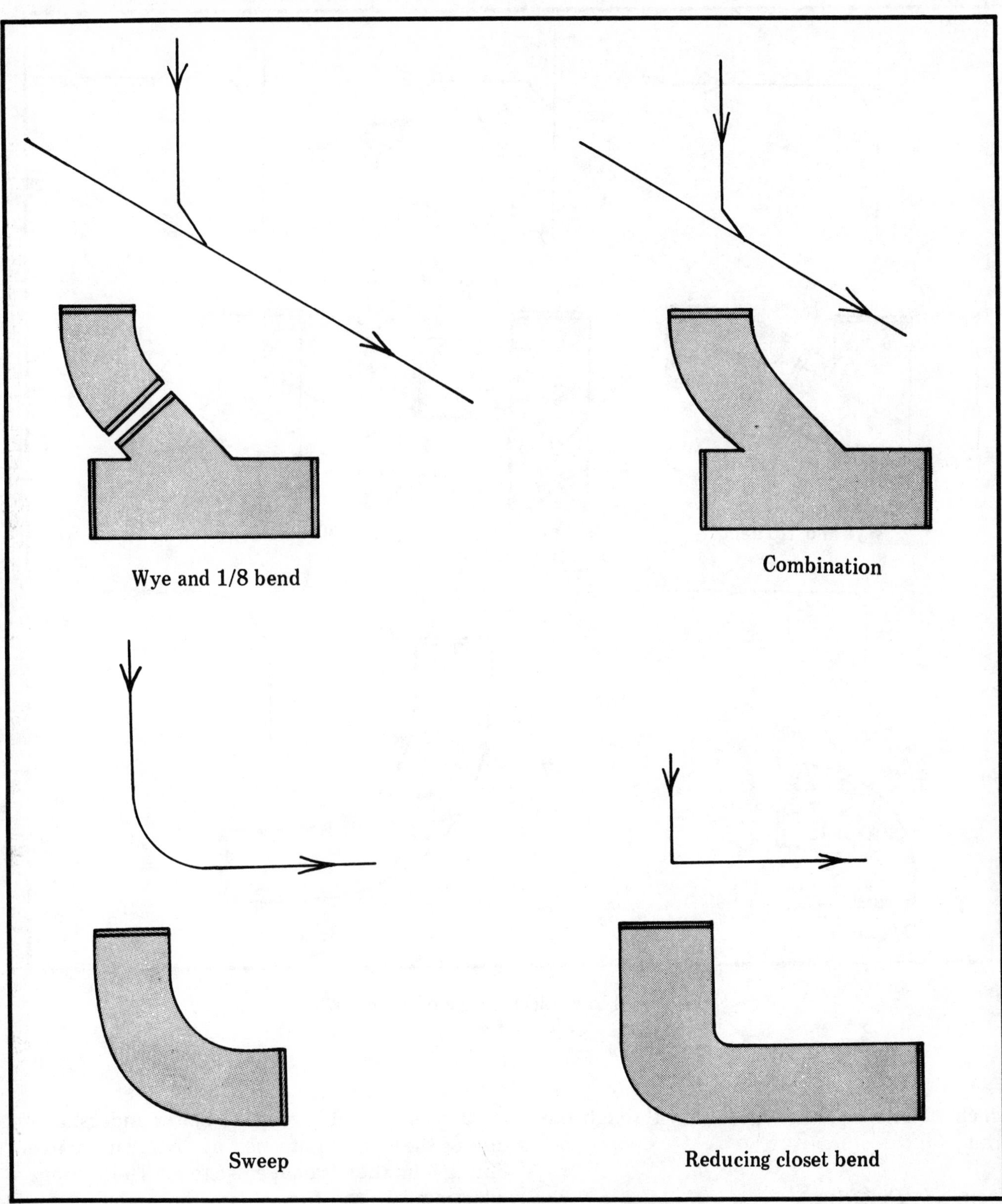

Vertical to horizontal change of direction
Figure 4-23

Increasers and Reducers

Where different sizes of pipes, or pipes and fittings are to be connected, the proper size increasers or reducers or reducing fittings should be used between the two sizes.

Prohibited Joints and Connections in Drainage Systems

You can not use any fittings or connections which have an enlargement, chamber or recess with a ledge, shoulder or reduction of pipe area

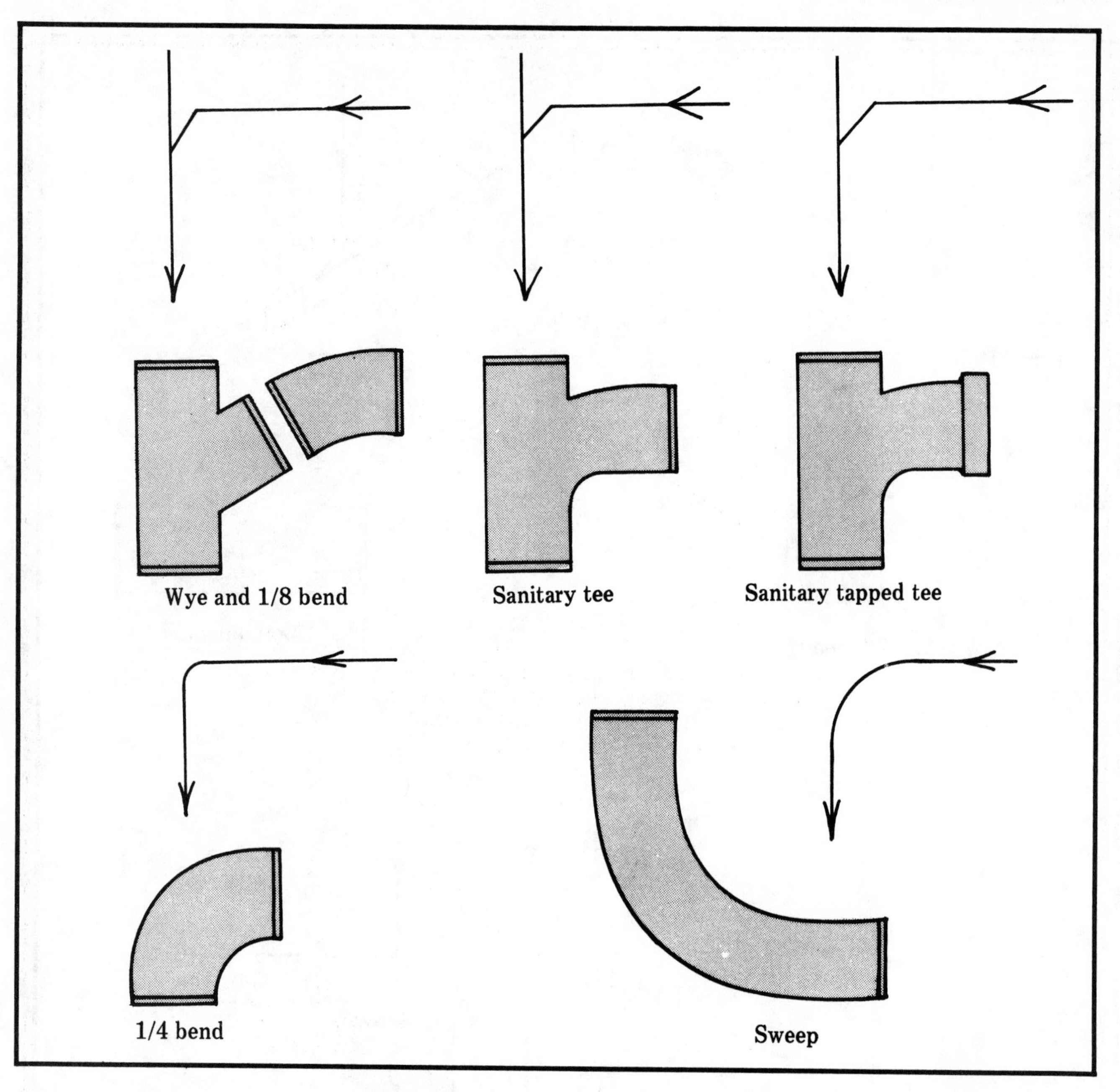

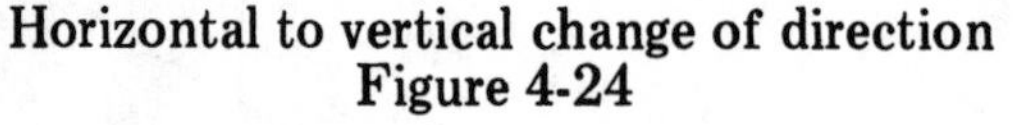
Horizontal to vertical change of direction
Figure 4-24

which offers an obstruction to flow through the drain.

Change in Direction

Changes in direction in horizontal or horizontal to vertical drainage systems must be made with 45 degree wyes, long or short sweep quarter bends, sixth, eighth, or sixteenth bends, or by a combination of these or other approved fittings. Where the direction of flow is from the horizontal to the vertical, sanitary tees, quarter bends and one-fifth bends can be used. Figures 4-22 to 4-24 should provide a better understanding of the appropriate use of direction change fittings in the drainage system. The fittings illustrated in the figures are the no-hub type.

Rough-In Clarifications

Figures 4-25 to 4-30 should help eliminate some of the confusion that is common when designing drain, vent and waste layouts. The intent of the code should be clear if you study these drawings and refer to the notes on each illustration.

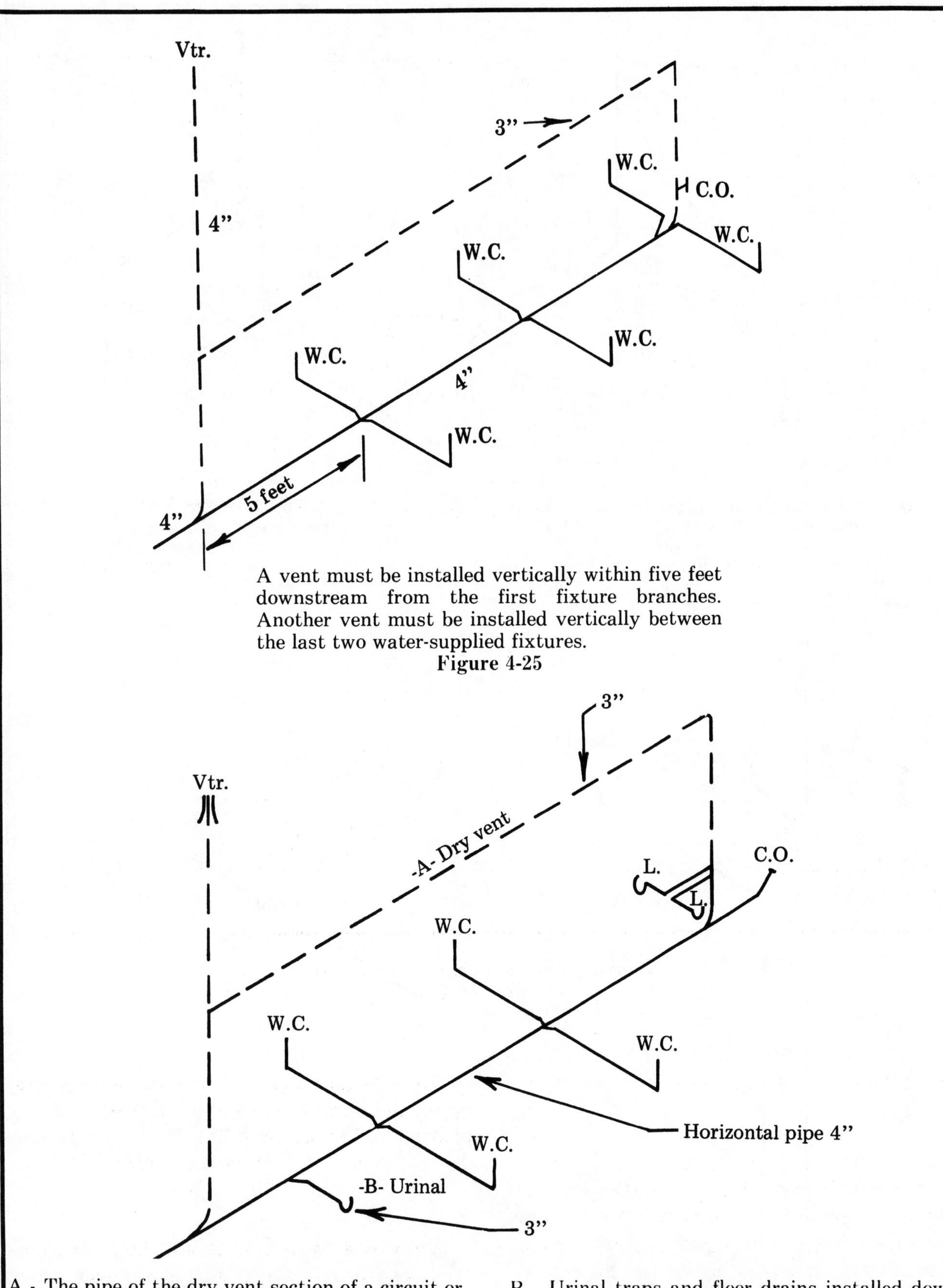

A vent must be installed vertically within five feet downstream from the first fixture branches. Another vent must be installed vertically between the last two water-supplied fixtures.

Figure 4-25

A - The pipe of the dry vent section of a circuit or loop vent may have a diameter of one pipe size less than the diameter of the pipe of the horizontal soil or waste drain it serves.

B - Urinal traps and floor drains installed downstream from a water closet in a circuit or loop vent group must be three inches in diameter where permitted by code.

Figure 4-26

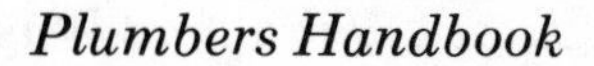

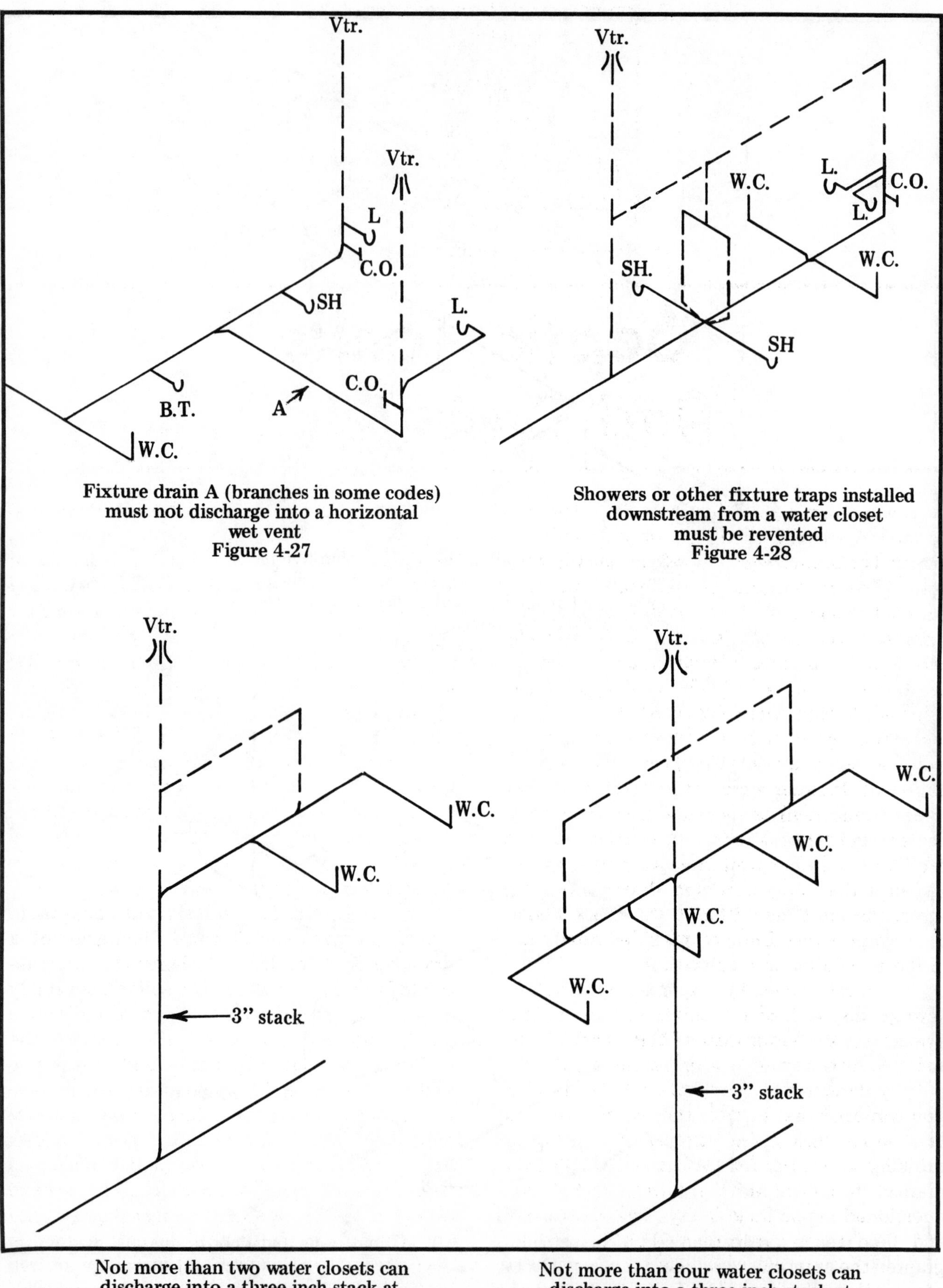

Fixture drain A (branches in some codes) must not discharge into a horizontal wet vent
Figure 4-27

Showers or other fixture traps installed downstream from a water closet must be revented
Figure 4-28

Not more than two water closets can discharge into a three inch stack at the same point
Figure 4-29

Not more than four water closets can discharge into a three inch stack at the same level
Figure 4-30

Septic Tanks And Drainfields

Not too many years ago cesspools and outhouses were the most common means of disposal of human waste in both urban and rural areas. The inadequacy of cesspools and outhouses became clear when population densities increased. Local authorities who enforce model codes will no longer accept outhouses or cesspools as an acceptable method of waste disposal. In rural areas where drinking water is taken from open or closed wells contamination from sewage is a real possibility. The contamination of drinking water by untreated sewage (cross connection) can spread diseases such as cholera and typhoid. The lack of basic sanitary facilities was responsible for many of the diseases that were common during the Dark Ages, the most feared being the Black Plague that swept across Europe and killed nearly half of the population in a short time.

The most proven and acceptable method for sewage disposal where public sewers are not available is the septic tank system. There is still some controversy as to whether the septic tank is truly danger-free. The Septic Tank Association contends that there is not one proven case of a septic tank contaminating any source of drinking water. But the Department of Environmental Resource Management (DERM) has questioned septic tank safety. The controversy and field testing continue. DERM has been granted the authority to regulate the installation of septic tanks and has established strict guidelines that have been adopted by most model code organizations.

Although septic tank and drainfield installations are an important part of plumbing work, few plumbing contractors handle this type of work. A licensed septic tank contractor usually does the installation and maintenance work for these systems.

The plumbing professional usually handles only the connection of the building drainage system to the inlet tee of a septic tank. However, since there are questions concerning septic tanks and drainfields on the journeyman's and master's examinations, you should learn the basic principles of these unique treatment systems.

Definition of the Septic Tank

A septic tank is a watertight receptacle which receives the sewage discharge of a drainage system. It is designed to separate solids from liquid wastes. The solids are usually about ¾ of a pound in each 100 gallons of water. The heavier portions settle to the bottom of the tank while the lighter particles and grease rise to the top of the liquid. The tank is sized to have a capacity equal to approximately 24 hours of anticipated flow. This retention period allows the solids to be digested through a biological process involving bacterial action. The sewage waste within the tank is thus transformed into gases and harmless liquids. As new sewage is discharged from the plumbing drainage system and enters the septic tank, the gases are forced up and through the drainage vent pipes and into the atmosphere above the building roof. An equal amount of treated liquid is forced out through the tee of the septic tank as effluent. This effluent enters a subsurface system of open-joint or perforated piping installed on a bed of washed rock. Further nitrification takes place in the form of oxidation and evaporation as the effluent seeps out between the joints or holes in the perforated piping. The treated effluent is thus returned to the soil outside the tank.

Bedrooms	Minimum liquid capacity required (gallons)	Inside length	Inside width	Air space	Liquid depth
2 or less	750	7'-2"	3'-6"	8"	4'
3	900	8'-0"	3'-9"	8"	4'
4	1050	8'-6"	4'-2"	8"	4'
5	1200	9'-0"	4'-6"	8"	4'

Table 5-1

When the bacterial process is complete within the tank, a small amount of solids remains in the tank and settles to the bottom as sludge. Lighter undigested particles rise and form a scum (crust) on top of the liquid content. Periodically the tank has to be cleaned of these undigested materials by a certified professional who has the equipment to perform this work.

Sizing Septic Tanks

The minimum capacity requirements for septic tanks in normal residential use are outlined in Table 5-1. The capacity is based on the number of bedrooms.

For each bedroom over five, add 150 gallons to the minimum liquid capacity. A six-bedroom house, for example, would be sized at 1350 gallons. These liquid capacities will provide the necessary retention period so that the septic tank can digest the solids and function properly.

Septic tanks for commercial establishments are generally sized by the Department of Environmental Resource Management or, in some cases, by the local health department. Table 5-2 shows common minimum tank capacities for most commercial establishments and should be helpful when planning these disposal facilities.

Minimum disposal facilities for eat and drink establishments
Liquid capacity of septic tank in gallons

Up to 25 persons	26 to 50 persons	51 to 100 persons	101 to 150 persons
1200 gal.	1500 gal.	2400 gal.	3600 gal.

Minimum disposal facilities for bars
Liquid capacity of septic tank in gallons

Up to 25 persons	26 to 50 persons	51 to 100 persons
750 gal.	900 gal.	1200 gal.

Minimum disposal facilities for stores
Liquid capacity of septic tank in gallons

1 store	2 stores	3 stores
750 gal.	1100 gal.	1500 gal.

Common minimum tank capacities for commercial establishments
Table 5-2

Septic Tank Construction

Blocks, brick, or sectional tanks are not permitted by most model codes. Metal tanks are approved by local authorities in some geographical areas. Fiberglass tanks are relatively new and have been approved in other areas as long as they are properly installed and protected. Precast concrete or cast-in-place septic tanks are the most common type and will meet practically all model code requirements. The interior wall of all septic tanks should be finished smooth and impervious. Voids, pits, or protuberances on the inner wall of a septic tank are prohibited.

Concrete septic tank inlet and outlet tees must be of terra cotta or concrete with a wall thickness of at least one inch and a cross-section area not *less* than the building sewer pipe size and not *more* than two times greater than the building sewer pipe. The outlet invert must be a minimum of one inch and a maximum of three inches lower than the inlet tee. See Figure 5-3. Inlet and outlet tees must be installed at opposite ends of the septic tank and be a maximum of 5 inches above and 18 inches below the liquid level line.

Septic Tank Installation

Cast-in-place septic tanks can be installed where they are subject to overburden loads such as parking lots. In this case, the tank lid must be

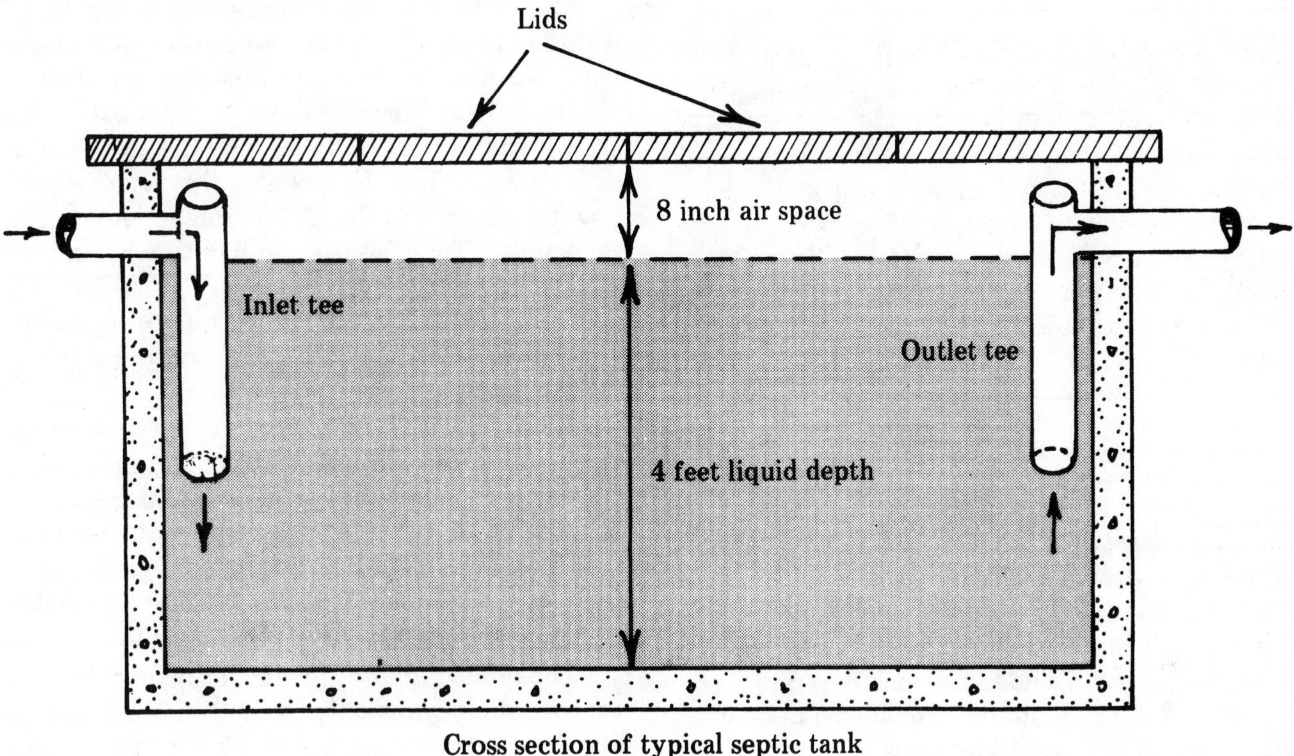

Cross section of typical septic tank
Figure 5-3

designed and installed to support an anticipated load equal to the weight of a 10 ton truck.

Precast septic tanks must not be used in parking lots or any area where vehicular traffic is anticipated unless it is designed with a traffic lid which is supported on the soil independent from the septic tank.

Where septic tanks are installed under any form of paving, all cast-in-place lids must have a 22 inch manhole located directly above the inlet and outlet tees. The manhole cover must be brought up to grade level. It is essential that all septic tanks be installed level.

Restrictions for Septic Tanks

- The tank can not be located under or within 5 feet of any building.
- The tank can not be located within 5 feet of any water supply lines.
- The tank can not be located within 5 feet of property lines other than public streets, alleys or sidewalks.
- The tank can not be located within 50 feet of the shore line of open bodies of water.
- The tank can not be located within 50 feet of a private water supply well which provides water for human consumption, bathing or swimming.
- The excavation for a septic tank must not be made within the angle of pressure as transferred from the base of an existing structure to the sides of an excavation on a 45 degree angle. See Figure 4-6.
- Circulation of air within the septic tank and drainfield must be through the plumbing system and then through the inlet and outlet tees of the septic tank. No other circulation is permitted.

Sizing Drainfields

It is essential that the soil surrounding a septic tank be able to absorb the effluent properly. Where soil porosity appears to be less than usual (as might be the case in heavy clay, sugar sand, etc.) a percolation test should be conducted. The percolation rate determines the time it takes water to be absorbed into the soil. The fastest rate given in Table 5-4 that is acceptable is three minutes per inch. Anything below or above the three minute level would constitute a special condition and would require that the drainfield be sized accordingly.

The drainfield absorption area per bedroom required for single family and duplex residences is to be determined from the percolation rate given in Table 5-4. For example, if the percola-

Percolation rate time in minutes for water to fall one inch	Absorption area in square feet per bedroom
1 or less	70
2	85
3	100
4	115
5	125
10	165
15	190
Over 15	Unsuitable for absorption field

Table 5-4

tion rate were 4 and you were sizing a drainfield for a five bedroom residence, you would multiply 5 times 115. The required drainfield size would be 575 square feet of trench bottom. Each linear foot of drainfield tile is considered to have one square foot of bottom trench even though the trench is at least 16 inches wide.

Type of establishment	Gallons per day
Residential	
Single family	450 (GPD for 2 bedrooms)
	150 (GPD for additional bedroom)
Townhouse	150 (GPD per bedroom)
Multiple apartments	150 (GPD per bedroom)
Trailer park - private baths	60 (GPD - 3 persons per trailer)
Trailer park - community baths	50 (GPD - 3 persons per trailer)
Commercial	
Barber shop	200 (GPD per chair)
Bowling alleys	100 (GPD per lane)
Full service restaurant	50 (GPD per seat)
Laundries, self-service	300 (GPD per machine)
Drive-in theaters	5 (GPD per space)
Industrial	
Churches	3 (GPD per seat)
Nursing, rest homes	150 (GPD per bed)
Airports	5 (GPD per passenger) and 20 (GPD per employee)
Warehouse - industrial	5 (GPD per 100/S.F.)

Table of sewage flow determination
Table 5-5

The minimum size drainfield for a residence with one or two bedrooms would be 200 square feet.

Drainfields for commercial establishments are usually sized by DERM or by the local health department. They are sized by the daily sewage flow based on a table of flow rates given in the code book. Table 5-5 is similar to what might appear in your code book. The following example uses the table and shows how to determine the size of a commercial drainfield when the daily sewage flow is known.

We will use as an example a 10 unit apartment building with one bedroom in each unit. The code assumes that each bedroom means an occupancy of two persons. The daily sewage flow would be 75 gallons per person. Multiply 20 (which is 10 units x 2 persons) times 75, which equals 1,500 gallons of sewage flow daily.

The following formula should be used to size the drainfield when only the daily sewage flow is known: Daily sewage flow x 43.5 ÷ 100 equals the area required in square feet. Using this formula, multiply 1,500 (from the example in the paragraph above) x 43.5 (65250.0), and then divide by 100. This will give the size of the drainfield required for a 10 unit apartment building serving 20 persons. The answer is 652.5 square feet.

Use the formula given when a commercial building is not listed in the "Table of Flow Determination" in the code book. To arrive at the amount of daily sewage flow, simply check the water bill for the month. This should show the amount of water used. Divide this amount by the number of days covered by the bill (probably 30). When this has been determined, you would proceed with the formula: daily sewage flow x 43.5 ÷ 100.

When drainfields are located under pavement the absorption area must be increased as follows:

Percolation rate	Area increased
0 to 5	10%
5 to 10	17%
10 to 15	25%

As an example, if the required drainfield for a three bedroom house is 300 square feet and the percolation rate is 3, the drainfield area would need to be increased by 10%, an additional 30 square feet.

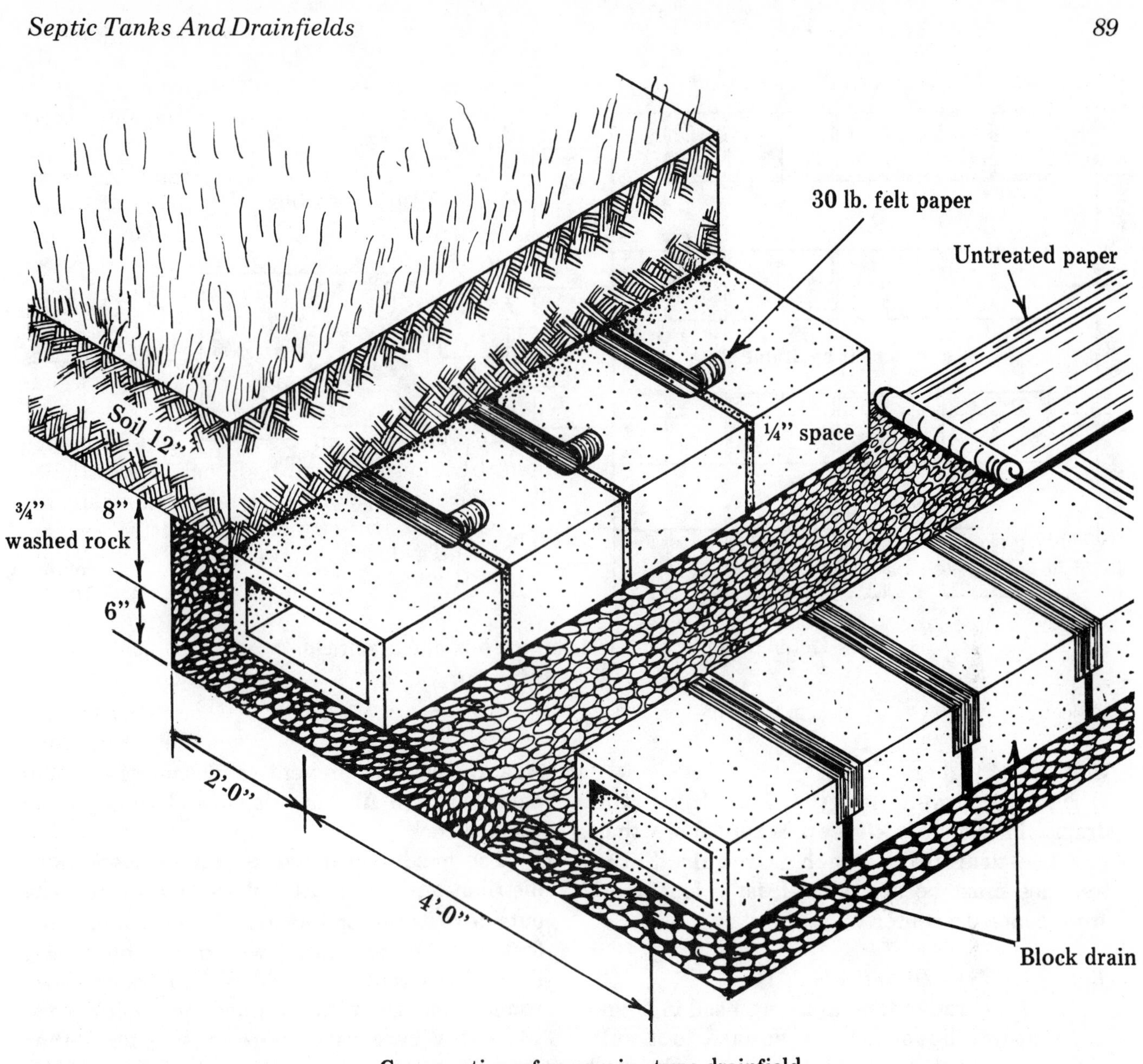

Cross section of reservoir - type drainfield
Figure 5-6

Drainfield Construction and Installation

Several materials can be used to distribute the effluent from septic tanks evenly throughout the drainfield bed. The most common are open jointed or perforated drain tile, block or cradle type drain units, and the newer corrugated plastic perforated tubing. Each has its own requirements and the particular installation methods for each will be considered separately. The ends of each distribution line must be sealed by capping or by cementing a block to the ends.

Tile Drainfields

Drainfield tile must have a minimum inside diameter of 4 inches and must be installed on a slope not more than ½ inch per 10 feet. They should be laid on a bed of ¾ inch washed rock extended 6 inches under the tile. Another 6 inches of rock is used to level the rock to the top of the tile, giving a total depth of 12 inches for the full width of the trench.

The tile must be laid with a space of ¼ inch between the tile ends. A strip of 4" x 16" 30-pound bituminous saturated paper must cover the ¼ inch space between tiles to prevent sand or other small particles from filtering into the openings.

Regardless of the width of the trench, each linear foot of drainfield tile is considered to be one square foot. Each trench must have a minimum width of 16 inches. The maximum

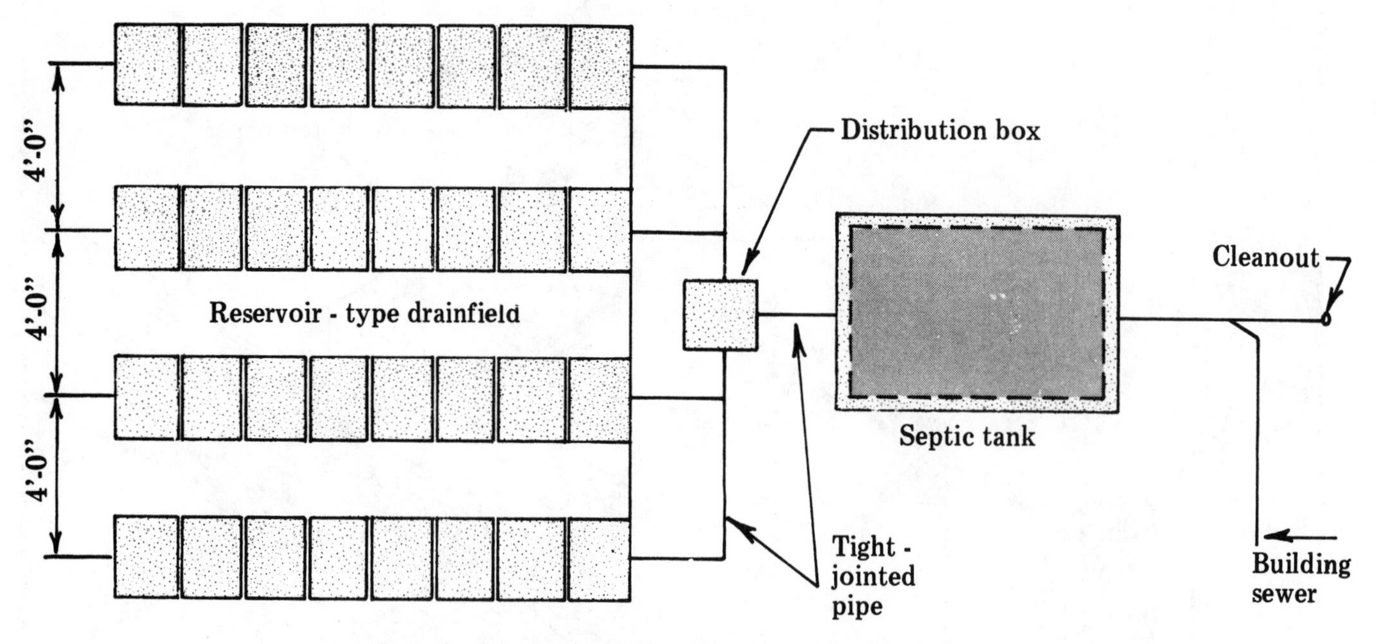

Detail of septic tank distribution box and drainfield
Figure 5-7

width is 24 inches.

The maximum length of a single tile drainfield trench is 40 feet. Where more than one tile drainfield trench is required, the trenches must be spaced at least 6 feet apart from center to center.

Reservoir-Type Drainfields

Block or cradle drain units are used in single excavations based on the square foot rule (length x width) rather than in individual trenches like drainfield tile.

Block or cradle drain units are installed to slope not more than ½ inch per 10 feet. They are laid on a bed of ¾ inch washed rock which extends from 6 inches under the drain units to 8 inches above the bottom of the units. This gives a total of 14 inches of rock for the full width of the drainfield.

Where drain units have a fixed opening to provide seepage, units may be butted tight against each other. Where a fixed opening is not provided, units must be laid with a ¼ inch space between the ends. The ¼ inch space is covered with a suitable length of 4 inch wide, 30 pound bituminous saturated paper to protect the top seam. The paper must extend down 4 inches on each side of the units. (See Figure 5-6.) The entire area of drainfield filter material (washed rock) must be covered with untreated paper. Again, this is to prevent sand and other small particles from filtering down and through the washed rock.

The maximum distance between centers of distribution lines must not exceed 4 feet. The outside distribution line must be a minimum of 2 feet from the excavated wall of the filter bed. Each drain unit is considered to occupy four square feet. Distribution lines must not exceed 100 feet. Where two or more lines are used, they should be as near the same length as possible and must be connected by a distribution box. A tight-jointed pipe must connect the septic tank's outlet tee to the distribution box and tight jointed pipes must connect the distribution box to the fixed reservoir distribution lines. (See Figure 5-7.)

Corrugated Plastic Perforated Tubing Drainfield

Plastic tubing drainfields are most often used in single excavations based upon the square foot rule (length x width) rather than in individual trenches. This material was approved for drainfields in the last few years and seems to work rather well.

It, too, must be installed to slope not more than ½ inch in each 10 feet. It must be laid on a bed of ¾ inch washed rock extending from 6 inches under the tubing to 6 inches over the tubing bottom, a total depth of 12 inches for the

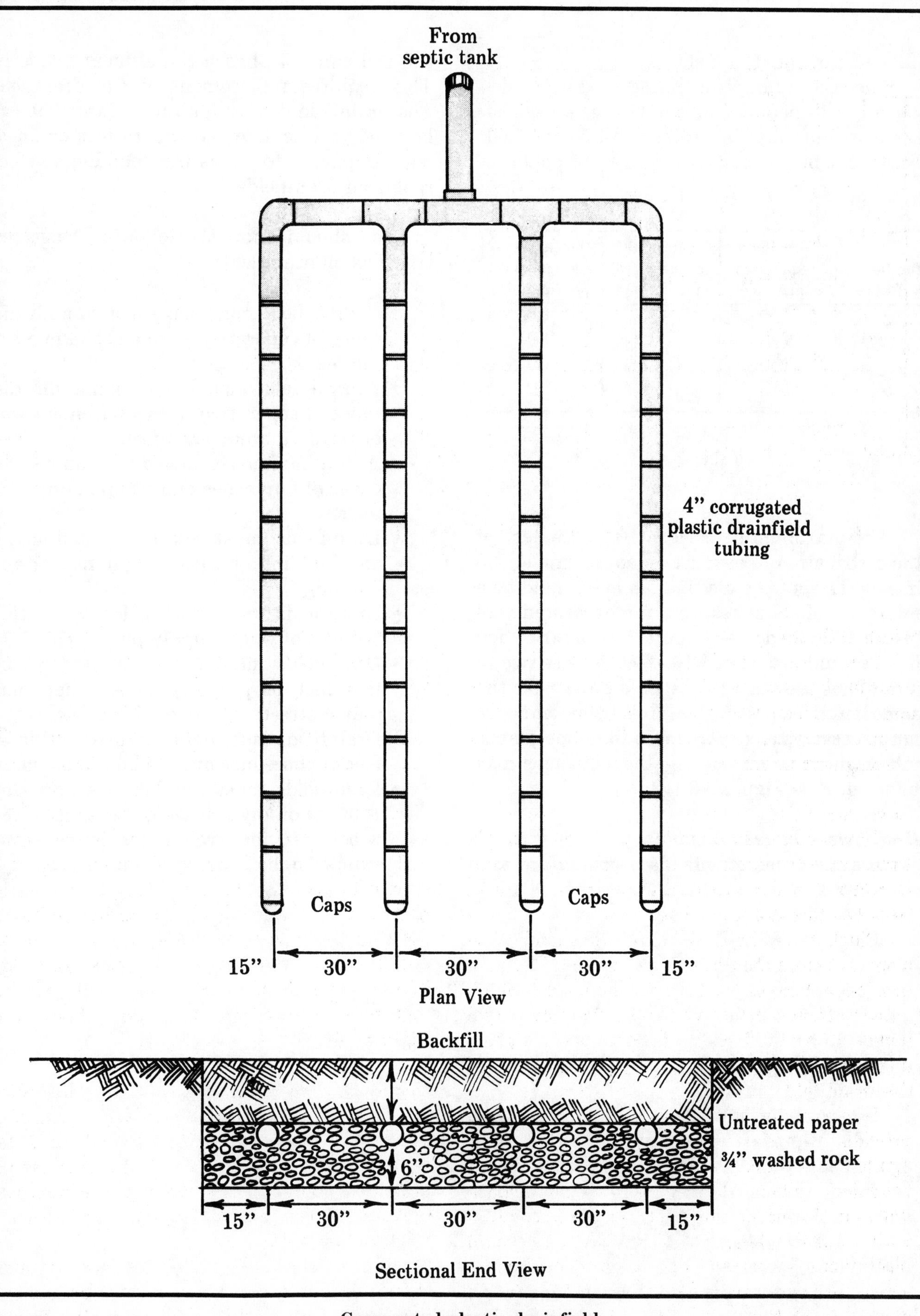

Corrugated plastic drainfield
Figure 5-8

full width of the drainfield.

Since the tubing is manufactured with adequate fixed openings, all joints are tight. The entire area of drainfield filter material (washed rock) must be covered with untreated paper.

Number of bedrooms	Drainfield required in square feet	Linear feet conventional block drain	Linear feet corrugated 4" plastic tubing
2	100	25	40
3	125	32	50
4	150	38	60
5	175	44	70

Relays
Table 5-9

The maximum distance between centers of the distribution lines must be approximately 30 inches. The outside distribution line should be a minimum of 15 inches from the excavated wall of the filter bed. See Figure 5-8. Distribution lines should not exceed 100 feet. Where two or more lines are used they should be as near the same length as practicable. A distribution box is not necessary to connect more than one plastic tubing distribution line together, but tees may be used. (See Figure 5-8.)

Drainfield Conversion Table

In some cases drainfields become *dead* and the soil can not absorb the effluent properly. This requires replacement of the drainfield. The drainfield conversion table, Table 5-9, can be used to calculate the square feet or linear feet required for existing buildings when replacing drainfields.

You should know the following basic restrictions on drainfields:

- A drainfield must have a minimum distance of separation from a soakage pit of ten feet.
- A drainfield must have a minimum distance of separation from a basement wall or terraced area of ten feet.
- A drainfield must have a minimum earth cover of ten inches and a maximum of 24 inches.
- Drainfields must not be located under any building or within eight feet of any building.
- Drainfields must not be located within ten feet of water supply pipe lines.
- Drainfields must not be located within five feet of property lines other than public streets, alleys or sidewalks.
- Drainfields must not be located within 50 feet of shorelines of open bodies of water.
- Drainfields must not be located within 100 feet of any private water supply well which provides water for human consumption, bathing or swimming.

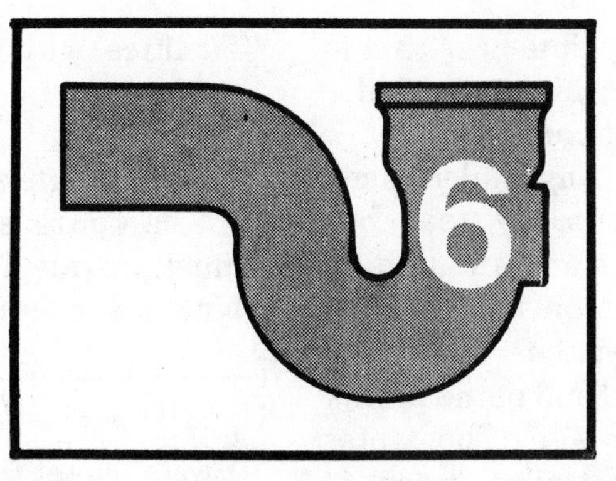

Mobile Home And Travel Trailer Parks

The popularity of both the towable mobile home and the travel trailer or motor home have forced local code adopting bodies to compile codes which provide adequate protection in trailer parks. Many if not most codes now establish minimum sanitary plumbing facilities and installation methods specifically for trailer parks. These standards vary considerably from requirements for conventional permanent structures. You should understand the standards established for designing and installing sanitary collection and water distribution systems in trailer parks. Most plumbers will eventually be required to do this type of work.

Travel Trailers

Travel trailers and motorized homes are generally defined in the code as *dependent trailers.* They do not have a plumbing system that is suitable for semi-permanent connection to the park sewage system. They are vehicles, portable structures built on a chassis for travel on public thoroughfares. Modern motor homes and travel trailers are usually self-contained units. They have a kitchen, bath, and living quarters and carry a water storage tank which operates the plumbing fixtures and a holding tank which retains the waste water.

Trailer parks are usually equipped with facilities which provide the sanitary services required by travel trailers. As the supply of potable water begins to run out and the waste holding tank becomes full, the park facilities are used to empty the waste holding tank and replenish the supply of fresh water.

Mobile Homes

Mobile homes are commonly defined as *independent trailer coaches.* They are used as permanent dwelling units on space leased in an approved trailer park. The operator of the trailer park provides each trailer coach space with water, electricity and a gas and water tight connection for sewage disposal. The owner of a trailer coach and the operator of the trailer park are jointly responsible for the sanitation of the trailer coach.

Definitions

There are a few code definitions that are unique to trailer park facilities:

Air lock The flexible hose connection from the trailer to the sewer connection may form a trap if not properly installed. Air trapped in the drain hose slows or completely stops the flow of liquid waste or sewage.

Branch service line That portion of the water distribution system extending from the park service main to a trailer site. This includes connections, devices and appurtenances.

Dependent travel trailer Any trailer coach used as a temporary dwelling unit for travel, vacation and recreation. It usually has built-in sanitary facilities but not a plumbing system suitable for connection to the park sewage and water supply system.

Drain hose An approved flexible hose that is easily detachable and used to connect the trailer drain to the park's sewer inlet connection.

Independent trailer coach Any trailer coach designed for permanent occupancy that has kitchen and bathroom facilities and a plumbing system suitable for connection to the park sewerage and water supply system.

Inlet coupling The terminal end of the park's water system to each trailer site. The water service connection from the trailer coach is made by a swivel fitting or threaded pipe end.

Intermediate waste holding tank An enclosed tank used on travel trailers for temporary retention of water-borne waste.

Mobile home or travel trailer park The site where one or more trailers are parked. The trailers may be used for temporary or permanent living quarters.

Park sanitary drainage system The entire drainage piping system used to convey sewage or other liquid waste from the trailer drain connection to a public sewer or private sewage disposal system.

Park water main The portion of the water supply piping that extends from the public water supply or other source of supply to the branch service lines.

Service connection The portion of the water distribution system which extends from the park branch service line to the inlet fitting at the trailer.

Service building A building in a trailer park which houses toilet or laundry facilities.

Trailer A mobile home, truck coach, travel trailer or recreation vehicle which can be used as a dwelling.

Trailer coach Any vehicle which can be licensed for use on public streets but is designed for permanent occupancy as a dwelling for one or more persons.

Trailer coach space A site or lot within a trailer park designated for use by one trailer coach.

Refuse All solid wastes, including garbage, rubbish and ashes, but excluding body waste.

Trailer Park Sanitary Facilities

Sites approved for either independent trailers or dependent trailers must have a service building with toilet facilities. Sites approved for use only by dependent trailers must have a service building with public toilet and bath facilities within 200 feet of the most distant trailer site.

Toilet Facilities for Dependent Trailers

The park's service building or buildings must provide a minimum number of fixtures for both sexes as follows:

	Women	Men
1 water closet for each	15	20
1 lavatory for each	20	20
1 shower bath for each	20	20
1 urinal for each	--	25

Parks completely sewered and intended to service both dependent and independent trailers must provide a service building or buildings with a minimum number of fixtures for each 100 trailer spaces or fraction thereof:

For Women	For Men
1 water closet	1 water closet
1 lavatory	1 lavatory
1 shower	1 shower
	1 urinal

All parks must provide laundry facilities as follows: For each 25 trailer coach spaces, an automatic washer and one 2-compartment laundry tray. Where wringer-type washing machines are used, one 2-compartment laundry tray must be provided for each machine. For example, if a park has 100 trailers, the park has to provide four automatic washers and one 2-compartment laundry tray, or four wringer-type washers and four 2-compartment laundry trays.

In determining the facilities required, each trailer is assumed to be occupied by three people.

Materials

All plumbing materials, fixtures, and appliances must comply with the standards in Table 4-2. Piping materials used in the park drainage system can be of any of the materials listed across from "Building House Sewer" in Table 4-1.

Sizing a Park Drainage System

Fixture unit load value which must be

Size of sewer (inches). Based upon slope, pitch of 1/8" per ft.	Maximum number of trailers, individually vented system	Maximum number of trailers, loop or circuit vented system
3	2	0
4	20	12
5	42	25
6	80	55
8	175	166
10	325	270

Installation method
Table 6-1

assumed for each trailer can be as few as six or as many as 15, depending on the code being used. Typically, codes use a value of nine fixture units per trailer. This procedure differs considerably from conventional buildings where each plumbing fixture is totaled separately. Because the number of fixture units assumed is only an estimate, most codes include a table for sizing the park drainage system based on the number and type of trailers. Table 6-1 is typical of what many codes require. It shows the maximum number of trailers that may be connected to each sewer size.

Installation

The installation of bedding and backfill are the same as for building sewers and depend on the type of materials used. Each trailer site must be provided a sewer lateral not less than 3 inches in diameter (4 inches in some codes) and the line must be capped when not in use. Vent pipes on building drainage systems must be located at least ten feet from adjoining property lines and must extend ten feet above ground level. Vent pipes should be securely anchored to

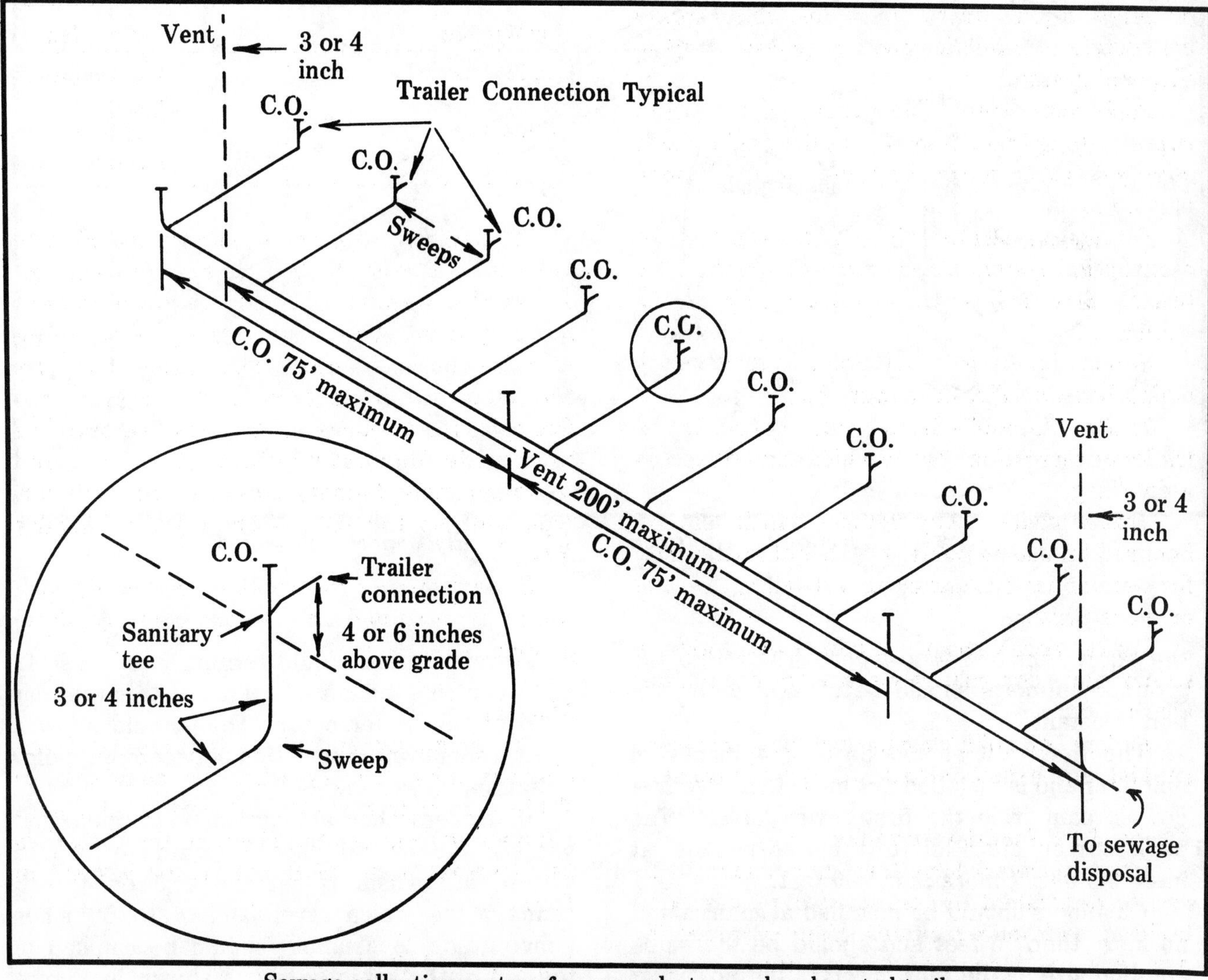

Sewage collection system for properly trapped and vented trailers
Figure 6-2

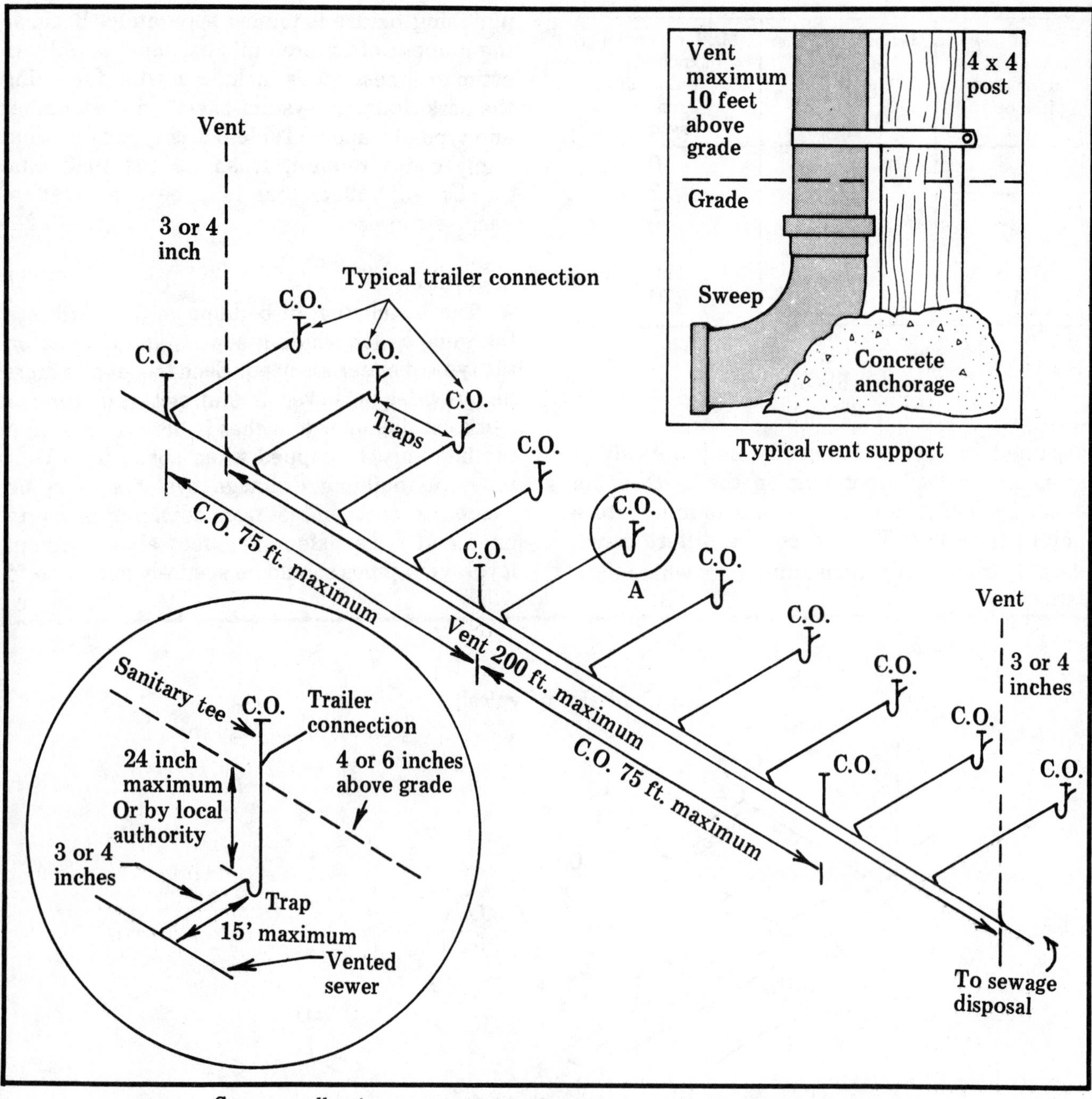

Sewage collection system for trailers not properly trapped and vented
Figure 6-3

the equivalent of a 4 x 4 post driven into the ground. Supports should be rot and deterioration resistant.

The first vent should be 3 or 4 inches in diameter and is installed not more than five feet downstream from the first sewer lateral. The park sewer main should be re-vented at intervals of not more than 200 feet.

Cleanouts should be installed at intervals of no more than 75 feet and should be the same nominal size as the pipe it serves but no larger than six inches.

Sewer laterals should terminate at least 12 inches outside the left wheel and within the rear third of the trailer coach. This should allow a short trailer drain connection between the trailer outlet and sewer inlet.

In trailers that are properly trapped and vented, the lateral should terminate with a sweep into which is caulked a 3 or 4 inch sanitary tee which terminates 4 to 6 inches above grade. A cleanout should be caulked in the top of the sanitary tee as shown in Figure 6-2. Trailers that are not properly trapped and vented should have a lateral that terminates with a 3 or 4 inch P-trap into which is caulked a sanitary tee terminating 4 to 6 inches above grade. A cleanout should be caulked in the top of the sanitary tee as shown in Figure 6-3. Since a P-trap is required at the end of the lateral (branch line), the measured horizontal distance from a vented sewer not having a re-vent must not exceed 15 feet. The P-trap must not be placed more than 24 inches below grade unless specifically allowed by local authorities.

Water Distributing System

A branch service line connected to the park main supplies potable water to each trailer site. The line terminates on the same side of the trailer site as the trailer sewer lateral.

The park's distributing system should provide a minimum pressure of 20 p.s.i. at each trailer site. The minimum size pipe in the park's water distributing system is ¾ inch in diameter. The service connection to each trailer must be a minimum of ½ inch diameter.

Trailers are connected to the park's water distributing system with a separate shutoff valve and a springloaded, soft-seat check valve on each branch service line. The valve must be located near the service connection for each trailer.

The service connection should be made with an approved flexible tubing with either end having fittings of a quick disconnect type. These fittings do not require any special tools or knowledge to install and remove.

Piping materials used in the park water distributing system should comply with Table 4-1. The materials and installation methods are the same as for water systems in Chapter 9.

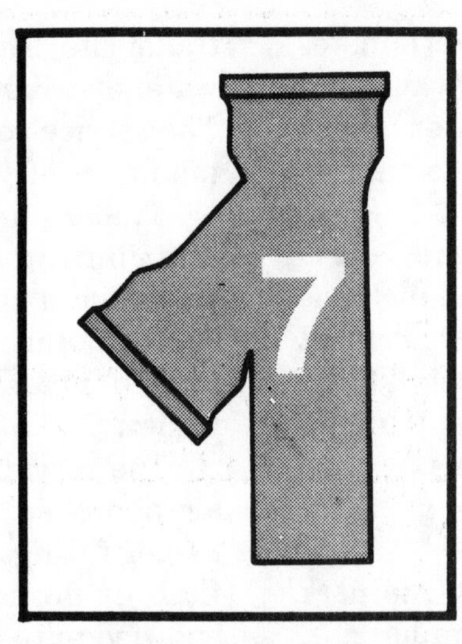

Public Water Supply And Distribution Systems

Two thousand years ago the citizens of Rome constructed overhead aqueducts and underground tunnels of masonry to bring water into the city from mountains fifty miles away. These aqueducts are ranked among the world's engineering triumphs. The aqueducts created a need for pipe to carry water to buildings within the city. Romans imported lead from the isles of Britain and the plumber ("plumbarius" or lead worker) of Rome developed the first lead pipe. These primitive pipes conveyed water by gravity flow to the private bathrooms of the upper class. Some sections of these lead pipes still exist today. Clearly, the workmanship of the Roman plumber was excellent. A recent hydraulic test was conducted on a section of ancient 4 inch diameter, ¼" wall thickness pipe. It did not fail until a pressure of 250 psi was reached.

Early American Water Systems

In 1652 in Boston, Massachusetts, the first gravity water supply system in America was built. Hollowed hemlock logs were used as pipes. About 1700 the city of New York installed a gravity water supply system with pipes built of wood. These were laid under streets, to street pumps or hydrants where the water could be sold. Prior to this, New York and other cities got their water from wells or ground cisterns in which rainwater was collected. With the exception of a few wooden sinks using hand pumps to draw water from shallow wells, plumbing fixtures were not installed within buildings until the 1800's. Plumbing fixtures required pressurized water systems and none were in use. The first U.S. city to install a water pumping station was Bethlehem, Pennsylvania, in 1754. Almost a hundred years later (1842), New York City installed a pressurized water system. This brought about a revolutionary change in construction because plumbing fixtures could be installed within buildings for the first time. The drudgery and inconvenience of having to carry water from street hydrants, wells, and cisterns was eliminated. A safe and abundant supply of water for all domestic purposes was available at the turn of a handle. Large cities and modern civilization would be nearly impossible without adequate water distribution and sanitary liquid waste disposal systems.

The sources of water for public water supply systems are lakes, rivers, and deep wells. This raw water is generally not acceptable for human consumption and must be filtered and treated until it is safe for drinking and other household applications. Unpleasant taste, odors, and impurities are removed through treatment before the water is distributed through mains.

Local authorities usually require that a connection be made to a public water supply

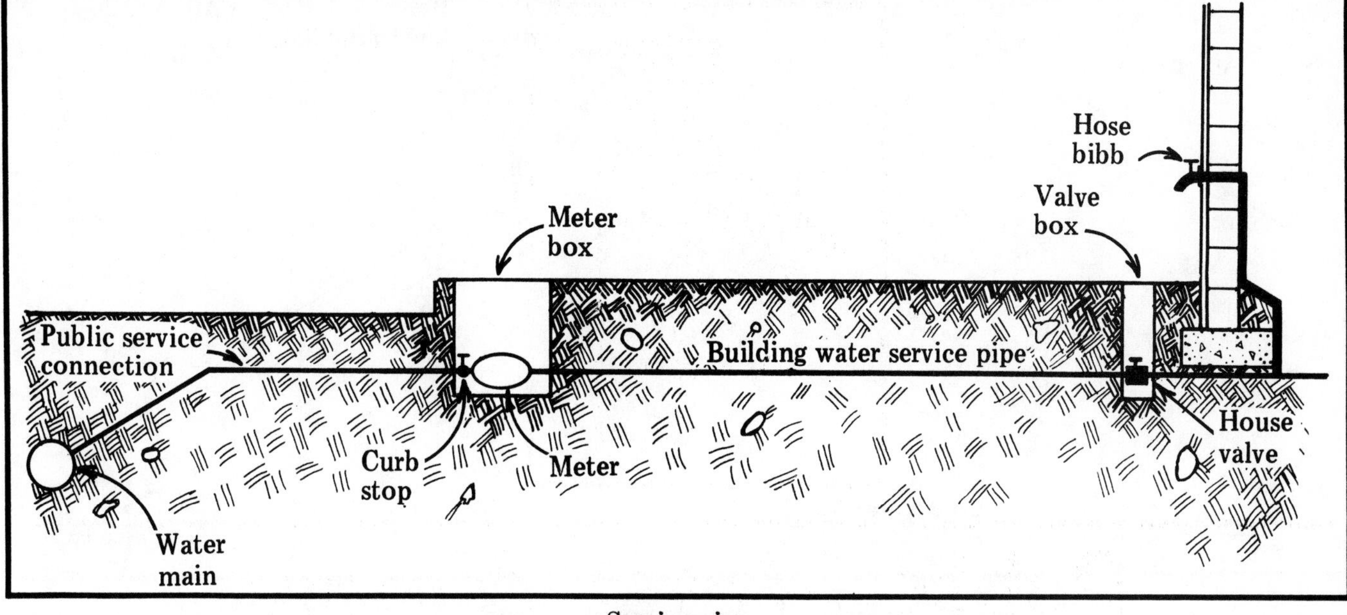

Service pipe
Figure 7-1

system if one is available. Local health departments monitor public water supply systems on a regular basis so that pure and wholesome water is always assured.

Definitions and Illustrations

The definitions in this chapter are similar to what you will find in the code. Every effort has been made to simplify the code's complex wording. Included here are specific water piping diagrams which should give you a better understanding of what the code requires.

Water main This public water distribution system is located in either the street, alley or a dedicated easement adjacent to each owner's parcel. The main carries *public water* for *community use*. This means that the main is a common pipe installed, maintained and controlled by local public authorities.

Property owners can tap into this public water system when the utility company approves the connection and has received the fee established by the local public regulatory agency. The service connection must have a curb stop (valve) and be connected to a water meter at the property line. (See Figure 7-1.)

Water service pipe A building water service pipe begins at the outlet side of the water meter at the property line and ends where the line reaches the first water distributing pipe. Service pipe installation and maintenance are the responsibility of each property owner. (See Figure 7-1.)

Water distributing pipe This is any pipe within a building (premises) which conveys water from the water service pipe to the plumbing fixtures, appliances and other water outlets. (See Figure 7-2.)

Water supply system This includes the water service pipe, the water distributing pipes, standpipe system, and the necessary connecting pipes, fitting, control valves and all appurtenances on private property.

Water outlet As defined in the code and used in connection with the water distributing system, the water outlet is the discharge opening for water:

- To a plumbing fixture.
- To atmospheric pressure (except into an open tank which is part of the water supply system).
- To a boiler or heating system.
- To any water-operated device or equipment not a part of the plumbing system but which requires water for operation.

Potable water It is any water which is satisfactory for drinking, culinary and domestic purposes and meets the requirements of the health authority having jurisdiction. It is water considered to be *purified* by being treated by one or several processes as may be required by its original untreated condition.

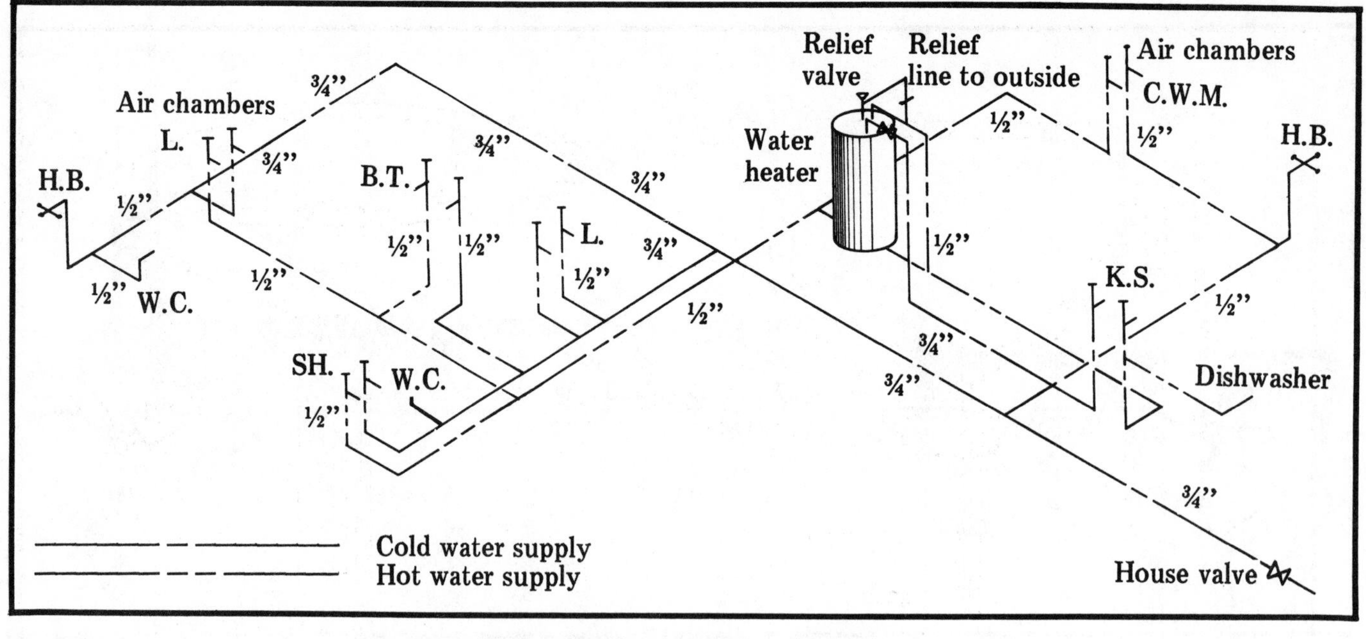

Water distributing piping diagram
Figure 7-2

Cross-connection Any physical connection or arrangement between two otherwise separate piping systems which may allow an exchange of potable water with anything that is not potable water if pressure differential is not maintained to create the assumed direction of flow. These cross-connections are prohibited. The most common cross-connection possibility specifically prohibited by the code is the physical connection between a private water supply system (domestic well) and a public water supply system.

Back-siphonage Any flow into a potable water distributing system from any source other than its intended source due to a *negative pressure* in the pipe. For example, assume a running garden hose is being used to dispense insecticide thru a sprayer. If the pressure in the public water system should suddenly fall (due, for example, to the demand fire engine pumper trucks can place on the system) a negative pressure could draw poison back through the garden hose and into the water supply system. When pressure was restored the contaminated water could be distributed to many homes.

Backflow preventer A device to prevent backflow of questionable water into the potable water system. A backflow preventer is generally known in the trade as a "vacuum breaker." There are four general types: pressure type, flush valve type, ballcock type, and hose-outlet type. The design of all "vacuum breakers" is the same: when a vacuum occurs (negative pressure) the breaker opens to admit air and relieve the vacuum. Air enters the pipe to replace water drawn from the pipe. This prevents back siphonage of questionable liquids at the ends of the distribution system.

Backflow connection (sometimes known as "*backflow condition*") Any arrangement that is likely to produce a backflow.

Backflow A flow into a potable water distributing system from any source other than its intended source.

Air gap (in a water supply system) The unobstructed vertical distance through the free atmosphere from the lowest opening from any pipe or faucet supplying water to a tank, plumbing fixture, or other device, to the flood level rim of the receptacle.

Codes today prohibit below-rim potable water supply outlets (faucets). The water supply outlets (faucets) must terminate at a suitable height above the flood level rim of the receptacle to provide an effective air gap. (See Figures 7-3 and 7-4.)

How the Water System is Sized

The maximum rate of flow or the demand in a building's water supply system cannot be determined exactly. You can't predict how many fixtures will be used at the same time. But you still have to estimate as accurately as possible the maximum demand to ensure that there will be an adequate water supply to all fixtures when

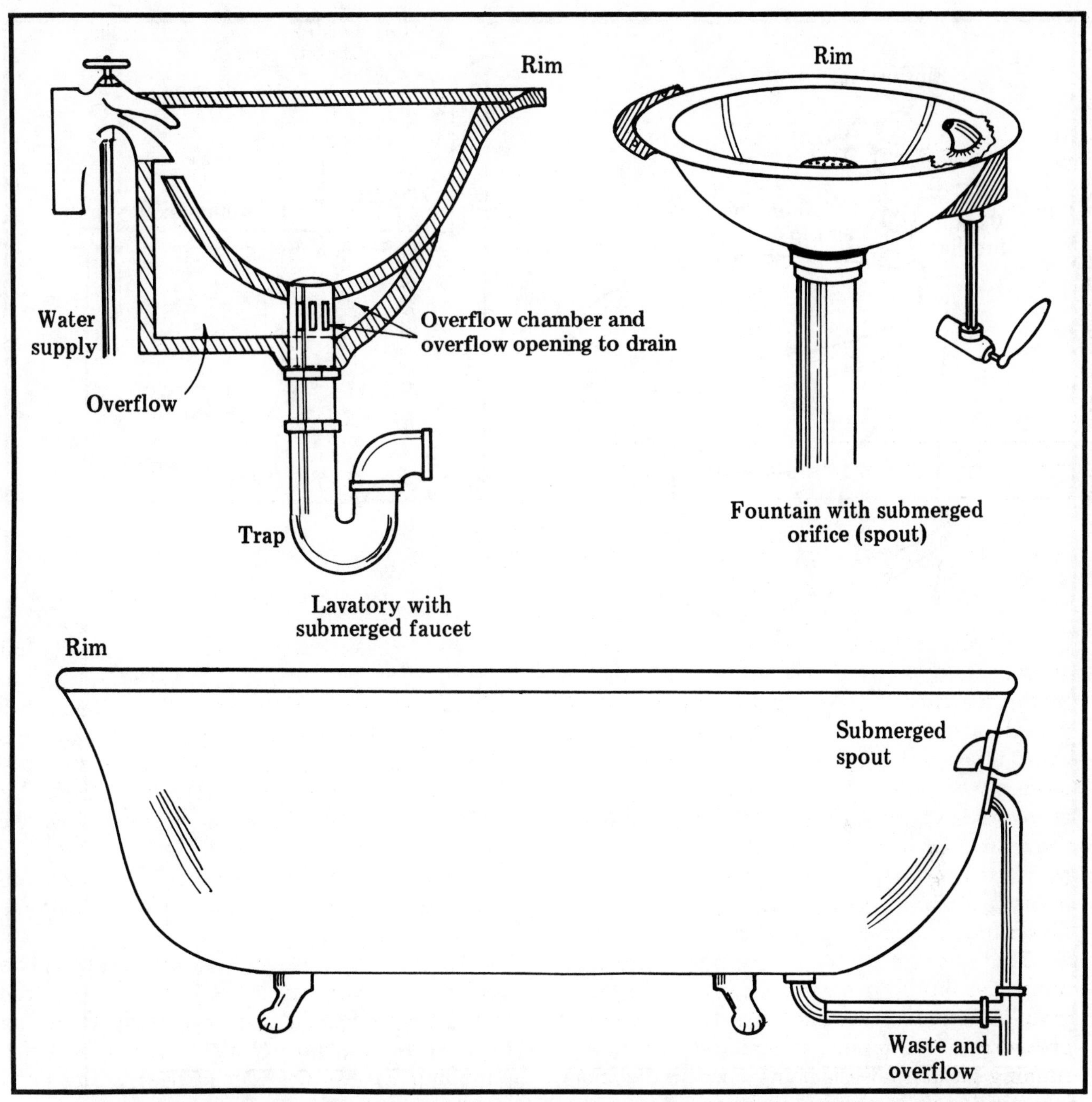

Fixtures with supply outlets below the overflow rim are prohibited. A stoppage in the waste line could cause a cross-connection. The liquid in the fixture bowl could rise to the flood-level rim and be siphoned into the water supply system through the submerged water outlets.

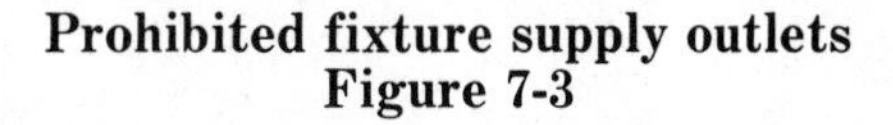

Prohibited fixture supply outlets
Figure 7-3

required. You want to avoid undersizing but still be as economical as possible in sizing water supply piping.

Complex calculations and great precision may be needed to avoid loss of pressure in elaborate water supply systems. You won't be expected to figure complex systems or account for all of the physical properties that can affect water in supply pipes: (1) density, (2) viscosity, (3) compressibility, (4) boiling point, (5) minimum available pressure, (6) friction loss, and (7) velocity flow. But you need to be familiar with the basic principles that are explained in the next few paragraphs.

There is a simplified method for sizing building water supply systems based on the

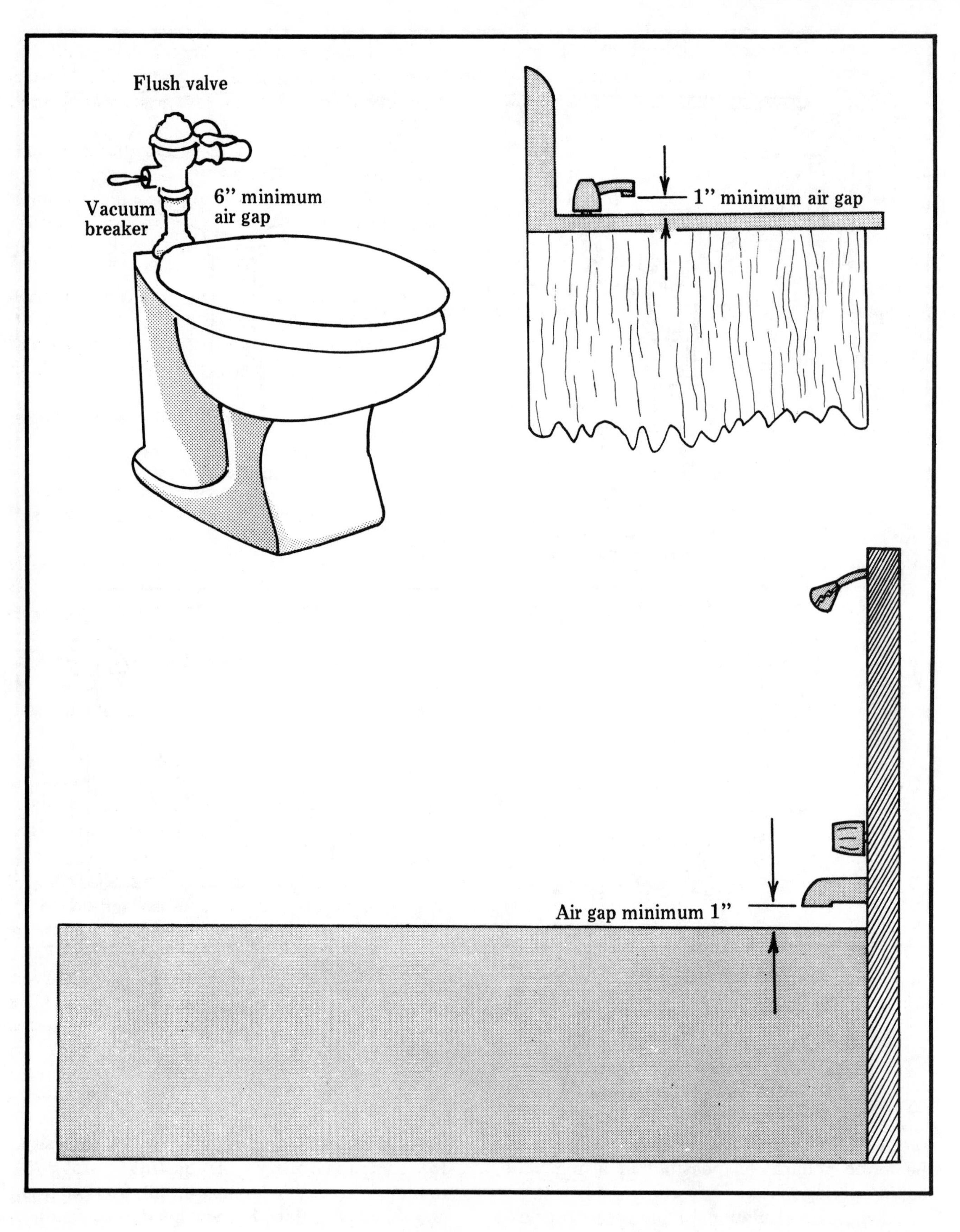

Approved fixture supply outlets
Figure 7-4

Fixture	Rate pressure- p.s.i.	Flow rate g.p.m.
Ordinary basin faucet	8	3.0
Self-closing basin faucet	12	2.5
Sink faucet - 3/8 inch	10	4.5
Sink faucet - 1/2 inch	5	4.5
Bathtub faucet	5	6.0
Laundry tub cock - 1/2 inch	5	5.0
Shower	12	5.0
Ball-cock for water closet	15	3.0
Flush valve for water closet	10-20	15-40
Flush valve for urinal	15	15.0
Garden hose (50 ft.) and sill cock	15	5.0

Rate of flow and required pressure during flow for different fixtures
Table 7-5

demand load. This method is expressed in terms of water supply fixture units and has been found to be an adequate and proper method for sizing the water supply system of many buildings. It works well for all buildings supplied from a source which has water pressure adequate to supply the highest and most remote fixtures during peak demand. The supply system is sized in accord with the velocity limitations in Table 7-5. These limits must be observed. You can use the method described below for almost all one and two family dwellings, most multiple family dwellings not over four stories high and many smaller one or two story commercial and industrial buildings if the minimum water pressure is at least 50 p.s.i. In buildings such as these, 50 p.s.i. is usually more than enough to overcome the static head and ordinary pipe friction losses. Pipe friction is not an additional factor to consider in sizing.

The method is based, first, on the application of velocity limitations that are recognized as good engineering practice and second, on the recommendations of manufacturers of piping materials.

The calculations take into account the seven physical properties listed previously and are considered adequate by authorities in most areas. This sizing method is easy enough for any professional plumber and is actually the method plumbing plan examiners and inspectors use to check building water supply pipe sizes. The tables supplied in this chapter are similar to what you will find in many codes. Illustrations have been included to explain how the tables are used.

The first thing you must find is the minimum pressure required for normal fixture use at the highest water outlets during periods of normal flow conditions. Minimum pressure requirements can be assumed to be as follows:

- For most water supply outlets at common plumbing fixtures, 8 p.s.i.
- For floor mounted ball-cock equipped water closets and urinals with flush valves, 15 p.s.i.
- For flush-valve equipped water closets, 10 to 20 p.s.i.

For other types of water supplied fixtures, refer to Table 7-5 for flow pressure in p.s.i. and flow rate in g.p.m.

The design of water supply systems using pipe sizes larger than those included in the following tables should be left to the professional engineer.

It is essential that you understand how to use Table 7-5 and the code sections that bear on sizing the piping within a water-supply system.

Sizing the Water Service Pipe

The water service pipe is the first part of the water supply system to size. It is the heart of the water supply system. This pipe must convey the necessary volume at an acceptable velocity to the water distributing pipes within a building if the fixtures are going to operate properly.

The water service pipe size must continue within the building to become the building water main, whether the pipe is horizontal or vertical as it would be in a high-rise building. After water distributing branch pipes are connected into the building main, using proper fittings, the main may be reduced in size. As the main progresses through the building, the demand likely to be placed on the line decreases.

Tables 7-6 and 7-7 show the correct pipe sizes for residential and commercial buildings. The tables are subject to certain limitations which are not difficult to follow when sizing water supply pipes. These tables are adequate if these restrictions are kept in mind:

- Tables 7-6 and 7-7 apply only where water

Number of bathrooms and kitchens		Diameter of water service pipe	Recommended meter size	Approximate pressure loss meter and 100' of pipe	Number of bathrooms and kitchens	
Tank-type closets					Flush valve closets	
Copper	Galvanized	Inches	Inches	P.S.I.	Copper	Galvanized
1-2	--	3/4	5/8	27	--	--
--	1-2	3/4	5/8	40	--	--
--	--	1	1	30	1	--
3-4	--	1	1	22	--	--
--	3-4	1	1	24	--	--
--	--	1-1/4	1	32	2-3	--
--	--	1-1/4	1	36	--	1-2
5-9	--	1-1/4	1	28	--	--
--	5-8	1-1/4	1	32	--	--
--	--	1-1/2	1-1/2	29	4-10	--
--	--	1-1/2	1-1/2	30	--	3-7
10-16	--	1-1/2	1-1/2	17	--	--
--	9-14	1-1/2	1-1/2	21	--	--
--	--	2	1-1/2	26	11-18	--
--	--	2	1-1/2	32	--	8-18
17-38	--	2	1-1/2	27	--	--
--	15-38	2	1-1/2	32	--	--
--	--	2	2	25	19-33	--
--	--	2	2	24	--	19-24
39-56	--	2	2	25	--	--
--	39-45	2	2	24	--	--
--	--	2-1/2	2	28	34-57	--
--	--	2-1/2	2	32	--	25-57
57-78	--	2-1/2	2	28	--	--
--	46-78	2-1/2	2	32	--	--
--	--	3	3	16	58-95	--
--	--	3	3	19	--	58-95
79-120	--	3	3	16	--	--
--	79-120	3	3	19	--	--

Residential Use
Minimum water service pipe size for one and two-story buildings
Hotels, motels, and residential occupancy only
Table 7-6

main pressure does not fall below 50 p.s.i. at any time.

- **In buildings exceeding two stories but not four stories, use the next larger pipe size.**
- **Buildings located where the water main pressure falls below 50 p.s.i. should use the next larger pipe size.**
- **Table 7-8 shows the minimum sizes for fixture supply pipe from the main or from the riser to the wall openings.**
- **According to a footnote in the code, only two fixtures can connect to a ½ inch water supply branch.**
- Table 7-5 gives the required rate of flow and pressure during flow for operating different types of fixtures.
- For buildings exceeding four stories or where residual pressure in the system is below the minimum required at the highest water outlet when the flow in the system is at peak demand, an automatic control pressure pump or gravity tank must be installed. This tank must have the capacity to supply sections of the building which are too high to be supplied directly from the public water mains.

Number of fixture units flush tank water closet		Size service	Size meter	Approximate pressure loss	Number of fixture units flush valve water closet	
Copper	Galvanized iron or steel	Size Inches	Size Inches	P.S.I.	Copper	Galvanized iron or steel
18	--	3/4	5/8	30	--	--
--	15	3/4	5/8	30	--	--
19-55	--	1	1	30	--	--
--	16-36	1	1	30	--	--
--	--	1	1	30	9	--
56-84	--	1-1/4	1	30	--	--
--	37-67	1-1/4	1	30	--	--
--	--	1-1/4	1	30	10-20	--
--	--	1-1/4	1	30	--	14
86-255	--	1-1/2	1-1/2	30	--	--
--	68-175	1-1/2	1-1/2	30	--	--
--	--	1-1/2	1-1/2	30	21-77	--
--	--	1-1/2	1-1/2	30	--	15-52
226-350	--	2	1-1/2	30	--	--
--	176-290	2	1-1/2	30	--	--
--	--	2	1-1/2	30	78-175	--
--	--	2	1-1/2	30	--	53-122
351-550	--	2	2	30	--	--
--	291-450	2	2	30	--	--
--	--	2	2	30	176-315	--
--	--	2	2	30	--	123-227
551-640	--	2-1/2	2	30	--	--
--	450-580	2-1/2	2	30	--	--
--	--	2-1/2	2	30	316-392	--
--	--	2-1/2	2	30	--	228-343
641-1340	--	3	3	22	--	--
--	581-1125	3	3	22	--	--
--	--	3	3	22	393-940	--
--	--	3	3	22	--	344-785

Commercial Use

Minimum water service pipe size for one and two-story buildings

Table 7-7

- You should have the assistance of a professional engineer if a tank is required.

Note that more fixtures can be installed on a copper system than on a galvanized iron or steel system of the same size. Note also that in most cases as the water service pipe increases in size, the pressure loss is reduced or remains stable. The load values assigned to fixtures are the same for sizing water supply pipes as for sizing drainage, waste, and vent pipes. Refer again to Tables 2-8 and 2-10.

Test your knowledge of Tables 7-6, 7-7, and 7-8 with several examples. Assume that you are sizing the cold water service piping in the small apartment building shown in Figure 7-9. The hot water piping is not considered when sizing the water service and distribution lines. The horizontal and vertical piping in Figure 7-9 serves 18 kitchens and 18 baths. The street level water pressure is 55 p.s.i. The water piping material selected is galvanized. Size the 20 sections of pipe lettered *A* through *T* on Figure 7-9 without looking ahead to find the right answer.

Begin with Table 7-6. Start at the most

Type of fixture or device	Pipe size (inches)	Type of fixture or device	Pipe size (inches)
Bathtub	1/2	Laundry tray - 1, 2, or 3-compartment	1/2
Combination sink and tray	1/2	Shower (single head)	1/2
Drinking fountain	3/8	Sink (service, slop)	1/2
Dishwasher (domestic)	1/2	Sink (flushing rim)	1
Hot water heater	3/4 minimum	Urinal (flush tank)	1/2
Kitchen sink, residential	1/2	Urinal (direct flush valve)	3/4
Kitchen sink, commercial, one or more compartments	1/2	Water closet (tank type)	1/2
Lavatory	1/2	Water closet (flush valve type)	1
		Hose bibbs	1/2

Size of fixture supply pipe
Table 7-8

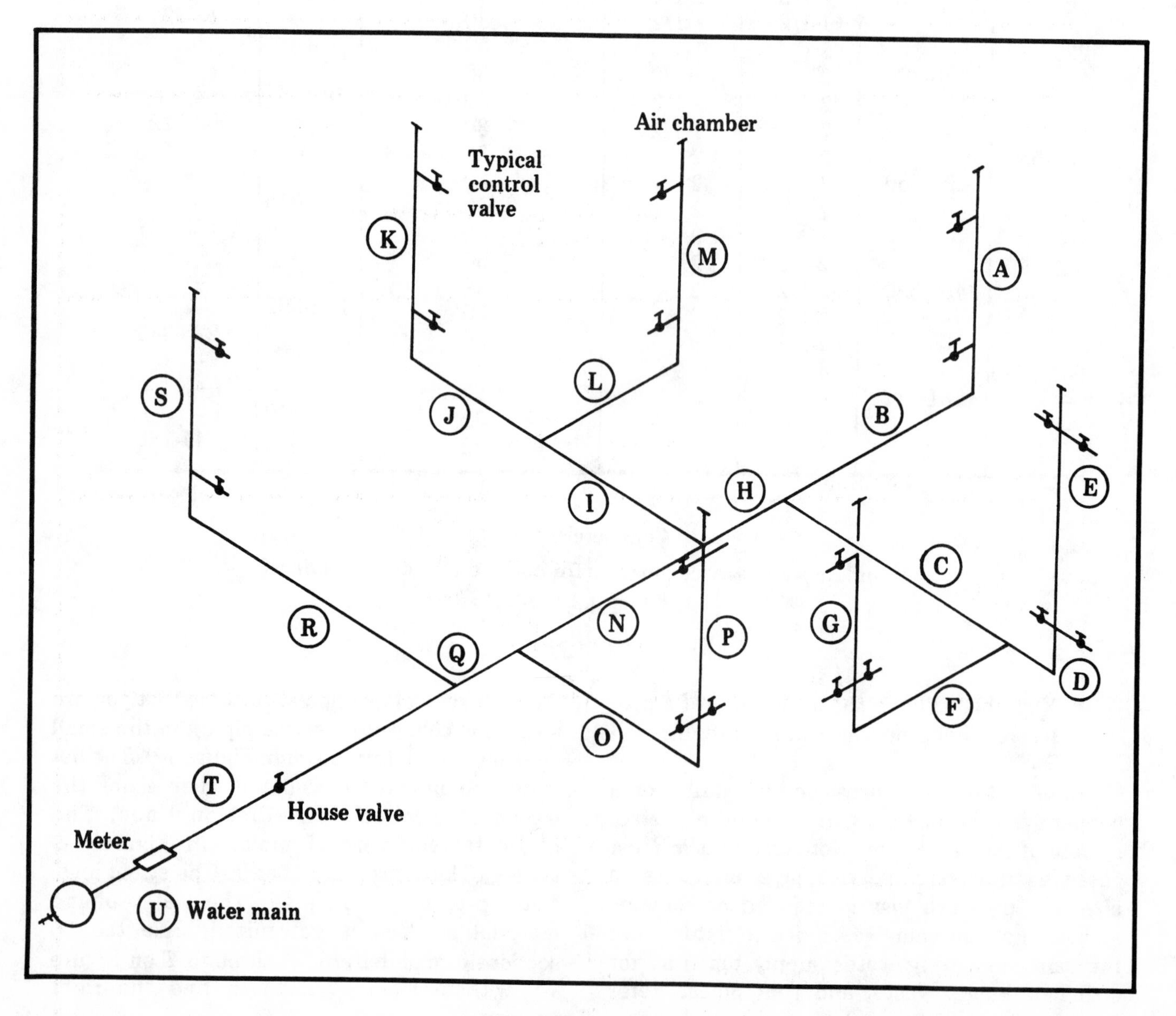

Piping sizing problem
Figure 7-9

remote outlet *A* and work back toward the water main *U*. Compute the cumulative demand for the apartment units. Remember that galvanized pipe is being used.

Answers For Figure 7-9

A -	one bath unit	¾"	K -	one bath unit	¾"
B -	two bath units	¾"	L -	two bath units	¾"
C -	seven bath units	1¼"	M -	one bath unit	¾"
D -	four bath units	1"	N -	thirteen units	1½"
E -	two bath units	¾"	O -	three bath units	1"
F -	three bath units	1"	P -	one bath unit	¾"
G -	one bath unit	¾"	Q -	sixteen units	2"
H -	nine bath units	1½"	R -	two bath units	¾"
I -	four bath units	1"	S -	one bath unit	¾"
J -	two bath units	¾"	T -	eighteen units	2"

The single water supply riser in Figure 7-10 serves eight kitchens and eight baths (two on each floor). The water piping material selected is copper. The street level water pressure is 55 p.s.i. There are four sections of riser pipe to size (*A, B, C,* and *D*) as water rises from the water main (*E*).

Without looking ahead to find the right answer, figure the sizes for sections *A* to *D*. Starting at the most remote outlet (*A*) and working back toward the building main (*E*), compute the cumulative demand for the apartment units on each floor. In column one of Table 7-6 ("copper"), you see that the next larger pipe size is to be used because the building is over two stories high.

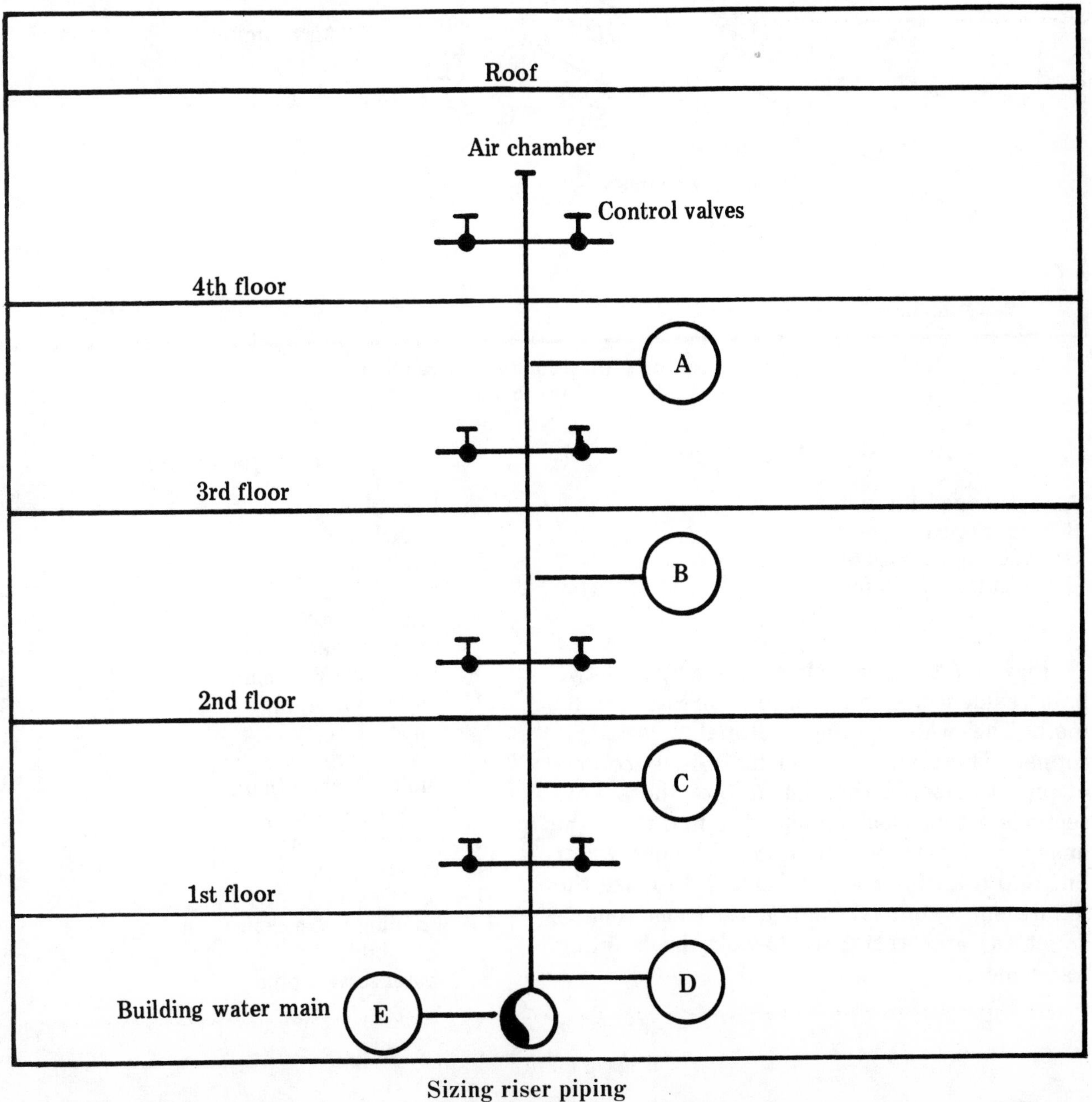

Sizing riser piping
Figure 7-10

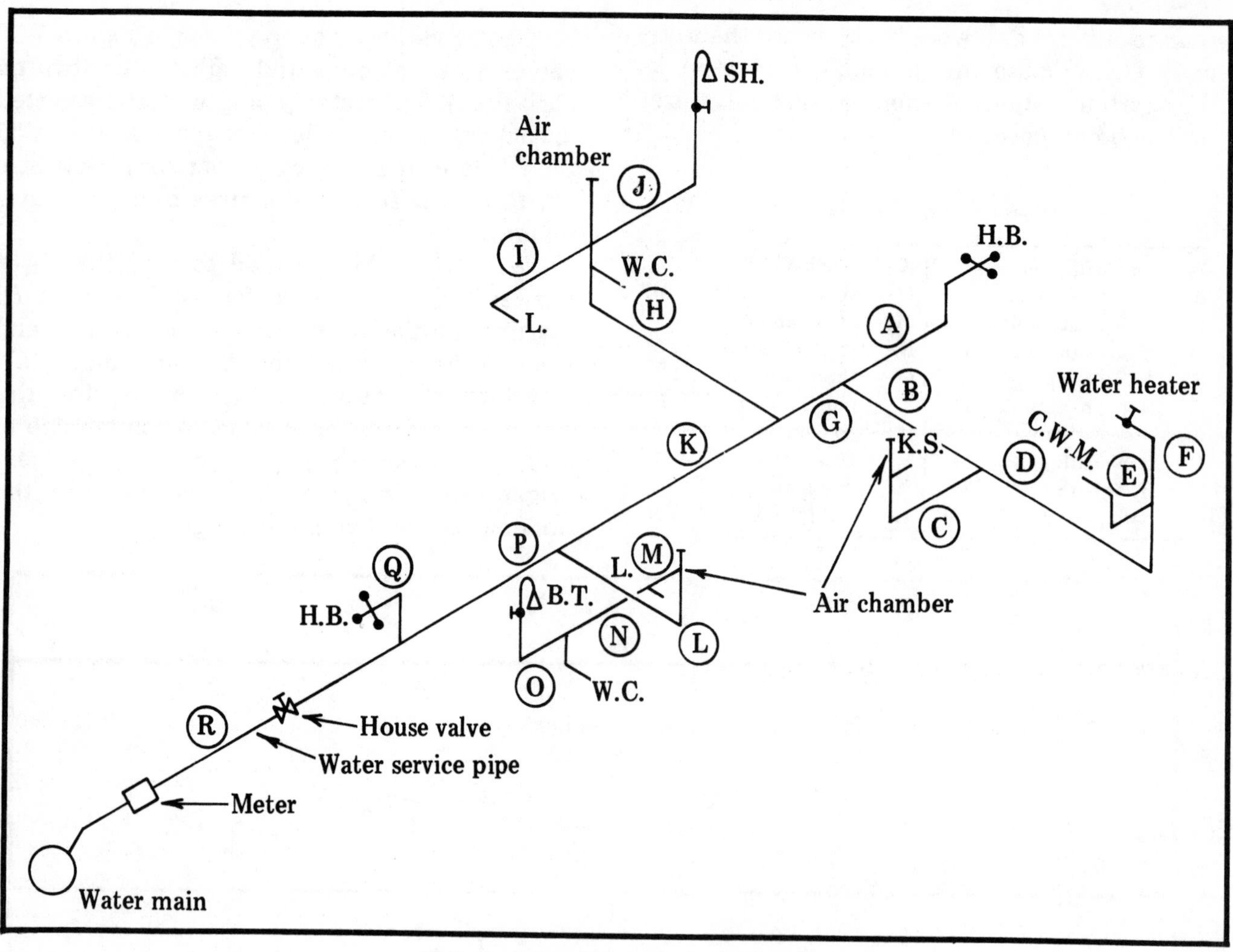

Sizing piping for typical residence
Figure 7-11

Answers for Figure 7-10

A -	two one bath units	1"
B -	four one bath units	1¼"
C -	six one bath units	1½"
D -	eight one bath units	1½"

Figure 7-11 shows piping for a typical two bath residence. The street level pressure is 50 p.s.i. The water piping material selected is copper. There are eighteen individual sections of pipe to size, *A* through *R*. Size these pipe sections without looking ahead to find the right answers. Use Table 7-6 to size the main water line and branch lines and Table 7-8 to size the fixture supply pipes. Start at the most remote outlet (*A*) and work back toward the building water meter.

Answers for Figure 7-11

A -	hose bibb	½" or ¾"
B -	branch line	¾"
C -	one fixture	½"
D -	branch line	¾"
E -	one fixture	½"
F -	water heater	¾"
G -	building water main	¾"
H -	bathroom supply	¾"
I -	one fixture	½"
J -	one fixture	½"
K -	building water main	¾"
L -	bathroom supply	¾"
M -	over two fixtures	¾"
N -	two fixtures	½"
O -	one fixture	½"
P -	building water main	¾"
Q -	hose bibb	½" or ¾"
R -	water service pipe	¾"

Hot Water Systems

Hot water systems are generally not required by plumbing codes. The exception to this is buildings designed for a special occupancy such as a restaurant or a rooming house. Apparently the writers of the code consider hot water a luxury and not a necessity. Even though the code does not require a hot water system, most buildings have one. The code is quite specific in requiring proper installation of safety devices if there is a hot water system. Some safety equipment is necessary to prevent damage to property and injury to persons using the facilities.

The code does not specify requirements for hot water distribution systems. It leaves the design of the system to professional engineers and plumbers. The code simply states: "The sizing of the hot water distribution system shall conform to good engineering practice.". In large commercial buildings a professional engineer usually designs the hot water supply systems. The plumber makes the installation from the plans furnished by the architect. In smaller commercial and residential buildings, design responsibility for the hot water system rests with the installing plumber. This chapter is intended to help you design safe and functional hot water systems for small commercial and residential buildings.

Design Objectives

There are two principal objectives in designing a good hot water system. First, the system must satisfy the hot water demand for a particular type of occupancy. Second, safety features must be built into the system to guard against the hazards of excessive pressure and temperature. Several sample diagrams are included at the end of this chapter to show you how simple hot water systems are planned.

The normal design temperature of hot water in most plumbing fixtures should range between 130 and 140 degrees. This is hot enough for most purposes but cool enough to prevent scalding of the skin. The heater thermostat is generally pre-set within the correct range at the factory and does not need additional adjustment.

Most direct heating units installed by plumbers are designed to burn gas or oil, or use electricity as a source of energy. This energy transfer rate must be adequate for the quantity of hot water stored within the heater tank. All commercial and residential storage tanks have insulation adequate to prevent excessive heat loss from the water stored.

Electric water heaters are clean and attractive and can be installed nearly anywhere within a building. Water heaters designed to burn gas or oil must be located in a well-ventilated area and have flues to carry away combustion gases. The possible locations within a building are limited. See Figure 8-1.

There is no need for constant circulation of

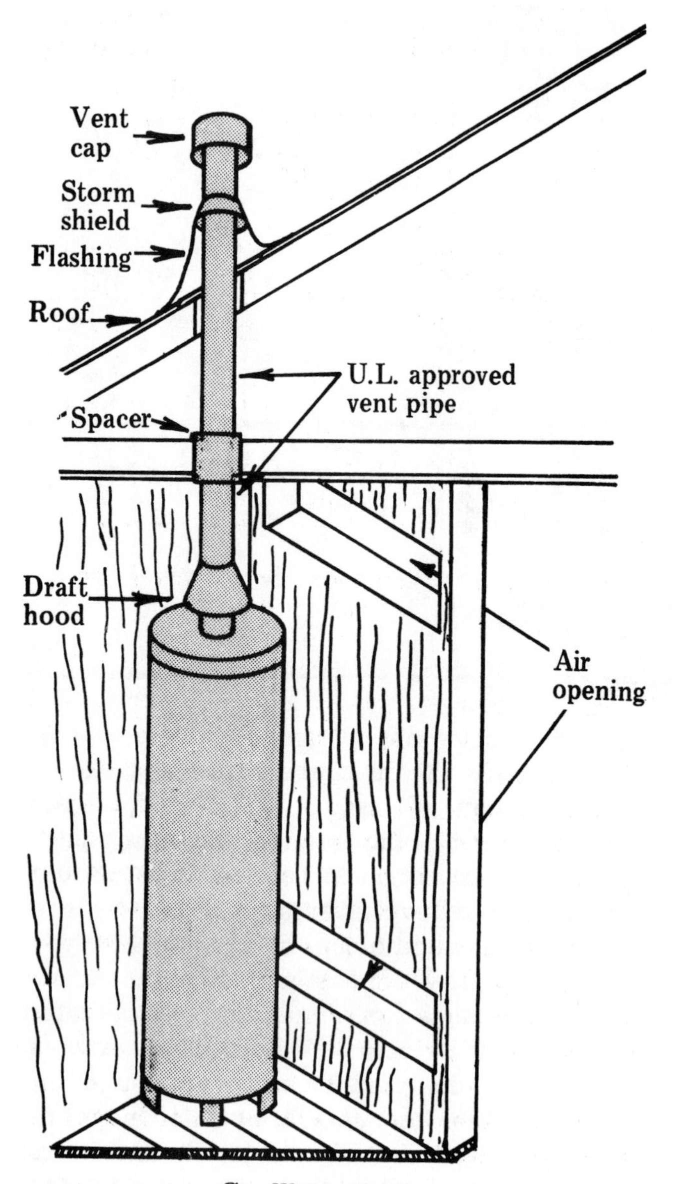

Gas Water Heater

The confined space must be provided with two permanent openings, one commencing within 12 inches of the top and one commencing 12 inches from the bottom. Adequate air for combustion, ventilation, and dilution of flue gases must be obtained.

Installation of fuel-burning water heaters
Figure 8-1

hot water in a system designed for a one or two family dwelling. A small system has small pipes and short runs. The hot water within these pipes will cool when water is not being used. But hot water is again available a short time after the valve is opened.

Circulating hot water is much more desirable in larger buildings with a central hot water system. Otherwise, there is a long waiting time and excessive waste of water each time hot water is required.

Water expands as it is heated but is otherwise a relatively incompressible liquid. Therefore, your design must allow for the expansion and contraction of hot water lines. In smaller dwelling units with short runs of pipe this is no problem. In larger units with hot water riser supply pipes and circulating lines, expansion and contraction can be a serious problem. Depending on the piping materials used, the expansion for 100 linear feet in a 100 degree F rise may be as little as ¾ inch or as much as two inches. An expansion variance of 1½ inches should be allowed for each 100 feet of piping materials. The common four elbow loop shown in Figure 8-2 should provide adequate developed length to prevent excessive stress in a hot water piping system.

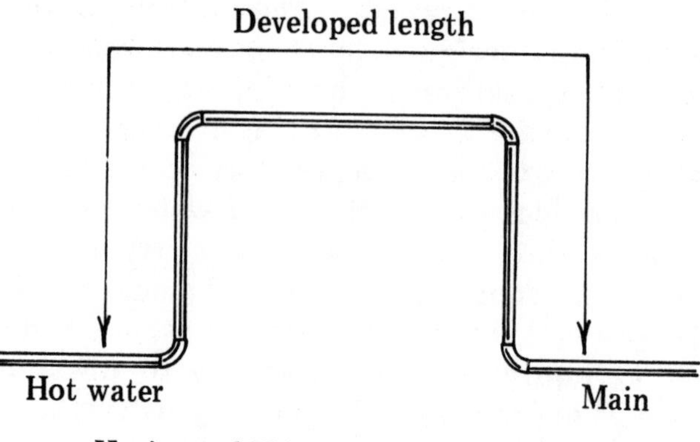

Horizontal U bend (four elbow loop)
Figure 8-2

Buildings large enough to use hot water riser pipes and circulating lines should have pipes that are insulated. The savings from insulating hot water pipes can be substantial.

Heater Capacities

The most common type of water heaters are the electric and fuel-burning storage tank type. These units are constructed with small storage tanks because they provide only the heating capacity necessary to maintain the water at 140 to 150 degrees Fahrenheit during the peak draw period. The peak draw period is usually assumed to be one hour though in most homes it is as little as 20 minutes.

When sizing hot water storage tank capacity,

assume that only 75% of the tank's hot water supply is available during the period of peak draw (one hour). If you think the peak demand rate may extend over a two hour period, assume that 37.5% of the tank's capacity may be used per hour. If the peak demand rate may last three hours, assume that 25% of the tank's capacity may be used per hour (75% over the 3 hour period).

For a peak draw period of one hour, a 20 gallon water heater should provide 15 gallons. (75% of the tank's capacity) at 140° to 150° water temperature. For a peak draw period of two hours, a 20 gallon water heater should provide 7.5 gallons, or 37.5% of the tank's capacity per hour. For a peak draw period of three hours, a 20 gallon water heater should provide 5 gallons or 25% of the tank's capacity per hour.

The heating capacity per hour for a storage tank type heater determines the amount of hot water which can be drawn from the tank per hour during the peak demand periods.

The following individual tank storage capacities have proved to be economical and satisfactory for the average dwelling unit *when the peak draw period does not exceed one hour.* Special demand requirements that may last more than one hour might indicate that the next larger size should be used.

Recommended Tank Sizes

Number of bedrooms	Hot water storage tank capacity
1	20 gallons
2	30 gallons
3	42 gallons
4	52 gallons

The number of bedrooms (not bathrooms) in the dwelling unit determines the size of both the septic tank and the capacity of the hot water storage tank. The number of bedrooms is the best indication of the quantity of water that will be used.

For each additional bedroom above four, use the next larger size hot water heater. You may want to split the system and install two water heaters, especially if long pipe runs are required.

For multi-family dwellings with a central hot water system for up to twelve living units, a packaged *high recovery rate* heater with a 75 gallon storage tank will generally be adequate. For buildings with up to eighteen living units, a packaged high recovery rate heater with a storage capacity of 100 gallons is usually adequate. Buildings with more than eighteen living units are usually designed by a professional engineer.

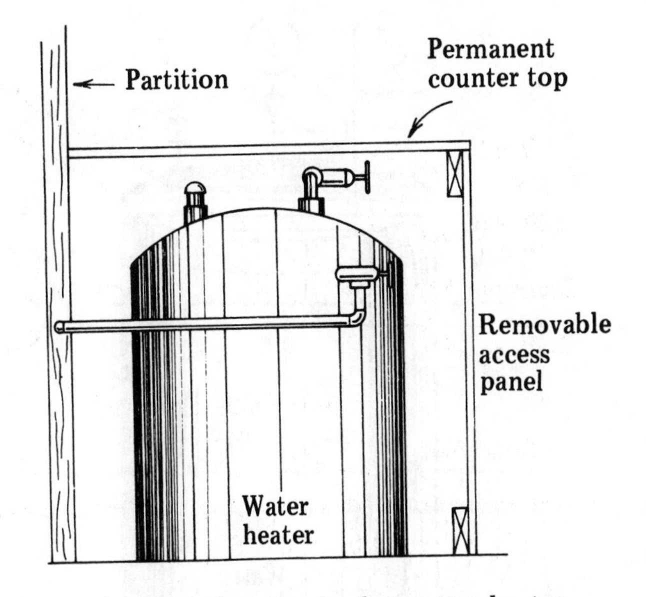

Installation of under counter hot water heater
Figure 8-3

Apartment buildings with a central hot water generating system should have a return line and circulating pump for greater efficiency. The hot water feed and hot water return lines should be well insulated to prevent heat loss.

Water heaters should be installed so that the maximum working water pressure plates and other data are visible. Locate the heater where it is accessible for servicing or replacement without removing any permanent part of the building. Note Figure 8-3.

Safety Devices

The code requires that all equipment used for heating or storage of hot water have a pressure relief valve installed to relieve excess storage tank pressure. The rate of discharge for the pressure relief valve must limit the pressure rise for any given heat input to within ten percent of the pressure at which the valve is set to open. The reason for this is that most standard hot water storage tanks are designed to withstand pressures up to 125 p.s.i. The pressure relief valve should open at 25 p.s.i. greater than the maximum working pressure, but under no circumstances should this be over 125 p.s.i.

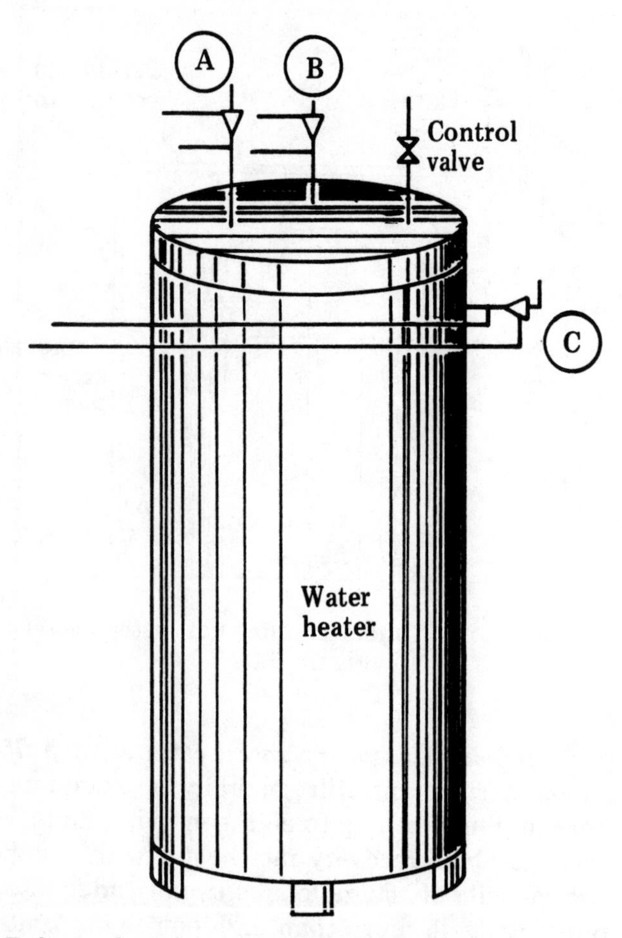

Relief valves are located in the top six inches of water heater. Positions illustrated at A, B, or C are acceptable
Figure 8-4

Temperature relief valves must be installed on all equipment used for the heating or storage of domestic hot water. The temperature relief valve must be the *reseating* type and be rated as to its B.t.u. capacity. The B.t.u. rating of all temperature relief valves must always be greater than the B.t.u. rating of the appliance it serves. This is to prevent premature opening of the valve.

In some cases the wattage of an electric water heater is known, but you do not have the B.t.u. rating. You need to choose the correct temperature relief valve and size the relief valve discharge line. If this should be the case, you can find the B.t.u. rating with the following formula: 1 watt = 3.4 B.t.u. Multiply 3.4 by the known wattage. This will give you the B.t.u. rating of the particular appliance or combination of appliances in question.

Separate pressure and temperature relief valves are not required on domestic packaged hot water heaters. A combination pressure and temperature relief valve is acceptable and is commonly used. Use only valves which meet the requirements of the American Gas Association, A.S.M.E., or other recognized authority.

The relief valves must be installed so that the temperature sensing element is immersed in the hottest water in the top six inches of the tank. See Figure 8-4.

The code prohibits the installation of a check valve or shut-off valve between the relief valve and the water heater or hot water storage tank it serves. The reason for this is simple. If the check valve should fail to operate or if someone should accidentally close the shut-off valve, the relief valve would become useless and could not serve its intended purpose. The tank could rupture, causing property damage and perhaps personal injury.

Relief valve drip pipes (*popoff lines*) should not connect directly to any plumbing drainage or vent system. This could cause contamination to the potable water system. It could also conceal from view any continuous discharge.

When a building covers the entire lot, the drip pipe can terminate in an indirect connection above a floor drain or other suitable fixture approved by local authorities. It must never terminate above a water closet, urinal, bidet, bathtub or shower stall where anyone would be scalded if the relief valve should discharge.

When hot water storage tanks are located above the roof of a building, the relief valve drip pipe may discharge on the surface of the roof.

The terminus of drip pipes for all other buildings should extend to an observable point outside the building and be carried down to within six inches of the ground. The end of relief valve drip pipes must not be threaded. This will make it difficult for someone to connect something to this pipe.

Relief Valve Discharge Lines

B.t.u.	Inside diameter pipe size
Up to 69,000 B.t.u.	⅜ inch
Over 69,000 to 127,000 B.t.u.	½ inch
Over 127,000 to 340,000 B.t.u.	¾ inch
Over 340,000 to 600,000 B.t.u.	1 inch

The sizing of relief lines is based on the B.t.u. rating of the appliance. Use the table above for sizing relief lines for single family

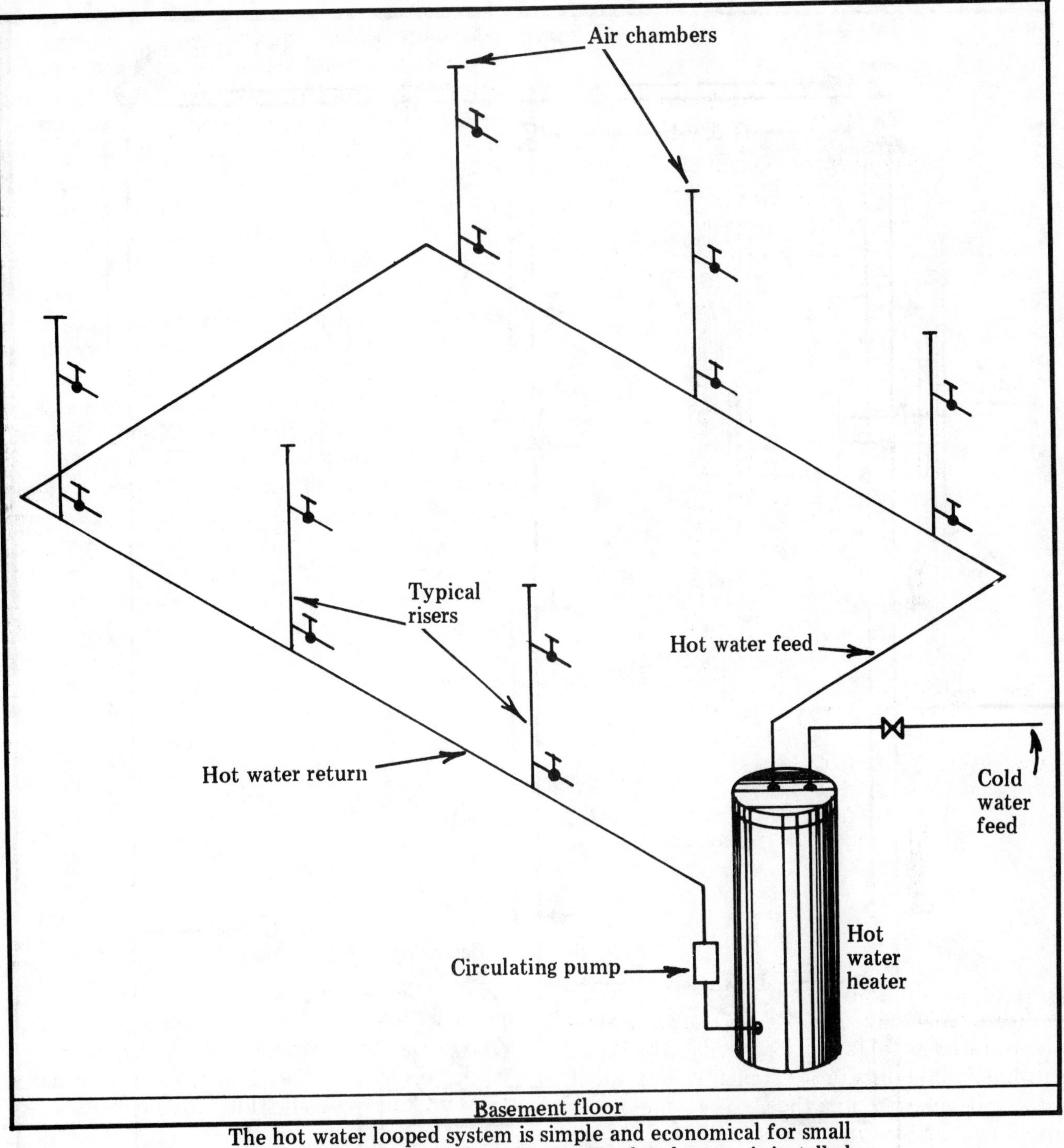

The hot water looped system is simple and economical for small apartment buildings. The hot water feed and return is installed directly beneath the bathrooms of each apartment. A circulating pump returns the water to a high recovery heater. This assures each apartment of adequate hot water

Hot water looped system
Figure 8-5

residences as well as multi-story buildings with manifold lines if you know the total B.t.u. rating. All relief pipes are sized by their inside diameter.

Hot water heaters and hot water storage tanks must have the drain cock in an accessible location. This is necessary both for flushing the tank of sediment and for emptying for repairs or replacements.

The cold water supply pipe to hot water heaters must have a minimum ¾ inch shut-off valve in an accessible location. The shut-off

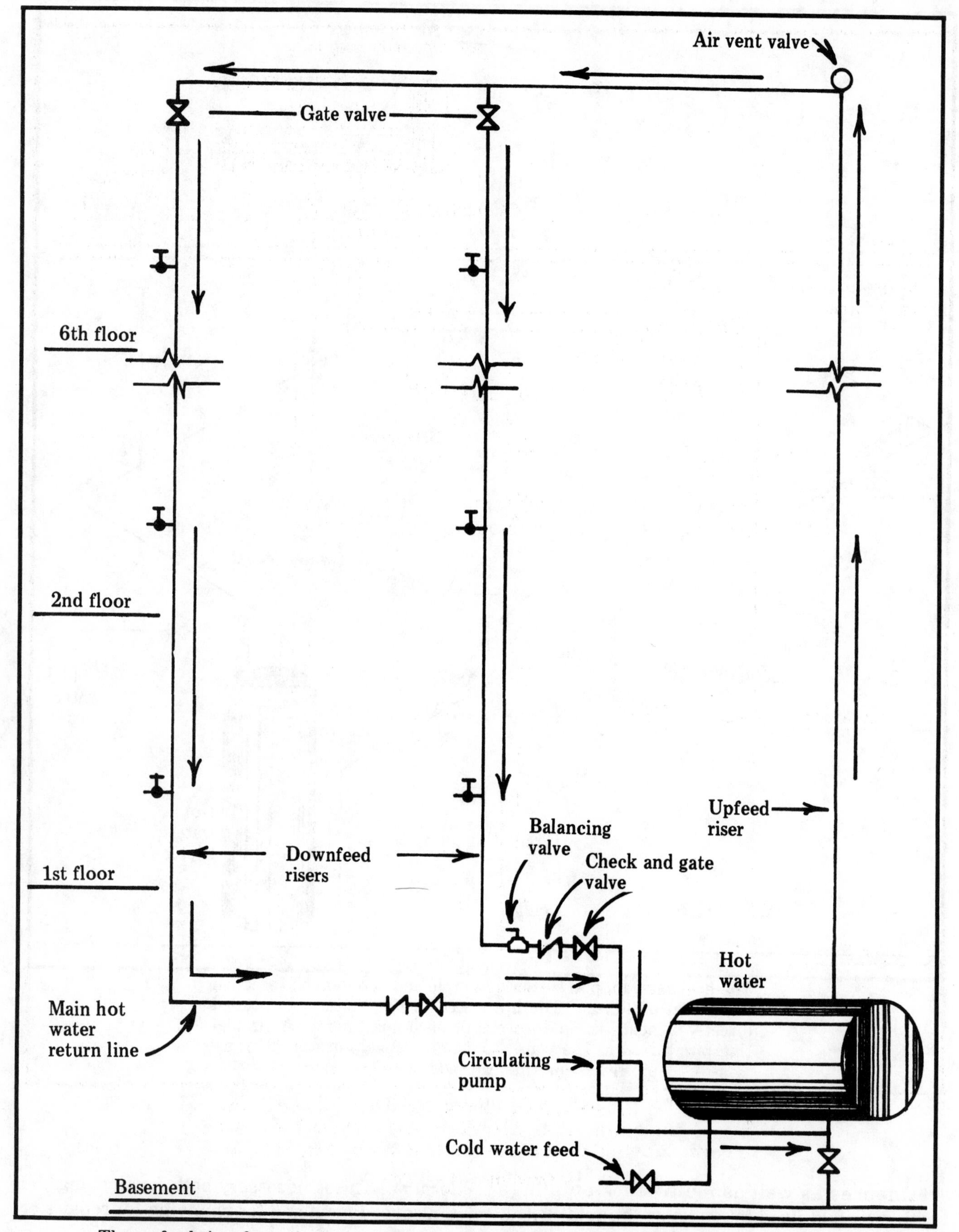

The upfeed riser from the hot water source conveys hot water to the downfeed risers. The circulating pump returns this water to the hot water generator.

Downfeed system
Figure 8-6

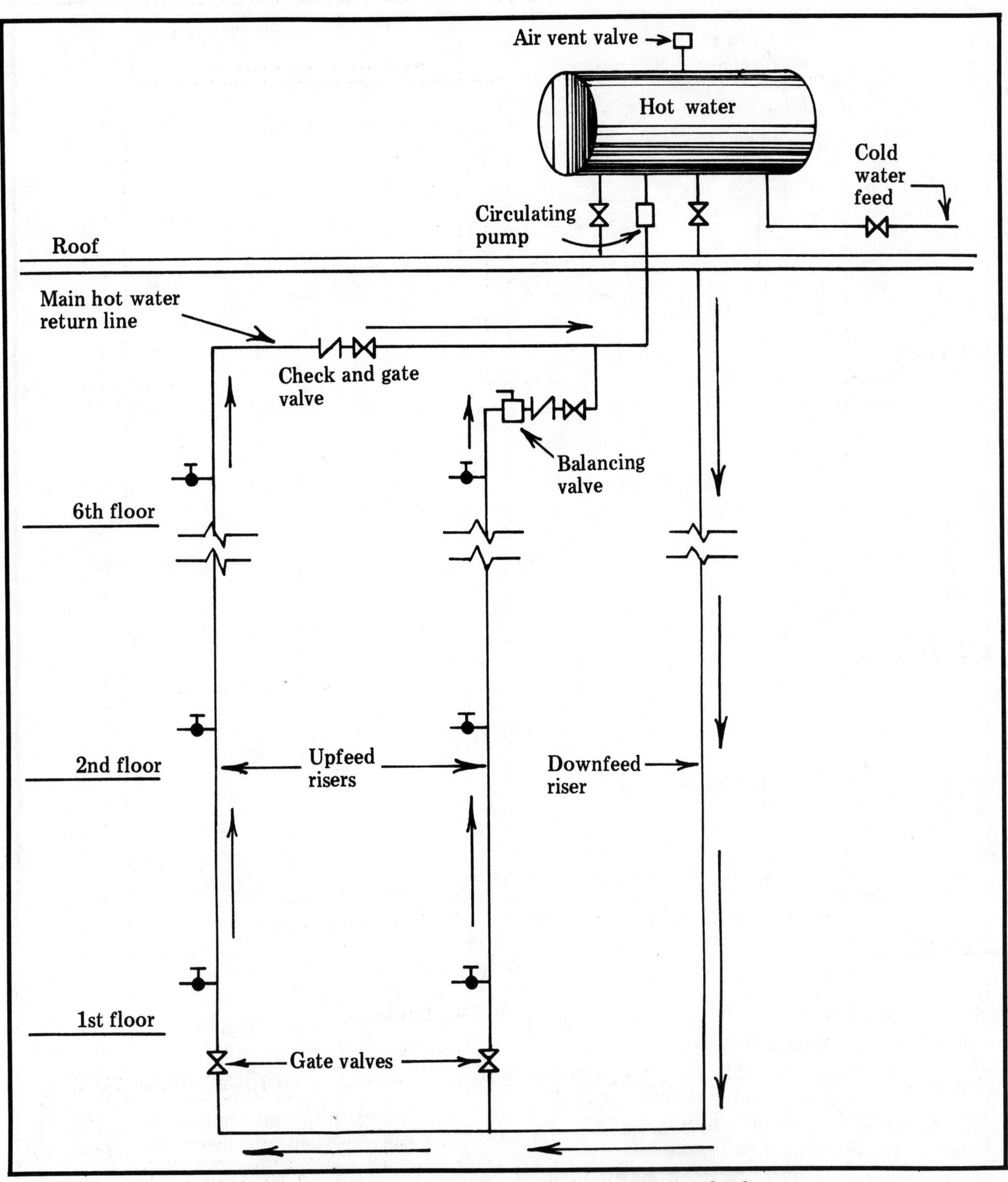

The inverted upfeed system conveys hot water to the lowest part of the system. The circulating pump returns the water through the upfeed risers to the hot water generator.

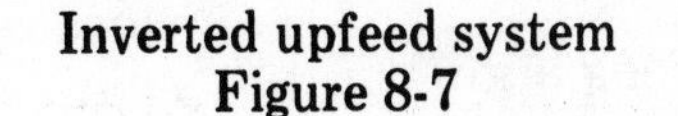

Inverted upfeed system
Figure 8-7

valve must have a cross-sectional area equal to 80% of the nominal size of the pipe in which it is installed.

The minimum size cold water supply pipe for any hot water heater, regardless of its capacity, is ¾ inch in diameter.

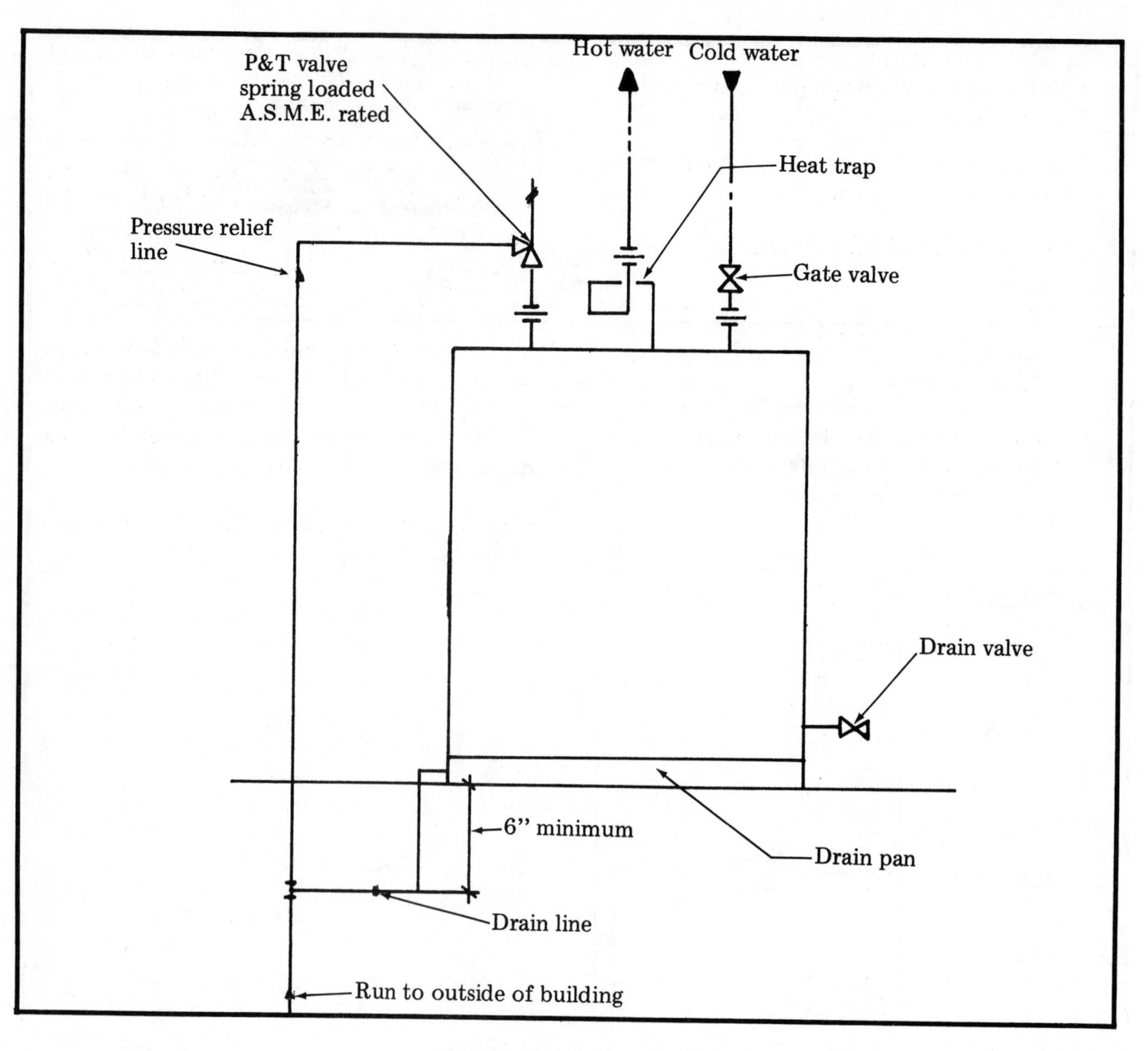

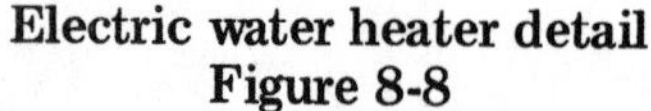
Electric water heater detail
Figure 8-8

Water Heater Drain Pans

A drain pan must be provided when water heaters or hot water storage tanks are located above the ground floor level of a building. This is to avoid injury to the building occupants or damage to the building. See Figure 8-8.

The following criteria are required by most codes and are enforced by local plumbing officials:

- Drain pans shall be constructed of a high-impact plastic of at least 60 mil. (1/16") thickness. Galvanized sheet steel or other corrosive resistant metal having a thickness of not less than 24 gauge is also acceptable.
- Drain pans shall be a minimum of 1½ inches deep (2 inches required by some codes).
- A 2-inch minimum clearance shall be provided between the drain pan sides and the heater.
- The drain pan shall have a minimum ¾ inch drain outlet (1 inch in some codes) located ½ inch above the bottom of the pan.
- In multiple vertical installations, the drain from pans shall run downward a minimum of 6 inches before connecting into the main riser. See Figure 8-5.

Sizing Drain Pan Risers

- A maximum of 3 drain pans may connect to a 1 inch riser.
- A maximum of 4 to 10 drain pans may connect to a 1¼ inch riser.
- 11 or more drain pans shall connect into a minimum 1½ inch riser.
- When receiving drain pan waste, horizontal pipes under a floor slab on grade shall be one size larger.
- Water heater relief lines and safe pan lines may use a common vertical riser if they are of material approved for such use.
- Water heater relief lines shall not discharge directly into the drain pan but may connect into the vertical riser or horizontal drain line. Some codes permit that a water heater relief line may discharge directly into drain pan. Check local code requirements.
- Both water heater relief line and drain pan line shall terminate over a suitable waste receptor or extend outside the building in a visible location 6 inches above grade.

Energy Conservation

Because of the rapid increase in the cost of electricity and fossil fuels, codes have adopted stringent methods to conserve energy. Water heaters account for approximately 25 per cent of the energy used in the average household.

Most codes enforce the following:

- Heat traps are required on hot water lines leading from water heaters. See Figure 8-8. The heat trap prevents hot water from creeping into the hot water line, thus saving approximately 2 per cent of the cost to generate hot water.
- Water heaters must be equipped with an energy shut-off device. This will cut off the supply of heat energy to the water tank and prevent temperatures from exceeding 210 degrees Fahrenheit.
- Water heaters are required to have more efficient heating devices and controls.
- More stringent standards for water heater insulation to prevent excess heat loss from hot water storage tanks are now required.

Figures 8-5, 8-6 and 8-7 are piping diagrams for hot water circulating supply systems and should help you when designing systems. The three most conventional systems used are illustrated. These are the looped system, the downfeed system and the inverted upfeed system.

Hot water piping materials and installation methods are the same as those described for water distribution in Chapter Nine. The only exception is that hot water piping supports must permit the expansion expected.

Materials And Installation Methods For Water Systems

The plumbing code regulates the materials, sizing and installation methods for water piping. It requires that there be a satisfactory supply of potable water to all fixtures so that they flush properly and remain clean and sanitary. The code establishes safeguards to avoid pollution of the water supply due to backflow or cross connection. There are many other requirements, limitations and restrictions and these will be discussed in this chapter.

MATERIALS

Consider the water supply in your area before selecting the material and size for water supply pipes, tubing or fittings. The water in many communities corrodes or leaves deposits on the interior walls of some pipe. Some types of soil and fill can corrode the exterior of the pipe. Remember that pipe that can leave toxic substances in the water supply must not be used for piping, tubing or fittings. Also, piping that has been used for other than a potable water supply system should not be reused in a potable water supply system. For example, pipe or fittings that were once used in a gas system should not be reused in a potable water supply system.

A wide variety of water piping materials will meet code requirements. Some materials are acceptable for use in underground installation; others can be used above ground only. Some materials are acceptable both above and below ground.

Water Service Supply Pipe

For a water service supply you can use cast iron water pipe, cast iron threaded pipe, wrought iron pipe or steel pipe. Wrought iron, steel pipe and fittings must be galvanized, zinc-coated inside and out. Other permissible materials for water service supply pipe under most codes are brass or copper pipe, type K, L, or M copper tubing, ABS plastic pipe, PVC plastic pipe, PE plastic pipe, and *polybutylene* plastic pipe and tubing.

All plastic pipe and fittings must carry the ASTM (American Society for Testing Materials) indentification numbers, or other recognized national standards of acceptance. All plastic pipe and fittings used in water service installations must have a minimum working pressure of 160 p.s.i. and carry permanent identification markings. Plastic pipe and fittings of like material listed above are accepted by most codes for water service supply piping use.

Water Pipe in a Building

Water piping which is permanently inaccessible in a building, such as piping installed under floor slabs, must be one of the following: cast iron water pipe, wrought iron pipe, steel pipe, brass, lead, type K, L, or M copper tubing. Wrought iron and steel pipe and fittings must be galvanized and only the appropriate approved fittings can be used. Many codes now approve the use of polybutylene plastic pipe, tubing and fittings under floor slabs inside a building.

Water piping installed above first floor slabs must be one of the following materials: brass, copper type K, L or M, lead, cast iron, wrought iron, block tin or steel. Wrought iron and steel pipe and fittings must be galvanized inside and out and only the appropriate approved fittings can be used.

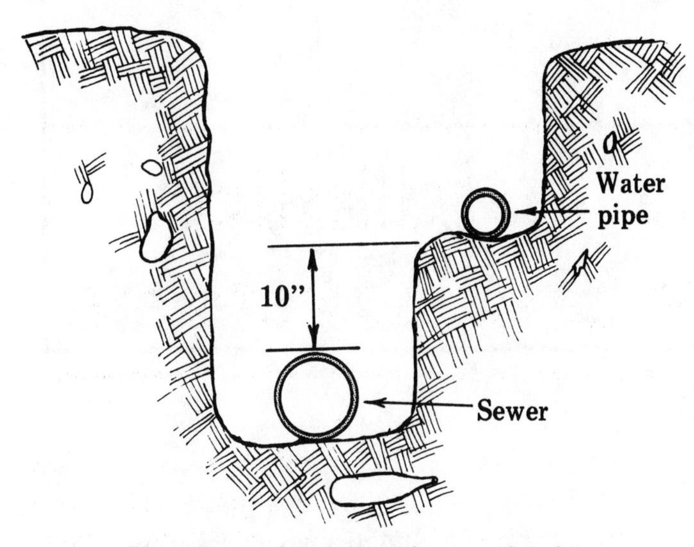

Placement of water service supply pipe in sewer trench
Figure 9-1

The use of plastic pipe and fittings below and above ground within buildings is fairly new.

The first plastic pipe and fittings approved by most codes in recent years was CPVC. Its use is limited to interior installation above the building floor slab.

CPVC is a high-temperature vinyl plastic designed for use in both hot and cold water systems, manufactured only in rigid lengths. It can withstand pressures up to 100 p.s.i. and temperatures up to 180 degrees Fahrenheit.

Many codes now accept for use (without limitations) the newer plastic pipe, tubing and fittings known as *polybutylene.* Polybutylene pipe and fitting resins are manufactured from butene-1 monomer. It is designed for use in both hot and cold water systems and is manufactured in both rigid piping and flexible coils. It can withstand pressures up to 100 p.s.i. and temperatures up to 180 degrees Fahrenheit. Polybutylene has a unique advantage over other approved plastic and metallic piping and fittings in the area of creep resistance (expansion). It is undamaged by freezing temperatures.

Plastic pipe and fittings have a distinct advantage over metallic pipe and fittings, as they resist corrosion, scala, sediment build-up, and are unaffected by soil conditions or electrolysis.

INSTALLATION METHODS

Water Service Supply Pipe

The water service supply pipe can be installed in the same trench as the building sewer. This avoids digging a separate trench. If a single trench is used for sewer and water service, the following conditions must be met:

- The water service supply pipe must be placed on a solid shelf excavated at one side of the common trench and above the sanitary sewer line. See Figure 9-1.
- The bottom of the water service supply pipe must be at least 10 inches (some codes require 12 inches) above the top of the sewer line. See Figure 9-1.
- The joints in the water service supply pipe must be kept to a minimum.

When metallic water service supply piping is installed on filled corrosive soil or where hydrogen sulphide gas is known to be present, the pipe must be protected by one or two coats of asphaltum paint or other approved coating. Fittings must also be protected by painting, wrapping with an approved material or applying other approved coatings.

Water service supply piping should be laid on a firm bed of earth for its entire length and installed in open trenches. It should be securely supported to prevent sagging, misalignment and breaking.

Cast iron water pipe, cast iron threaded water pipe, wrought iron pipe and galvanized steel pipe are considered superior as building materials because of their characteristic strength, durability and resistance to trench loads. They are especially desirable for outside use in water service supply piping. Because of their resistance to trench loads, the depth of placement is not critical except as protection against freezing. Avoid backfilling with large

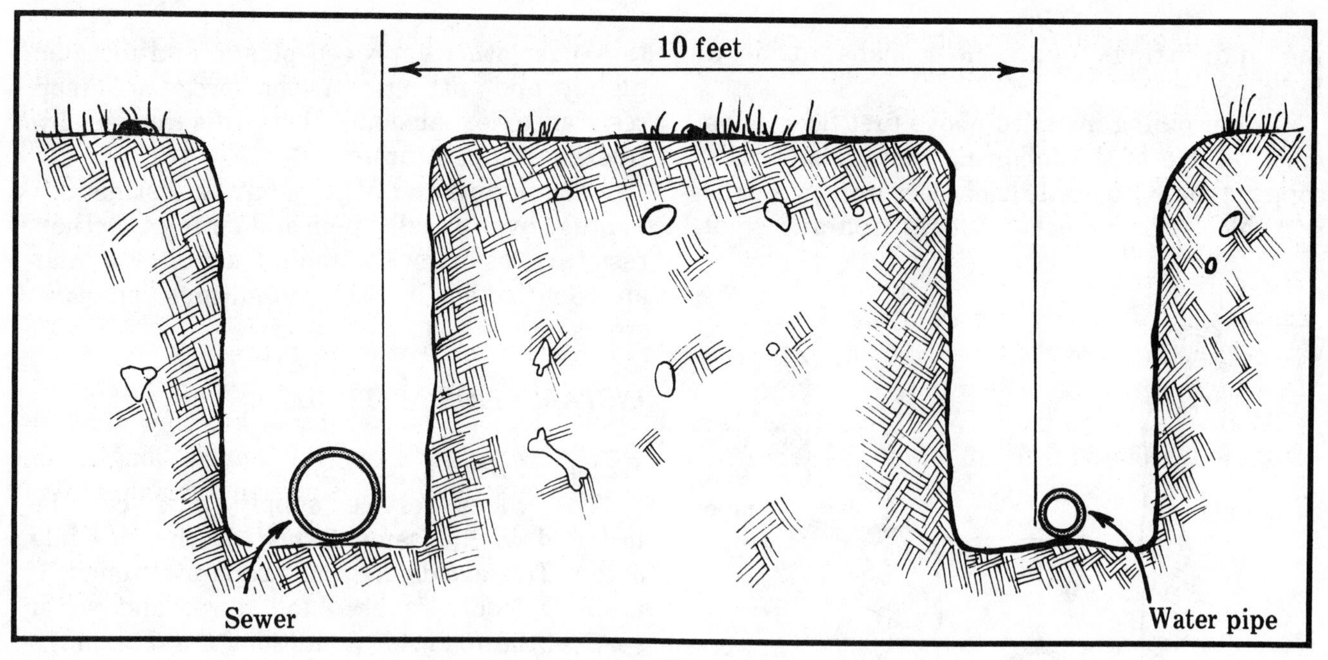

Placement of water service pipe in separate trench
Figure 9-2

boulders, rocks, cinder-fill or other materials which might physically damage or support corrosion of the pipe.

The depth for brass or copper pipe, type K, L, or M copper tubing is not established in most codes. Naturally, the pipe must be placed deep enough to avoid freezing. These materials are soft and are damaged easily. Select the backfill material with care. The backfill must be fine rather than coarse and should contain nothing that promotes corrosion of the pipe exterior.

ABS plastic pipe, PVC plastic pipe, PE plastic pipe and polybutylene plastic pipe are considered fragile. They require installation methods when used in open trenches similar to installation methods for plastic pipe used in building sewers. See Figure 4-4.

The water service supply pipe installed in a separate open trench usually must have a separation of ten feet from a sewer line. See Figure 9-2. Some codes require some other distance for separation.

The water service supply pipe must have a separation of at least ten feet from any septic tank or drainfield.

In climates where buried water service supply pipe is subject to freezing, the trench with the pipe should go below the frost line. Water service supply pipe entering a building above ground or in areas not protected from the cold must be thoroughly insulated to prevent freezing which could burst the pipe.

Water pipe in climates where freezing is not a problem should still be buried deep enough to avoid damage from edgers or other sharp tools. Always bury the pipe deep enough so that it is not heated by the sun on warm days.

Water service supply piping passing under the foundation of a building must have a clearance of two inches from the top of the pipe to the bottom of the foundation or footing. Refer back to Figure 4-5. Water supply pipe which passes through cast-in-place concrete such as a basement wall must be sleeved to provide ½ inch clearance around the entire circumference of the pipe. This avoids damage or breakage due to settlement of the building or normal expansion and contraction of the pipe. This also gives protection from the corrosive effects of concrete.

Where a lawn sprinkler system is connected to the potable water service supply pipe, there must be an approved backflow preventer on the discharge side of each valve. The backflow preventer must be at least six inches above the highest sprinkler head, and not less than six inches above the surrounding ground. This should eliminate the possibility of a cross-connection. See Figure 9-3.

Each building must have a separate water control valve installed in the water service supply pipe. This must be independent of the water meter valve. The control valve must be accessibly located at or near the foundation line outside the building, either above ground or in a separate approved box with a cover. See Figure 7-1. Some codes require an accessible control valve near the curb and a second control valve with a drip valve near where the water supply pipe enters the building. See Figure 7-1. The drip valve is required in cold climates to drain off water during cold weather to prevent freezing and bursting the pipes.

Water service supply piping must be electrically isolated from all other pipe, conduit,

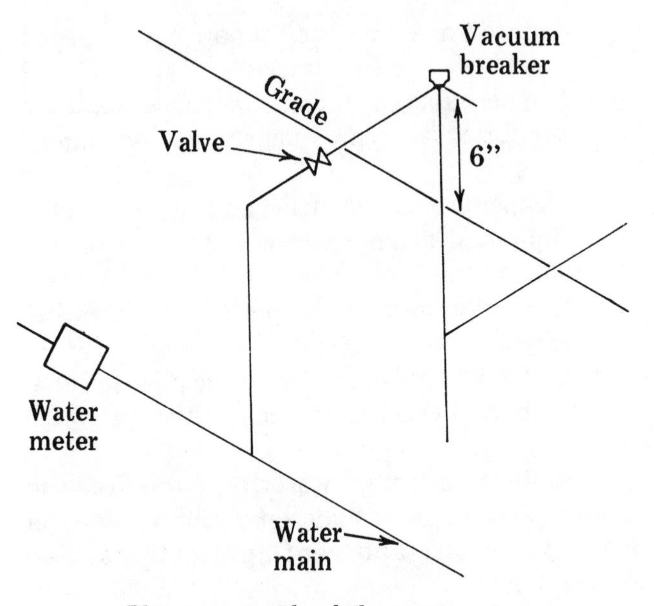

Placement of backflow preventer
Figure 9-3

soil pipe, building steel and steel reinforcing. An exception may be made where an electric ground is required by the code.

No private water supply, such as a well, can be interconnected with the water service of any public water supply.

If a swimming pool water supply is connected to the potable water service pipe, a positive air gap must be provided to prevent a cross connection. This will be examined further in Chapter 12, Swimming Pools.

Water Distribution Piping

Most code sections that cover water distribution piping make good sense if you take the time to understand them. Unfortunately, it is easy for you or your tradesmen to overlook fairly important points in the process of getting the job done on time and within the estimate. The code is written to ensure that the owners and occupants of a building have an effective, trouble-free plumbing system. If you comply with the code, you can be fairly sure that the owner of the building is getting good value for each dollar he spends on plumbing. Think about each of the points in this section and you will see why each is or can be important.

Underground water distribution piping installed under a slab within a building must be firmly supported for its entire length on well-compacted fill. This prevents misalignment and sagging and the formation of traps or depressions in the pipe that could collect sediment or mineral deposits. Some mineral deposits solidify over a period of time, reducing the flow of water and causing premature failure of the system. Use fine rather than coarse backfill in the trench so that the pipe is not damaged when the soil is compacted.

The entire system should be installed so that it will drain dry. This can be important if you have to make repairs or replace lost air in air chambers.

When you install the pipe, remember that the building will settle and the pipe will expand and contract in use. Make sure this movement doesn't damage the pipe. Water supply piping which passes through cast-in-place concrete floor slabs or bearing walls must be sleeved to provide ½ inch clearance around the entire circumference of the pipe. The sleeve provides crawling insects with access to the interior of a building unless the space is caulked with an approved flexible sealing compound.

You must protect water pipe installed under a concrete slab from corrosion. You also have to protect pipe in any location inside the walls of a building constructed on filled ground if hydrogen sulphide gas or other corrosive substances are known to be present. Protection can be provided by painting the pipe and fittings with two coats of asphaltum paint or other approved coating. It is better still to install water pipe overhead in the attic if you are working on filled ground. Some codes require this latter procedure.

In climate zones where water pipe is subject to freezing, pipe should not be installed in crawl spaces or other unheated areas unless it is protected with adequate insulation.

In restaurants, packing houses and other commercial buildings with walk-in cold storage facilities, don't install the water pipe through the cold storage room.

The code says that water distributing pipe must be electrically isolated from all ferrous pipes, electrical conduit, building steel and reinforcing steel. This means that the pipe can not come into physical contact with any good conductor of electricity.

The code usually prohibits water supply piping under elevators or in elevator hoistways.

Hot and cold water pipe must not contact each other when installed underground or within partitions. Contact or close installation tends to transfer the heat from the hot water pipe to the cold water pipe.

Domestic cold water piping installed above a roof or within ten inches of the roof must be adequately insulated with approved materials.

Other than polybutylene, plastic piping cannot be used underground within the walls of a building.

Each separate apartment or store in a building must have its own independent control valve or individual fixture control valve controlling all the fixtures in that apartment or store.

Each water closet and urinal supply must have an independent control valve installed above the floor.

Hotels, office buildings, hospitals, clinics, places of public assembly and manufacturing plants require either separate fixture control valves or a single control valve for each group of fixtures in a single room.

Residential buildings with more than two units and a single water service must have a separate control valve for each hose bibb (sill cock). This allows you to make repairs without interrupting the water supply to the residential units.

No more than two fixtures can be connected to a ½ inch cold water supply pipe.

Water pipe installations must be protected from water hammer with properly located air chambers or other approved devices. Air chambers must be installed so they drain by gravity without disconnecting the fixture water supply. An air chamber constructed of pipe must be at least 12 inches long and one pipe size larger than the pipe it serves.

Where water pressure within a building is in excess of 80 p.s.i., an approved water pressure regulator with strainer shall be installed to reduce the pressure in the building water distribution system to 80 p.s.i. or less.

Pipe Supports

Water piping inside walls and partitions must be secured by pipe straps or some other approved method to prevent movement of pipes as they contract or expand. Water supply piping installed above ground or in a vertical position must be securely supported by the building structure. The maximum distances between supports for horizontal water supply piping above ground are as follows:

- Screwed pipe must be supported at approximately twelve foot intervals.
- Copper tubing 1½ inches or smaller must be supported at approximately six foot intervals.
- Copper tubing two inches and larger must be supported at approximately ten foot intervals.
- Lead pipe must be supported for its entire length.
- CPVC and polybutylene plastic pipe must be supported every four feet.

Note that horizontal water pipe requires the same type of support and separation as used in drainage, waste and vent applications. See Figure 4-7.

Vertical water pipe requires the following support:

- Screwed pipe carrying cold water must be properly supported at the base and at every other story height.
- Screwed pipe carrying hot water must be properly supported to provide for expansion.
- Copper tubing carrying cold water 1¼ inches and smaller must be supported at intervals of approximately four feet.
- Copper tubing carrying cold water 1½ inches and larger must be supported at each story height.
- Copper tubing carrying hot water must be properly supported to provide for expansion.
- Lead pipe must be supported each four feet.
- CPVC and polybutylene plastic pipe carrying cold water must be supported at each

story height.

- CPVC plastic pipe carrying hot water must be properly supported for expansion.
- Polybutylene plastic pipe carrying hot water needs no special provision for expansion.

Note that vertical water piping requires the same type of support and separation as the material would if used in drainage, waste and vent applications. See Figure 4-8.

Give good support to the base of water pipe risers, particularly in high-rise construction where the pipe is expected to last for the life of the building. Don't let the riser weight fall on the smaller water pipe branches. This would crack the joints of the smaller pipes.

Cross Connection

The code requires that the potable water supply outlet terminate above the overflow rim of each fixture. This provides an air gap which will prevent siphoning of the fixture contents back into the water outlet, or faucet.

All water outlets equipped for hose connections (other than clothes washers) must have approved backflow preventers. Hose connections are common on restaurant faucets and service sinks.

Codes prohibit the use of below-rim water supplied fixtures. Where special fixtures cannot function properly otherwise, the plumbing official will require that a backflow preventer be installed in the water supply pipe and that individual backflow preventers be built into each individual unit or piece of special equipment.

Copper tubing must be connected to threaded pipe or fittings with a special brass or copper converter fitting. Unions in the water supply system must have metal-to-metal joints and ground seats.

The entire water distribution system must be tested, inspected and proved to be tight when completed. It must be pressurize tested to a water pressure of not less than the maximum working pressure under which it is to be used.

Remember the importance of air chambers and securing all water supply pipes. A system with water shock or hammer won't pass the final inspection.

Threaded Pipe, Fittings, and Valves

All standard pipe and fitting threads must conform to the standards adopted by the American Standards Association. The standard pipe thread has a taper of ¾ inch per foot. The

Size of pipe (inches)	Number of threads per inch	Approximate total length of threads (inches)	Approximate thread engagement (inches)
1/4	18	5/8	3/8
3/8	18	5/8	3/8
1/2	14	13/16	1/2
3/4	14	13/16	1/2
1	11-1/2	1	9/16
1-1/4	11-1/2	1	5/8
1-1/2	11-1/2	1	5/8
2	11-1/2	1-1/16	11/16
2-1/2	8	1-9/16	15/16
3	8	1-5/8	1
4	8	1-3/4	11/16

Data for standard pipe threads
Table 9-4

last column in Table 9-4 shows the number of threads that must be engaged on screwed pipe. Use the right size wrench when joining valves or fittings on threaded pipe. A wrench too small for the job will require unnecessary effort for your hands, arms and back. A wrench too large will force the fitting too far onto the threaded pipe. This can result in a bad joint or a cracked fitting. The following wrenches listed in Table 9-5 should be adequate for the pipe sizes given.

Cutting, Reaming, and Threading Pipe

Heavy duty pipe cutters are used to cut iron and steel pipe. The most common pipe cutter is the single wheel cutter. Begin each cut by *lightly* rotating the cutter completely around the

Pipe Size	Wrench Size
1/8, 1/4, 3/8 inch	6 or 8 inch pipe wrench
1/2 and 3/4 inch	10 inch pipe wrench
1 inch	14 inch pipe wrench
1-1/4 and 1-1/2 inch	18 inch pipe wrench
2 inch	24 inch pipe wrench
2-1/2 and 3 inch	36 inch pipe wrench
4 inch	48 inch pipe wrench

Larger sizes usually use chain tongs or special equipment.

Table 9-5

pipe. This will give a "bite" or groove for the cutter wheel to follow.

After each turn of the cutter wheel, tighten the handle slightly. Tightening the handle too rapidly will break the cutter wheel or spring the cutter frame, ruining the cutter. Thread cutting oil should be used occasionally on the cutter wheel and rollers.

The pipe cutter wheel leaves a bur on the inside of the pipe. See Figure 9-6. Mineral deposits can collect at the bur and cause premature failure of the line. Use a pipe reamer to remove the bur on threaded pipe up to 2 inches. On larger pipe use a coarse half round file.

Modern pipe threaders make plumbing work much less exhausting than it was formerly. Threaders cut the required standard threads and then disengage the dies.

Use plenty of thread-cutting oil on the threader dies when cutting pipe. This will prevent overheating. Occasionally check to be sure you are cutting clean threads. Chipped or torn threads mean that the pipe at this location s too soft, too hard, has impurities in it, or that the dies are worn and need replacing. Always cut new threads when this happens. Do not use pipe that has bad threads.

Only the plumber who makes the fitting knows whether the bur has been removed and whether the pipe had a good thread. The contractor paying his salary does not know, the plumbing inspector checking the job does not know, and the people relying on his workmanship do not know. You and your crew demonstrate your integrity, good workmanship, and personal pride on every joint made up on your jobs.

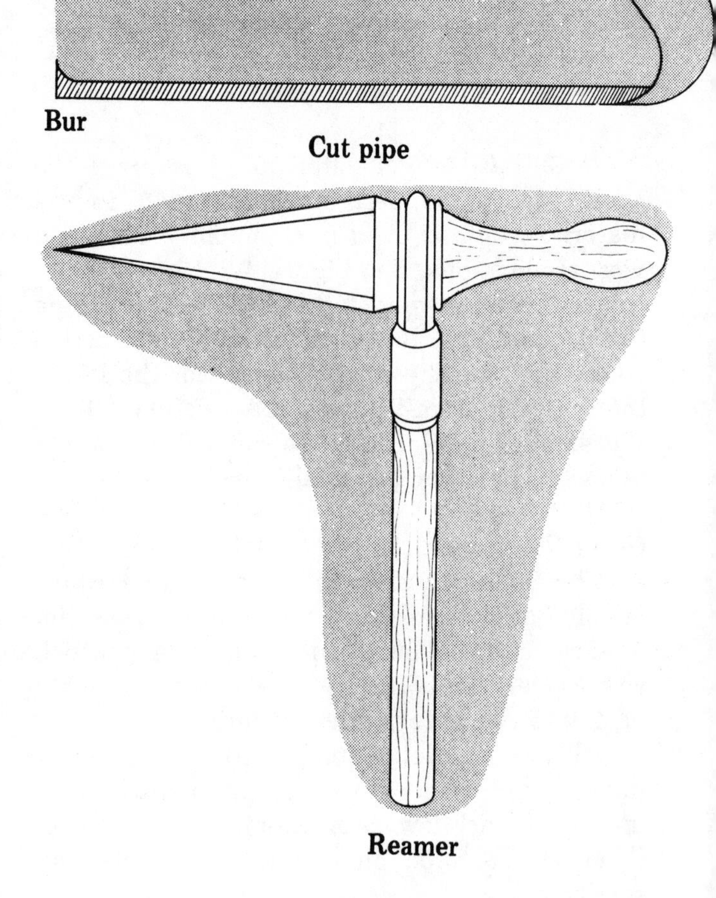

Figure 9-6

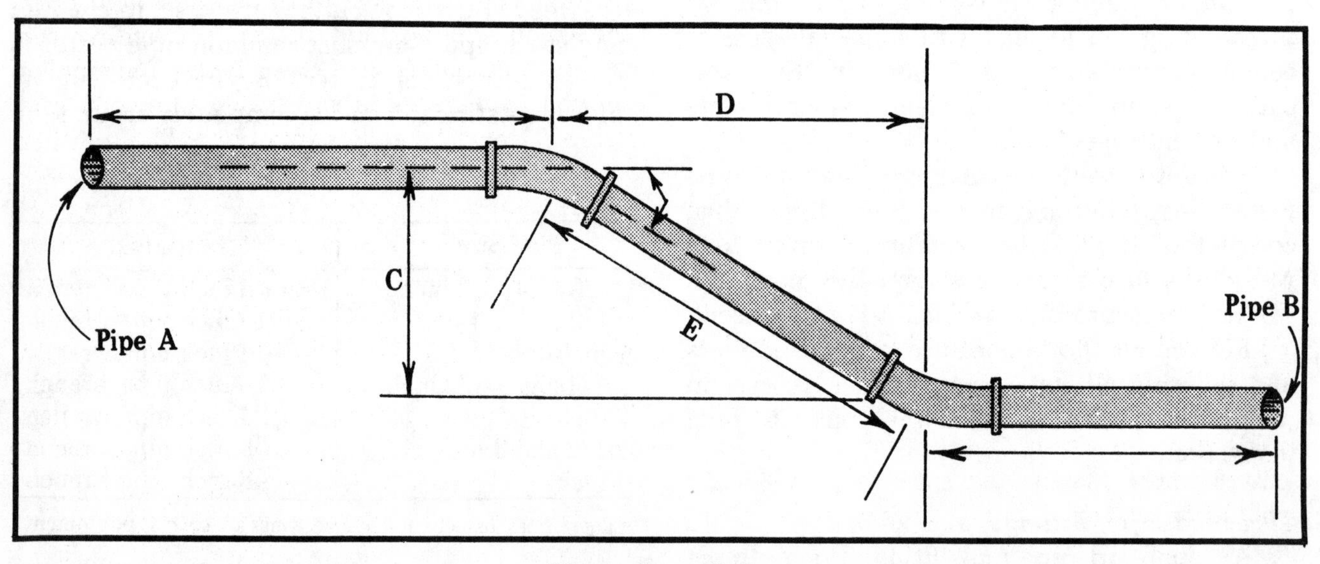

Measuring offsets
Figure 9-7

When you are satisfied that a good joint can be made, apply pipe compound (known as *pipe dope*) and screw the fitting snugly but firmly onto the threaded pipe. When installation is complete, open the house valve to test for leaks.

Measuring Offsets

In the layout and dimensioning of piping arrangements, the only place offsets are not a problem is where 90° elbows are used. You can find the exact distance between centers of fittings of offsets by careful calculation.

Figure 9-7 shows a common pipe offset between parallel runs of pipe. The center to center offset (C) is 10 inches, the distance between the centers of parallels A and B. Assume that pipes A and B are to be connected with 45° elbows. You need to determine the length of pipe required for E. Use Table 9-8. The letters refer to the letters on Figure 9-7. Find the figure 1.4142 opposite the 45° offset. Multiply this figure by distance C, in this case 10 inches. The length of pipe required would be 14.14 inches.

Degree of offset	When C = 1, D =	When D = 1, C =	When C = 1, E =
60°	0.5773	1.732	1.1547
45°	1.000	1.000	1.4142
30°	1.732	0.5773	2.000
22-1/2°	2.414	0.4142	2.6131
11-1/4°	5.027	0.1989	5.1258
5-5/8°	10.168	.0983	10.217

Finding the length of pipe to connect two parallel runs of piping
Table 9-8

Cutting, Reaming, and Joining Copper Tubing

Copper tubing, both rigid (hard) and flexible (soft), can be cut with a tubing cutter or a hacksaw. A clean square cut is made more easily with a tubing cutter. Copper tubing cutters are similar to pipe cutters, though much smaller. Apply oil to the cutter wheel and rollers sparingly as excessive oil will cause trouble when you are preparing the copper tubing for soldering. Be careful to remove the bur the cutter wheel leaves in the tubing. A reamer should be attached to the tubing cutter frame and be used for copper tubing up to 2 inches in diameter. In larger tubing use a file to remove the bur.

The most common joint for copper pipe is the sweat joint. Clean the inside of the copper fittings with a wire brush designed for this purpose or with emory cloth. The pipe end should be polished bright with emory cloth. Cleaned parts should be used as soon as possible and should not be handled with dirty or sweaty fingers. The bright surface oxidizes very quickly and the cleaning process then has to be repeated.

A thin coat of non-corrosive soldering flux should be applied to the inside of the cleaned fitting and the outside of the cleaned pipe. Fitting and pipe are then joined together, ready for soldering.

Soldering joins the two metallic surfaces together with a fusible alloy (solder) that has a melting point lower than that of the pipe to be joined. The solder recommended for use on copper pipe and fittings in a water supply system is known as "50-50" or "*half-and-half.*" The solder consists of 50% tin and 50% lead with a melting point of 370°. This type of solder is sold in fine or medium gauge wire rolled on a spool that weighs approximately one pound.

The soldering torch is fired from a tank of gas that is easily moveable from one area to another. The tank and gas weigh only about 20 pounds when full but provide enough gas for 3 days of continuous soldering. The tank has a regulator to control the flow of gas, a six foot long rubber hose, and another valve to control the flame at the burner tip. Select a tip appropriate for the size of copper tubing you are joining. Too large a tip will overheat and burn the copper tubing and fitting, preventing the solder from joining the two surfaces. Too small a tip will heat the pipe and fitting unevenly and will not draw the solder into the joint. In either case, a bad joint is the result.The joint may leak either when it is first tested or months later after the piping has been enclosed by building walls.

A system that has been pressurized is much more difficult to work with because there is usually some water present. The heat from the torch will turn the moisture in the tubing to steam. This will make pin holes in the newly applied solder before it has hardened. Heat applied to the tubing several inches on each side of the fitting to be soldered may dry the pipe long enough to make a good joint if you work quickly.

If you can not keep water away from the joint

while you are working, stuff a portion of plain white bread (not the crust) as far into the pipe as possible in the direction from where the water is flowing. The bread will absorb the water, giving you time to make a good joint. When water pressure is returned to the system, the bread flushes out easily.

If the joint is properly cleaned and heated, surface tension will spread the solder to all parts of the joined surfaces. This results in a sound joint that will last as long as the tubing.

Soldering is an art in itself. It looks easy but requires care and precision. Apprentice plumbers should solder only under the supervision of an experienced journeyman. It takes a lot of practice in soldering before you have the knowledge and skill necessary to solder copper water systems properly.

Copper tubing is measured the same way as threaded pipe. The only exception is that the smaller copper tubing is more easily bent when heated at the point where an offset is needed. In many minor offsets, fittings are unnecessary if you are using copper tubing up to one inch diameter.

Occasionally you will have to make a flare joint. This type joint is used on soft flexible copper tube when connecting certain types of fixtures and appliances. For example, a connector for a gas range might be joined with a flare fitting. The tubing is cut to the desired length. Then the flare nut that will make the connection is slipped on the cut tubing. A flaring tool is used to flare the tubing ends so that you get a perfect fit. Slide the nuts to each end of flared pipe and gently bend or shape the tubing by hand so that the ends fit together. The flare nuts must be screwed firmly by hand on the male thread of each fitting. Tighten with the proper type and size wrench until the fitting is snug. It should then be tested for leaks.

Cutting, Reaming, and Joining Plastic Pipe

Plastic pipe and fittings are common plumbing materials today. Plastic pipe is used for drainage, waste, vent and water service piping. More recently vinyl (CPVC) has been approved for interior (above slab) piping for hot and cold water. These plastic pipes are lightweight, easily handled by one person, and rigid once cemented in place.

Plastic pipe is usually cut with a hacksaw blade, but a special cutter which is similar to the copper tubing cutter may be used. A square end cut is essential with plastic pipe. Use a hacksaw and a miter box. A "freehand" cut is an invitation for problems. Use a fine tooth hacksaw blade when cutting plastic pipe.

Remove the bur with a reamer if a pipe cutter is used. A hacksaw does not leave a bur, but the interior and exterior of the pipe are left with a rough ridge. Remove this roughness with a knife or file.

You can measure offsets for plastic pipe and fittings the same way as for threaded pipe. Refer back to Figure 9-7 and Table 9-8.

Special plastic cement is always used to join plastic pipe and fittings though you may hear the joint created referred to as a "welded" joint. First, check the fittings and pipe. Don't use any pipe or fittings that have gouges, deep abrasions or cracks. After the pipe is cut to the proper length, check the dry fit before cementing. The pipe must enter the fitting socket smoothly but must not be so loose that the two surfaces do not make good contact.

Liquid cleaner (developed for plastic pipe and fittings) or fine sandpaper should be used to remove impurities and gloss from the surfaces to be joined. Use only the cement recommended by the manufacturer of the pipe and fittings you are using. Don't use cement that will not pour from the can or that has a rusty or dark brown color.

Plastic cement sets very quickly once it is applied to plastic surfaces. For this reason, only one joint can be cemented at a time. You have less than one minute to do the following:

- Brush a light coating of cement on the plastic pipe with the brush supplied with the can.
- Brush a thin coat in the fitting socket and quickly brush the plastic pipe a second time.
- Push the pipe fully into the fitting and then give the pipe a quarter turn. If a direction is required (a tee facing up, etc.), adjust the pipe at once. This is your last chance to make corrections.
- Wipe excess cement from the fitting and close the cement can immediately to prevent drying.
- Hold the pipe and fitting together for approximately 15 seconds until the cement sets.

Since a cemented joint looks like a dry fit joint, check all joints to make sure that they have been cemented.

Wait at least one hour after cementing the last joint before testing the system. If time is not critical, it is best to let the joints harden over-

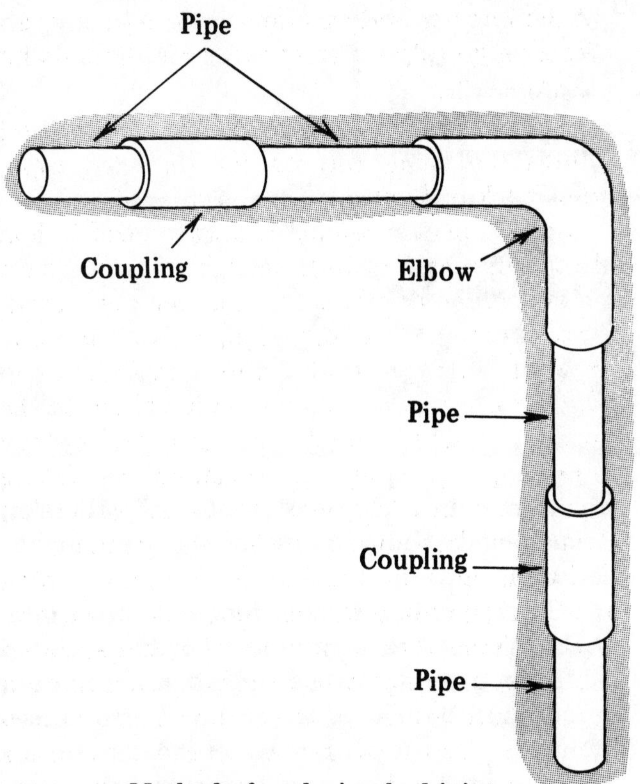

Method of replacing bad joint
Figure 9-9

night. Check for leaks after turning the house valve on and releasing trapped air from the ends of each branch.

If a leak occurs in a plastic system, cut out the bad joint and replace it with a new fitting. Remove enough pipe to allow room for two couplings and the fitting. Figure 9-9 shows how much pipe should be cut out if a joint at the elbow leaked. In this case three fittings, two short lengths of pipe and 6 joints are required. Do a first class job when cementing plastic pipe and fittings. Plastic pipe fittings can't be reused if they don't work the first time.

Approved male-threaded plastic adapters should be used if you have to connect plastic pipe or valves. Only the thread compound recommended by the manufacturer should be used.

Polybutylene is the newest plastic pipe approved for use in water distribution systems in many states and by local codes. Its use for both hot and cold water piping is unrestricted.

The piping may be cut into desired lengths using the tools and methods for other type plastic piping mentioned earlier in this chapter. It may be joined together by compression type fittings, crimp type fittings or by special heat fusion methods. It can also be flared (like soft copper) and joined together with flare fittings. When installation is complete, there is no waiting for joints to cure as in other plastic systems. The water may be turned on immediately to check out the system.

When the water system is complete, open a hose bibb on each section to flush out sand, pieces of pipe shavings and other impurities that may have collected inside the pipe during installation. This will save time when fixtures are set and prevents the damage sand and grit can do to washers.

Private Water Supply Wells

A recent survey by the Water Quality Association found that nearly one home in five is not connected to a public water supply. This means that about 40 million people depend on a private system for pure and wholesome water. Many suburban homes and practically all rural homes use water from private wells. In some rural areas, lakes and streams make wells unnecessary.

The rapid spread of urban communities around large metropolitan areas has often outstripped the installation of public water distribution systems. Many smaller towns and cities have lower population densities which make the cost of public water systems prohibitive.

Approval and inspection of domestic wells is controlled by either the local health department or the Director of Environmental Resource Management (DERM). These authorities set guidelines on depth and distance separation of the well from sources of contamination. A permit is required for any drilled or driven well, regardless of whether the water is intended for domestic or irrigation purposes. Each well is inspected to ensure the owner has complied with the regulations that apply.

Well water is generally classified as hard water because of its high mineral content. Many people have to acquire a taste for this untreated water because it has a distinctly different taste than "city water." It does not contain the chlorine or other chemicals that city dwellers are accustomed to but requires more soap to make a lather.

The minerals in untreated well water often stain the surface of plumbing fixtures a dark reddish brown. These stains are virtually impossible to remove. Minerals can also cause a buildup of scale within water heater storage tanks and the building water distribution pipes. Premature failure may be the result. Scale buildup can be greatly reduced and the taste and smell of the water improved considerably by installing a water softener on the building water service line. See Figure 10-1.

Source of Well Water

Well water is ground water that has filtered down through the soil to the water table level. It is known as meteoric water and makes up most of the estimated two million cubic miles of ground water in the upper crust of the earth.

Rain water soaks into the ground and moves slowly down to the underground water reservoir. The underground water table may be only a few feet below the surface of the ground or hundreds of feet below the surface. Since it has been filtered through sand and rock, it is usually cool, low in harmful bacteria and high in dissolved minerals.

When a well is driven or drilled, the bottom of the well casing must extend into the dry weather water table. Otherwise, during pro-

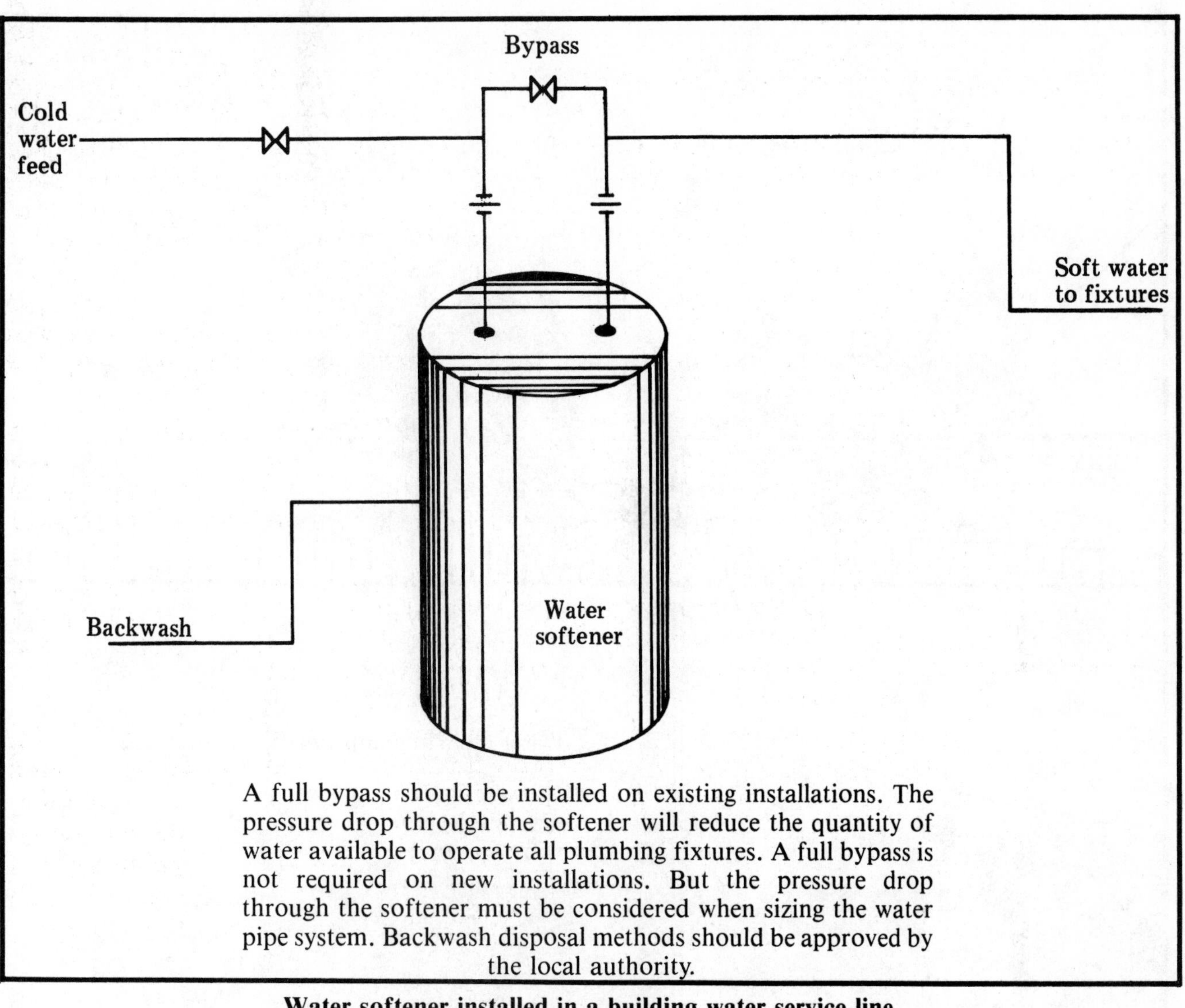

A full bypass should be installed on existing installations. The pressure drop through the softener will reduce the quantity of water available to operate all plumbing fixtures. A full bypass is not required on new installations. But the pressure drop through the softener must be considered when sizing the water pipe system. Backwash disposal methods should be approved by the local authority.

Water softener installed in a building water service line
Figure 10-1

longed droughts, the water table may fall below the level of the well. See Figure 10-2.

Well Installation

Generally, professional well drillers install wells, suction lines, pumps, and water pressure tanks. They are certified and licensed as a specialty contractor and are permitted to do only this type of work. Plumbing contractors are also licensed to do this work, but rarely become involved. If your contract includes a complete well supply system, you will usually subcontract this part of the job to a professional well driller. You should, however, have some knowledge of well systems because many of the people you deal with will assume you know something about wells. Also, the plumber's examinations include questions on well systems.

The depth of potable water supply wells is determined by the local authority and by the depth of the water table. Even though the underground water table is within a few feet of the ground surface, the local authority may require a 30 foot minimum depth. The authority may also require a separation of a domestic well from a source of contamination (such as a septic tank, drainfield, soakage pit or discharge well) of 100 feet. Some codes may require more, some less.

Wells which are dug or driven are classified as shallow wells. Dug or driven wells are used where the water table is within 22 feet of the ground surface. In some parts of the country,

Well extended to dry weather water table
Figure 10-2

plumbers install these shallow wells. The well must penetrate deep enough below the existing water table to assure a dependable supply of water, even in very dry seasons.

Wells which must be drilled are classified as deep wells and may penetrate hundreds of feet into the earth. The water from deep wells is more desirable because there is smaller chance of contamination from a well 100 feet or more deep. The water level in these wells is little affected by seasonal rainfall or dry years.

There are two types of driven wells, those with an open end casing and those with a casing equipped with a well point.

An open end casing is commonly used for domestic as well as irrigation purposes in areas where the water table is close to the ground surface and in a good rock formation. In some

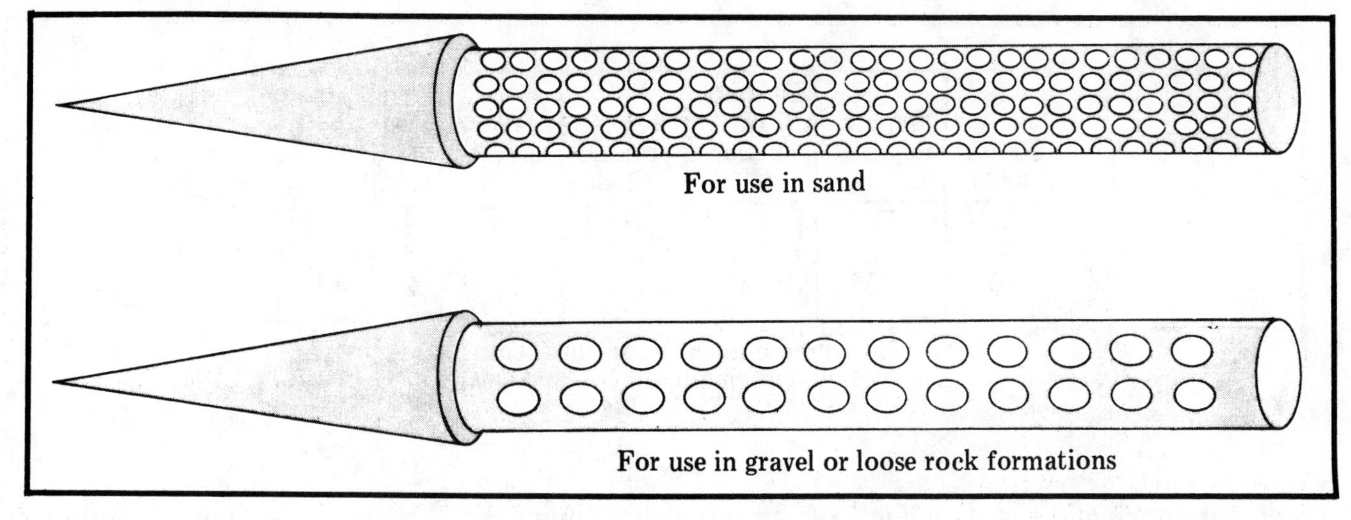

Well points
Figure 10-3

areas, rock or corrosion will clog the protective screening of a well point in a short time. When this happens, the well will not draw water and is useless. The open end well casing would be preferable under these conditions.

If an open end casing is used, the pipe is driven to the desired depth and then loose soil and rock are flushed from the driven section. Flushing is done with a smaller pipe with a sharpened point at one end for chipping into the rock. A garden hose is attached to the other end. The smaller pipe is inserted into the open end casing and water pressure is used to flush loose debris out of the casing. A water collecting pocket is formed in the rock at the lower end of the well casing during the flushing.

When you are sure that a good well has been installed, connect a three horse power gasoline driven centrifugal pump to the top of the well casing. Pump water out of the well until the water is free of rocks and sand.

A well point is used in areas where the water table is in loose shale or sand. The well point is attached to the well casing and is driven to the desired depth. There are two types of well points. One has a screen or fine perforations and is used in sand. The other type has larger openings and is used in gravel or loose rock formations. See Figure 10-3.

You are expected to know the installation methods and local code requirements for well drilling even though you may never have to do this work. Questions on wells are included in journeyman's and master's examinations. Be aware of the following requirements:

- Unless specifically approved by the local authority, a well must not be located within any building or under the roof or projection of any building or structure.
- The well casing must be conunuous, of new pipe, and must terminate in a suitable aquifer. Well casing pipe six inches or less in diameter must be of galvanized steel.
- A concrete collar a minimum of four inches thick and 36 inches wide must be poured around the top of the well casing. The concrete pad must be placed immediately below the tee and suction line. This collar must slope away from the casing to prevent surface water from carrying pollutants down the well casing to the water reservoir.
- A tee the same size as the well casing must be installed at the top of the well. This gives access to the casing for inspections, the introduction of disinfecting agents, measuring well depth and testing the static water level.

Suction Line

The suction line must be large enough to provide the water volume and pressure required to operate the plumbing fixtures in the building. The suction line or water service pipe from the well to the pump must not be smaller than one inch. It must be installed with a pitch toward the well. If the well requires a suction line longer than 40 feet, the suction pipe must be increased to the next pipe size shown in Table 10-4.

As an example, consider a residence having

Fixture units	Supply required G.P.H.	Diameter of suction pipe	Diameter pressure pipe	Diameter service pipe	Size of tank	HP	Well size
23	720	1	¾	¾	42	½	1½
30	900	1¼	1	1	82	¾	2
40	1200	1½	1	1	120	¾	2

(Predominantly for flush tanks)
Tank and pump size requirements
Table 10-4

30 fixture units and a suction line 65 feet long. The next pipe size in Table 10-4 would be 1½ inches. Therefore, a 1½ inch suction line pipe should be used.

A soft seat check valve rated at 200 pounds water test must be installed as close as is practical to the well. Check valves should be of brass up to a two inch diameter. They may be either spring-loaded or flapper-type. The suction line must have a union or slip coupling installed just before the pump.

Materials and installation methods for the suction line may be the same as for water service piping as described in Chapter 9.

The water piping from the pump to the hydropneumatic tank should not be smaller than the discharge outlet of the pump. A gate valve with the handle removed should be installed in the piping between the pump and the tank if the tank has a capacity larger than 42 gallons. A minimum ¾ inch gate valve must be installed on the water piping on the discharge side of the tank.

The hydropneumatic tank should be large enough to prevent excessive cycling of the pump. It should provide a draw-down of six gallons of water while maintaining an operating range of 20 to 40 p.s.i. water pressure. The minimum size tank for a single family residence would be 42 gallons. See Table 10-4.

The type and capacity of the pump, equipment, suction line, pressure line, and water tank must conform to Table 10-4.

When installing the pump and pressurizing system, be aware of spacing so that equipment may be reasonably accessible for repair or replacing.

Interior water piping, materials, and installation methods are the same as described in Chapter 9 for water distribution pipes.

Fire Protection

Fire protection equipment is an important part of many buildings and every plumber should know the code requirements for installing fire protection systems. Code sections relating to fire protection equipment appear in many parts of the code. This chapter will bring together everything the code requires in fire protection systems and explain what the code language means.

Standpipe Systems

Many larger buildings must have emergency fire hose connections on the site so firemen will have an adequate supply of water immediately available in the event of a fire. The hose connections are called *standpipes.* Standpipe systems can connect to a public water main. The main can provide the quantities and pressures required by the code. A public water supply should be used if the street water main is at least 4 inches in diameter and is located within 150 feet of the nearest point of the building. Sometimes the public water main can not provide the required quantities and pressures. In this case the system will have to use another approved method such as a fire pressure pump (Figure 11-1) or a pressure tank.

Fire Flow On-Site Well System (Figure 11-2 and 11-3)

Where a standpipe system is required and adequate public water is not available, an on-site well system may be the best solution. The system must meet the following requirements:

- All wells must be cased and sized for a flow of 500 gallons per minute (g.p.m.).
- The well must be of ample diameter and depth, and must be sufficiently straight to receive the pump.
- All casings must have a wall thickness of at least 3/8 inch.
- The well must be cased to a proper depth and properly sealed to prevent loose or foreign material from entering.
- The well must be properly developed to be free of sand or loose gravel.
- The well must be test-pumped at 150% of the capacity of the pump to be installed for two hours after it is free of sand.
- The drawdown in the well must not exceed four feet during pumping at 150% of pump capacity. See Figure 11-2.
- The flow rate must be 500 g.p.m. at 20 p.s.i. where the pump discharges.
- Hook-up must be done according to the requirements of the fire department having jurisdiction. Each fire department connection on the discharge side of the pump must be equipped with a shut-off valve with a diameter not smaller than the

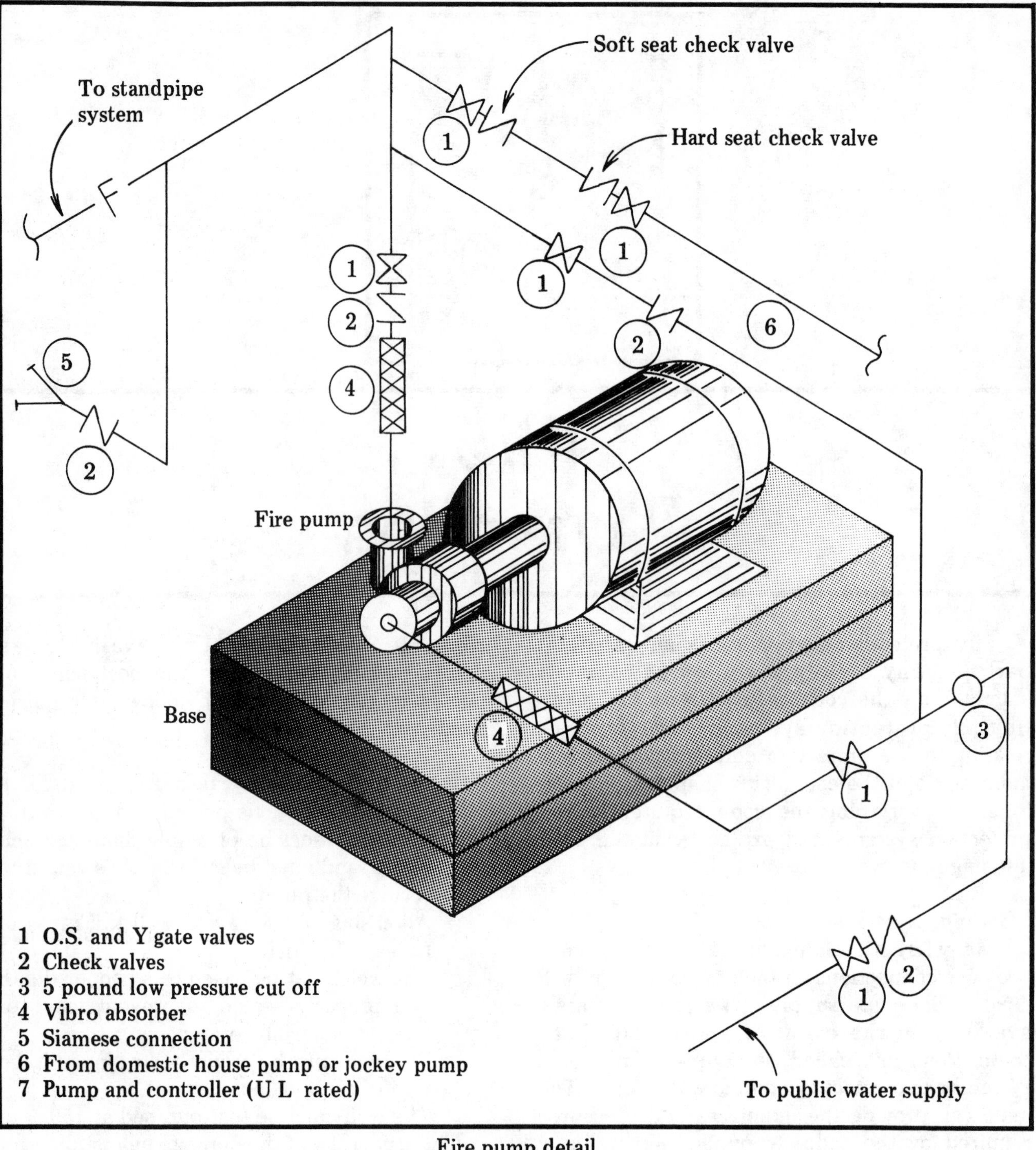

Fire pump detail
Figure 11-1

size of the discharge opening.

- Fire department connections rated by g.p.m. are required, and must be National Thread Standard (N.T.S.). These connections should have a flow of 500 g.p.m. with one 4½ inch connection and one 2½ inch connection. See Figure 11-3.
- All on-site systems must be tested before final approval by the fire department having jurisdiction.
- All on-site systems must be located a minimum of 50 feet from the buildings being protected whenever physically possible. Fire department access to the

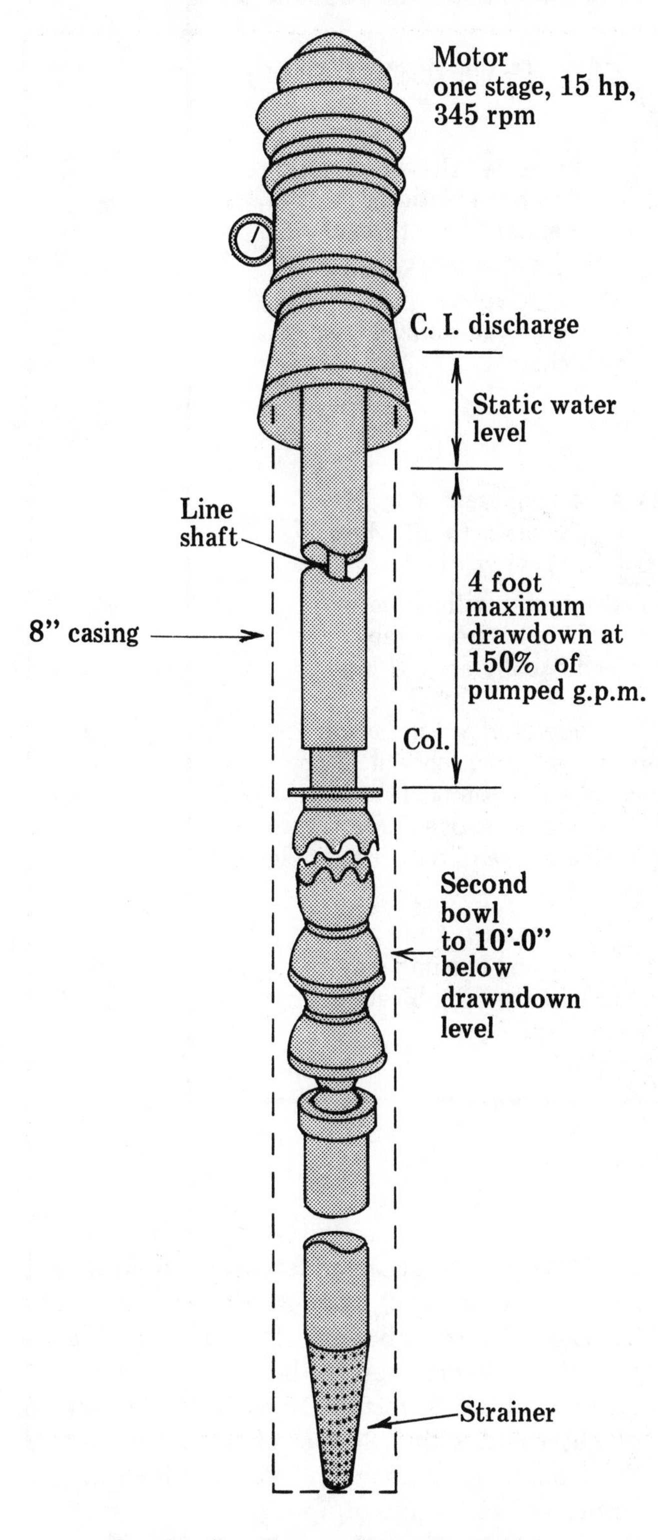

On-site fire flow well sectional view
Figure 11-2

on-site system must be by a roadway suitable for fire equipment. The fire department connections must be no more than 8 feet from the roadway.

- Direct connections between on-site fire protection systems and the potable water supply are not permitted.

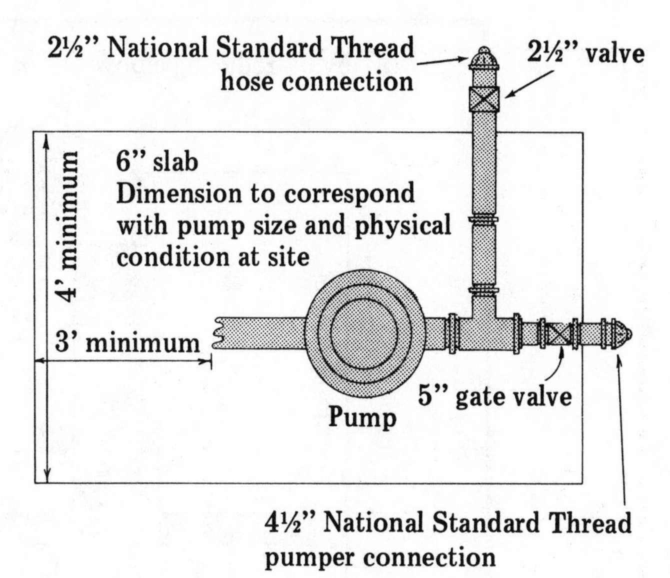

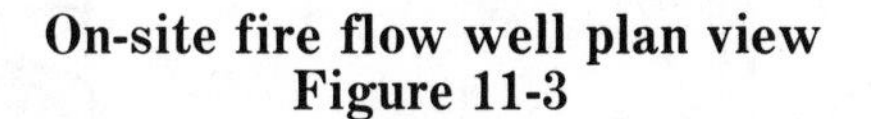

On-site fire flow well plan view
Figure 11-3

Even where a standpipe system is not required, a standard fire well may be required for certain occupancies if a public water supply is not available. See Figure 11-4. The well installation is nearly the same as the on-site well system with the following exception. No pump is required and the hose connection can be a single 4½ inch American National Standard hose connection.

Yard and Street Hydrants

Yard or street hydrants are usually required in commercial, industrial and residential areas. The code lists other areas that must include yard hydrants because of their size, isolation or type of occupancy. These include mobile home and trailer parks, marine terminals, marinas, boat yards, oil-storage tank yards, lumber yards and exhibition parks. To comply with the code, there must be one yard hydrant and hose station for fire protection for each 20,000 square feet of area in use.

Hydrants must have two 2½ inch connections with National Standard threads similar to those of the local fire department. The fire department must approve the location of all yard hydrants. Figure 11-5 shows a typical hydrant installation.

Standpipe Requirements

Where standpipes are required, the system must be pressurized ("wet") with a primary

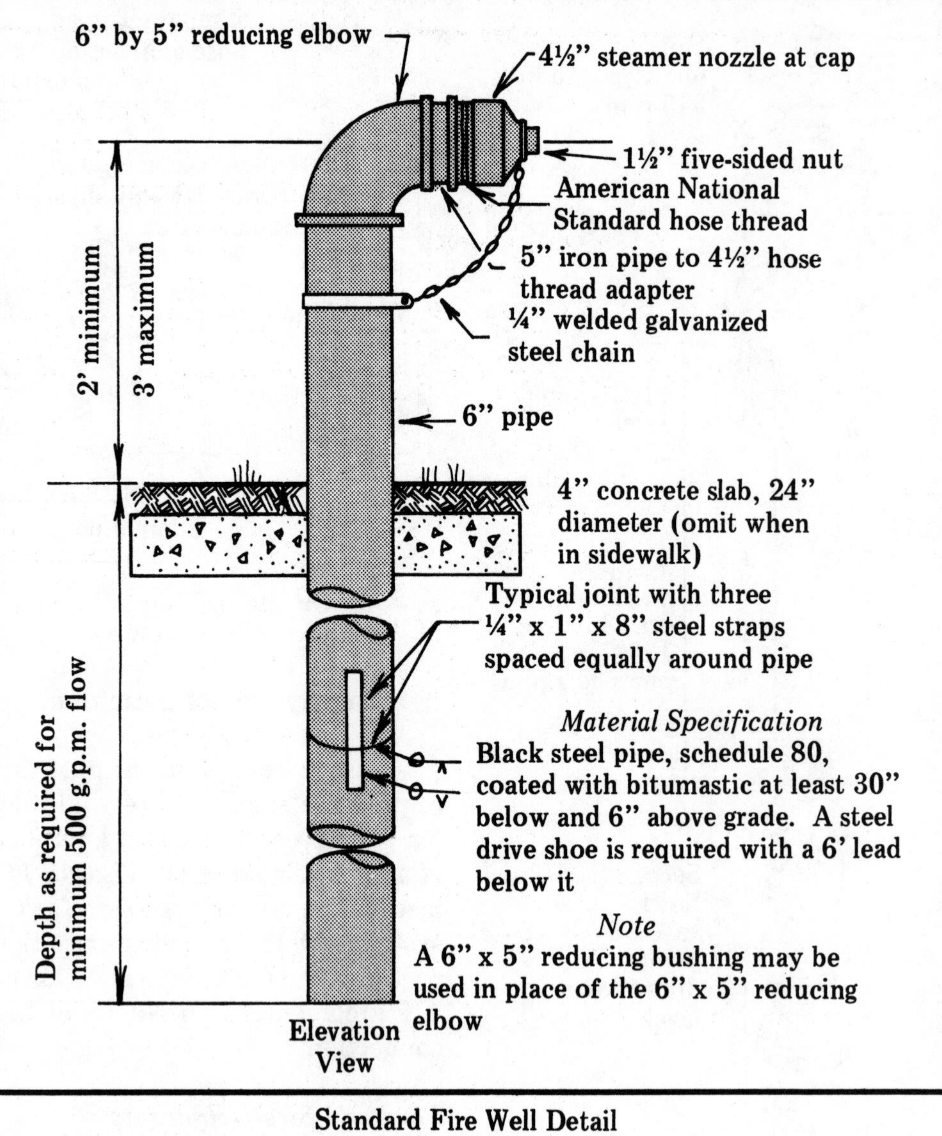

Standard Fire Well Detail
Figure 11-4

water supply constantly or automatically available at each hose outlet.

In buildings under construction where standpipes are required, the standpipes must be constantly available for fire department use as the construction progresses. A fire department connection must be provided on the outside of the building at the street level and at each floor up to the highest constructed floor.

A fire standpipe system is required in buildings over 50 feet high. That means that every seven story building must have standpipes, assuming there are 8 feet between floors.

Buildings designed for theatrical, operatic or similar performances must have a 2½ inch diameter standpipe on each side of the stage. A hose not over 75 feet long must be located at each standpipe hose station.

Standpipe locations must be arranged so that they are protected from mechanical and fire damage. The number of standpipes and hose stations is determined in the following way. All parts of all floors must be accessible to a stream of water and within 30 feet of the nozzle end of the hose when a hose not over 100 feet long is connected to the standpipe.

Standpipes must be located within an enclosed stairway. If the stairway is not enclosed, the standpipe must be within ten feet of the floor landing of an open stairway. Valve or hose connections can not be located behind any door.

Sometimes additional standpipes and hose stations are required to provide the protection on each floor. Additional stairways may not be provided as they may not be required. In this

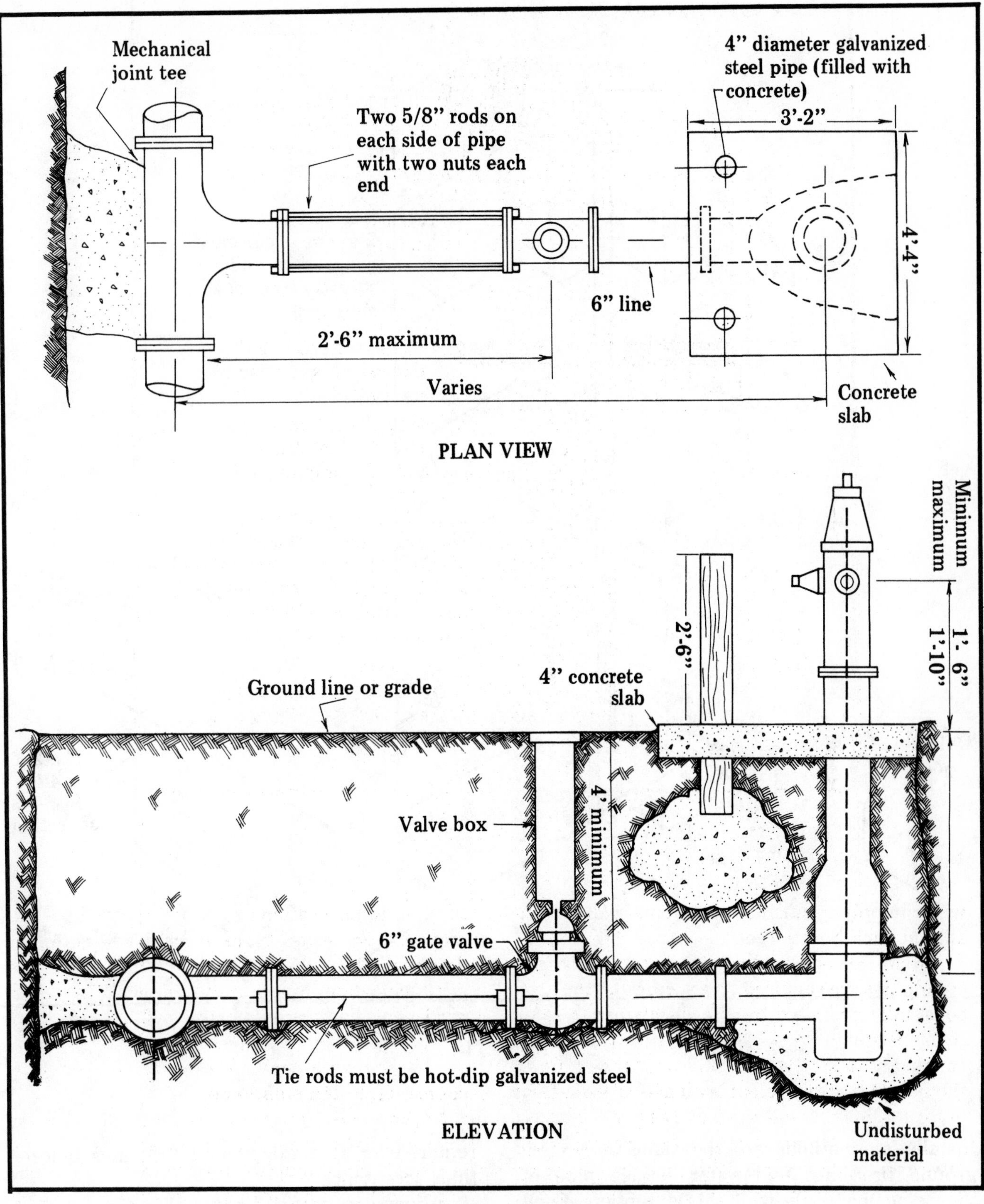

Underground fire line thrust blocks
Figure 11-5

case, the extra standpipes and hose stations on each floor can be located in hallways or other accessible locations approved by the authority.

Materials and Installation Methods

Underground fire lines should be of the same materials as water service pipe. (See Chapter 9.)

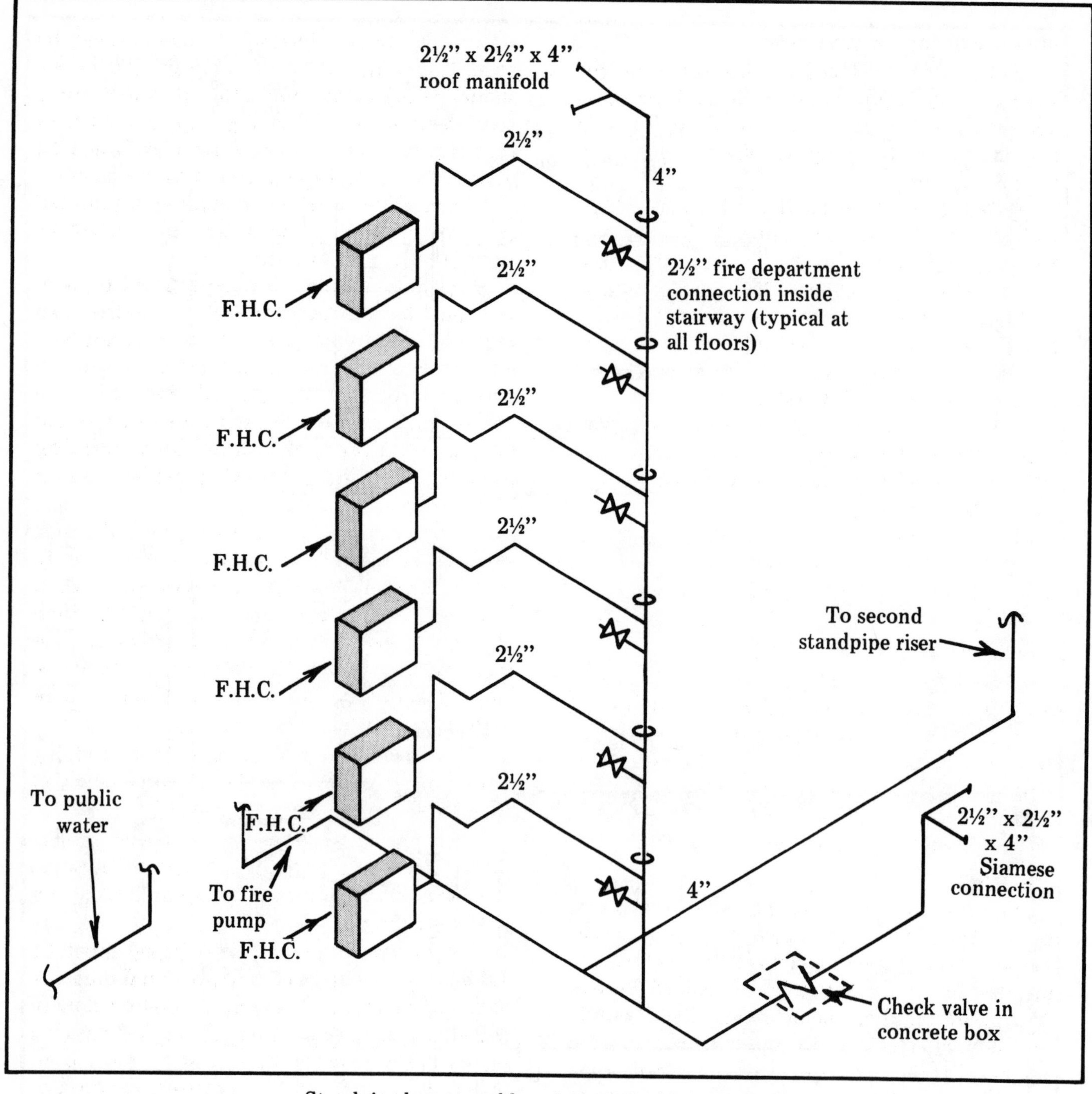

Standpipe layout and location of fire hose cabinets
Figure 11-6

Installation methods are the same as underground fire lines and water service pipe of the same materials with this exception: Each change of direction in underground fire lines must have poured concrete thrust blocks resting on undisturbed soil. See Figure 11-5. The thrust blocks prevent the pipe and fittings from blowing apart under the 200 p.s.i. pressure test required for all underground fire lines before their acceptance.

Above-ground fire lines within the exterior wall of a building must be black steel pipe, hot-dipped galvanized steel pipe or copper pipe. The only above-ground fire line is usually the standpipe. These lines, the fittings and all connections must be able to withstand 100 p.s.i.

pressure at the topmost outlet.

Buildings not exceeding 100 feet in height shall have standpipes sized at least 4 inches in diameter. Buildings in excess of 100 feet in height shall have standpipes sized at least 6 inches in diameter.

Standpipes shall be limited to 275 feet. Buildings exceeding a height of 275 feet must be zoned accordingly. Each zone requiring pumps shall be provided with a separate pump. Where two or more zones are supplied by pumps located at the same level, each zone shall be supplied by separate lines no smaller than the risers which they serve.

When it is necessary to pump the water supply from a lower zone to a higher zone, the risers shall not be smaller than the largest riser which they serve.

For systems with two or more zones, in which portions of the higher zones cannot be supplied with residual pressure of 65 p.s.i., an auxiliary means acceptable to the jurisdictional authorities must be provided.

Standpipes in buildings 50 feet or more high must extend above the roof a minimum of 30 inches. This extension must be with the same diameter pipe as the rest of the standpipe. An Underwriter's Laboratory approved duplex roof manifold with 2½ inch fire department connections must be installed on each standpipe.

Standpipes located in stairway enclosures must also have a 2½ inch fire department outlet. These outlets must have 2½ inch valves adapted for 2½ inch National Standard Thread fire department hose connections. Outlets must be accessible on each floor at the stair enclosure and at the stair enclosure in the basement if one exists. Hose outlets must not be located farther than ten feet from the standpipe or hose station, and must not be installed lower than five feet or higher than six feet above the finished floor.

A hose station must be provided within ten feet of the standpipe. The pipe connecting the hose station to the standpipe is usually 2½ inch pipe. Hose stations must never be located within any stair enclosure. See Figure 11-6 for a typical standpipe layout.

An approved wall mounted hose reel, cabinet or rack must be provided for each hose station. The station must be located so that it is accessible at all times. Each hose must be able to withstand 100 p.s.i. working pressure and be equipped with an adjustable nozzle that can be used to turn the water on and off. The nozzle should be adjustable from a fog spray to a strong stream of water. Where the pressure may exceed 100 p.s.i., pressure reducers must be installed to control the pressure to the hose.

When more than one standpipe is required to serve a building, each standpipe must be interconnected at its base.

A Siamese (duplex) fire department connection must be provided for each of the first two required standpipe risers. If more than two standpipes are required, the Siamese connections must be remotely located. Each Siamese connection must have the same diameter as the largest standpipe. For example, a building requiring a four inch standpipe must have a four inch Siamese connection.

An approved Underwriters' Listed check valve must be installed between the Siamese connection and the fire standpipe system. Each Siamese connection must have a 2½ inch National Standard Thread fire department hose connection, and must be installed on the street side of the building at least one foot but no more than three feet above grade.

The Siamese connection or its related piping must not project over public property (such as sidewalks) more than two inches.

Near the fire department Siamese connection a permanent "standpipe" sign with letters one inch high should be attached to the exterior of the building.

The water supply for standpipes must be sufficient to maintain 65 p.s.i. residual pressure at the topmost outlet, giving a required flow of 500 g.p.m. If more than one standpipe serves a building, the required flow must be 750 g.p.m. In most cases this additional pressure and water flow must be supplied by fire pumps. Refer back to Figure 11-1.

Fire pumps required to supply the 500 gallons per minute must be UL Listed. The pump controllers must also be UL Listed and may use limited service motors of 30 hp or less. Fire pumps must operate automatically with compatible controls. Pumps must be supplied with a separate electric service or be connected through a separate automatic transfer switch to a standby generator.

Here are the fire pump planning requirements. Refer back to Figure 11-1 to understand these requirements.

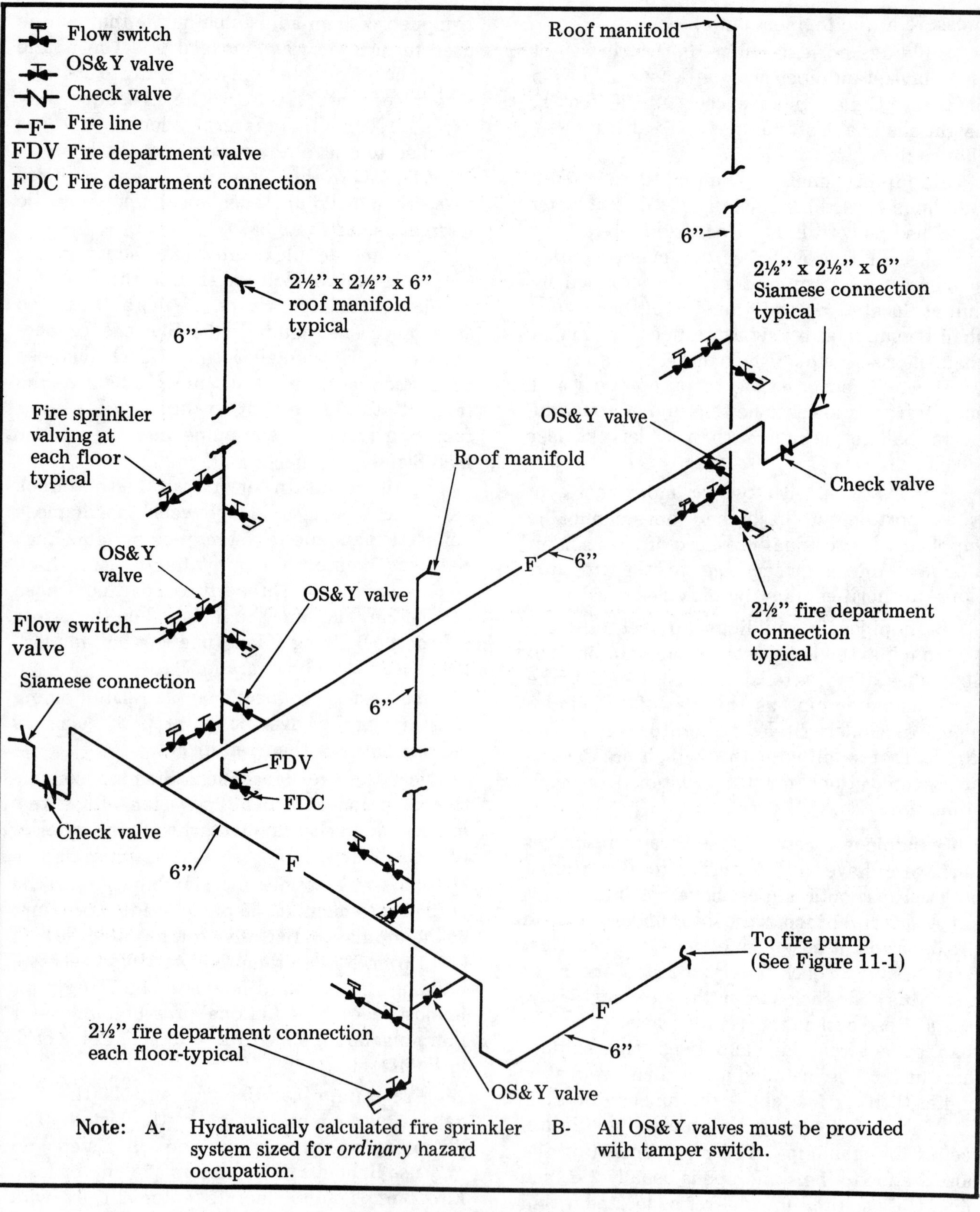

Automatic sprinkler-feed risers with sprinkler branch lines
Figure 11-7

- A 15 p.s.i. minimum pressure on a standpipe system at the roof must be maintained by a jockey pump actuated by a pressure switch or by connection to a suitable

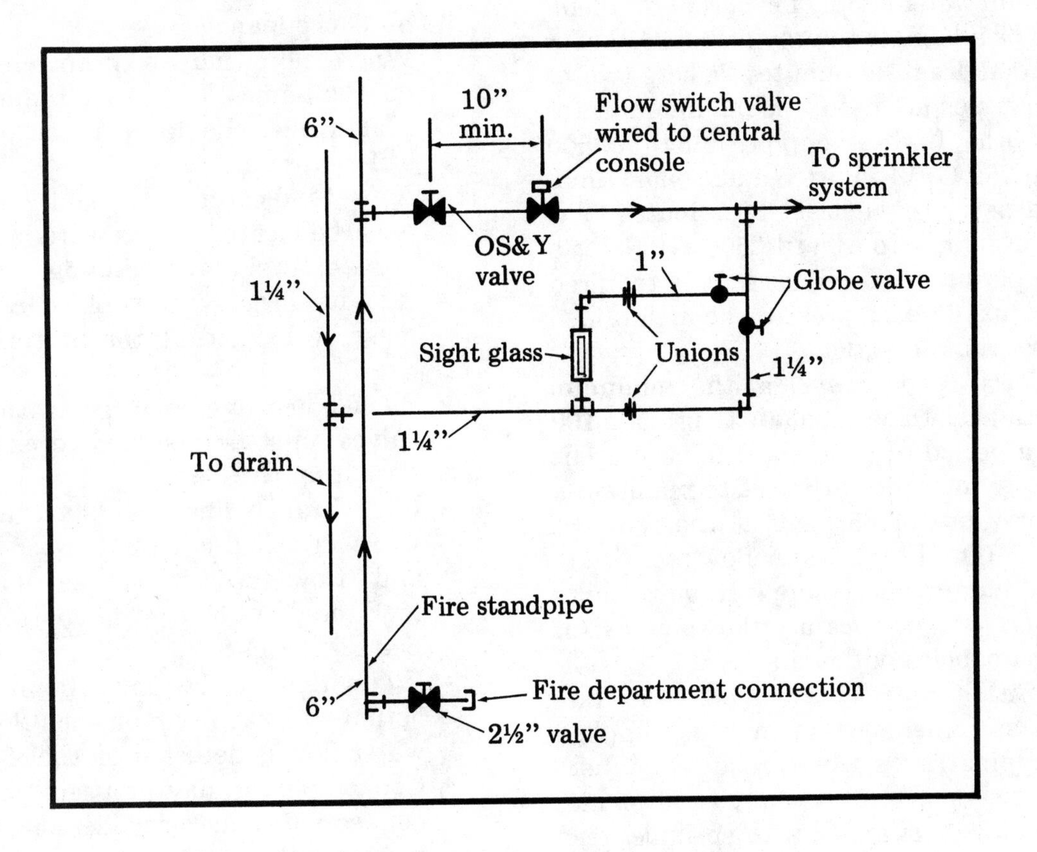

Typical fire sprinkler valving at each floor
Figure 11-8

domestic system through two 170 p.s.i. check valves. One of these valves must have a soft seat and one must have a hard seat.

- A full size bypass must be provided with an approved gate and check valve.
- The fire pump must have flexibly coupled drives.

Combined and Automatic Sprinkler Systems

In recent years, new construction requiring fire protection systems is using what is known as a "combined system" or an "approved automatic sprinkler system."

Plumbing contractors with local or state certification can install a fire protection system consisting of standpipes and fire hoses.

However, many states now require that plumbing contractors be certified by the state fire marshall to install fire protection systems having automatic sprinklers.

Let's review code requirements for these two systems.

Combined systems A combined system is one in which water-supplied risers serve the 2½ inch outlets for fire department use and also as outlets for automatic sprinklers. The minimum size risers for a combined system must be a 6 inch diameter pipe.

The water supply for a combined system is determined largely by the class assigned to certain type occupancies. There are three classes:

Class I: Includes occupancies where combustibility is of low quantity. These include bakeries, canneries, laundries, to name a few.

Class II: Includes occupancies where combustibility is of moderate quantity. These include cereal mills, distilleries, mercantiles, and the like.

Class III: Includes occupancies where combustibility is considered high. Feed mills, paper process plants and tire manufacturing plants are a few examples of high combustibility.

Minimum water supply for *Class I occupancies* must be adequate to provide 500 GPM for a period of at least 30 minutes. Where two or more standpipes are used, the minimum water supply to the first standpipe must be 500 GPM, and 250 GPM for each additional standpipe for a period of at least 30 minutes. The total supply is not to exceed 2500 GPM for a minimum period of 30 minutes. It is required that a residual 65 p.s.i. pressure be maintained at the most remote outlet.

For *Class II occupancies,* the minimum water supply has to be adequate to provide 100 GPM for a period of at least 30 minutes. The water supply must be sufficient to maintain a residual pressure of 65 p.s.i. at the topmost outlet with 100 GPM of water flowing.

The minimum water supply requirements for *Class III occupancies* are the same as for Class I occupancies outlined above.

Automatic sprinkler systems The automatic sprinkler system may be supplied from standpipe risers where required. Where standpipe risers are not required, automatic sprinklers may be supplied with sprinkler-feed risers sized by hydraulic calculations by a professional engineer. Fire Department connections of 2½ inches shall be located on each floor, similar to the standpipe and hose system. See Figure 11-7.

Other requirements for automatic sprinkler systems are set forth below:

- Where more than one standpipe riser is required, risers shall be looped at the lowest floor.
- Loop lines shall be the same size as the risers. For example, if the riser pipe is 4 inches, the loop piping must also be 4 inches.
- Branch lines shall be taken off the riser at the floor they serve.
- Where more than one riser is required, branch lines on alternate floors shall be taken off from different risers.
- Branch lines shall not supply areas exceeding 4000 square feet in apartment or hotel occupancies.
- Where more than one branch line is used on the same floor, each branch, where practicable, should be taken off from a different riser.
- A post-indicator valve and check valve must be located on the water supply line located outside the building.
- An outside screw and yoke (OS&Y) valve must be located at the bottom of each riser.
- When a loop system is used, other OS&Y valves must be installed to isolate each riser from the loop.
- Each branch line shall have an OS&Y valve installed with a tamper indicator and flow switch. These should be monitored at the central control station. See Figure 11-8.
- Water flow devices must be installed to actuate a local alarm on each floor, when water flow is detected on that floor.
- Copper piping may be used, but copper connections must be soldered with solder having a thermal strength of not less than 95 percent tin and 5 percent antimony.
- Water supply lines are not required to be graded.
- Where two or more fire pumps are used, each pump shall operate independently.
- Fire pumps usually must have electric motors. Engine-driven fire pumps are acceptable if first approved by local jurisdictional authorities.
- 1½ inch fire hose and cabinets may be omitted, provided each standpipe on each floor is equipped with a 2½ inch hose valve, a 2½ inch by 1½ inch reducer and cap with attachment chain for fire department use. See Figure 11-8.
- Where a secondary supply of water is required, an on-site supply of water must be provided. The supply has to be available automatically when the principal supply fails. See Figures 11-2 and 11-3 (alternate water sources).

Swimming Pools And Spas

Swimming pools and spas are big business for some plumbers. There are hundreds of thousands of swimming pools and spas in the United States, and about three-fourths of these are privately owned residential pools and spas. Approximately 75,000 pools and thousands of spas are built each year.

Specialty swimming pool contractors with a certificate of competency in that trade usually install the necessary piping and equipment in new pools and spas. These specialists also repair and maintain existing pools and spas.

You should know that your plumbing contractor's license also qualifies you to perform all work regularly done by a swimming pool contractor. Some plumbing contractors act as subcontractors on pool work. Everyone who wants to be licensed as a plumber in most parts of the country must be knowledgeable in this field to pass the journeyman's and master's examination.

Swimming Pools

The only common type of swimming pool is the *recirculating type*. This pool is equipped with a pump to recirculate water from the pool through a filter system. Some pool equipment for private use may have an automatic feeding device for adding chlorine or fluorine. Many private pool owners add their own chemicals, or use a professional pool company to maintain the quality of pool water.

The recirculating type of swimming pool can accommodate heavy use and yet use a minimum of water. Fresh water is added as needed when it is lost by evaporation, splashing or backwashing. Good filtration equipment assures the bather of water that is clean, clear of organic matter and safe from harmful bacteria.

Code Definitions

Swimming Pool: any constructed pool over 24 inches in depth or with a surface area exceeding 250 square feet and suitable for swimming or bathing.

Private Pool: a swimming pool located at a single family residence available only to the family of the household and its guests.

Public Pool: a pool used collectively by a number of persons for swimming or bathing. It may be operated by an owner, lessee, operator, licensee or concessionaire, whether a fee is charged or not. A public pool may be one of four types:

1) *Competition Pool:* a pool used for competitive swimming events.
2) *Public Pool:* a pool intended for public recreational use.
3) *Semi-public Pool:* a pool operated in conjunction with buildings such as hotels, motels and apartments.

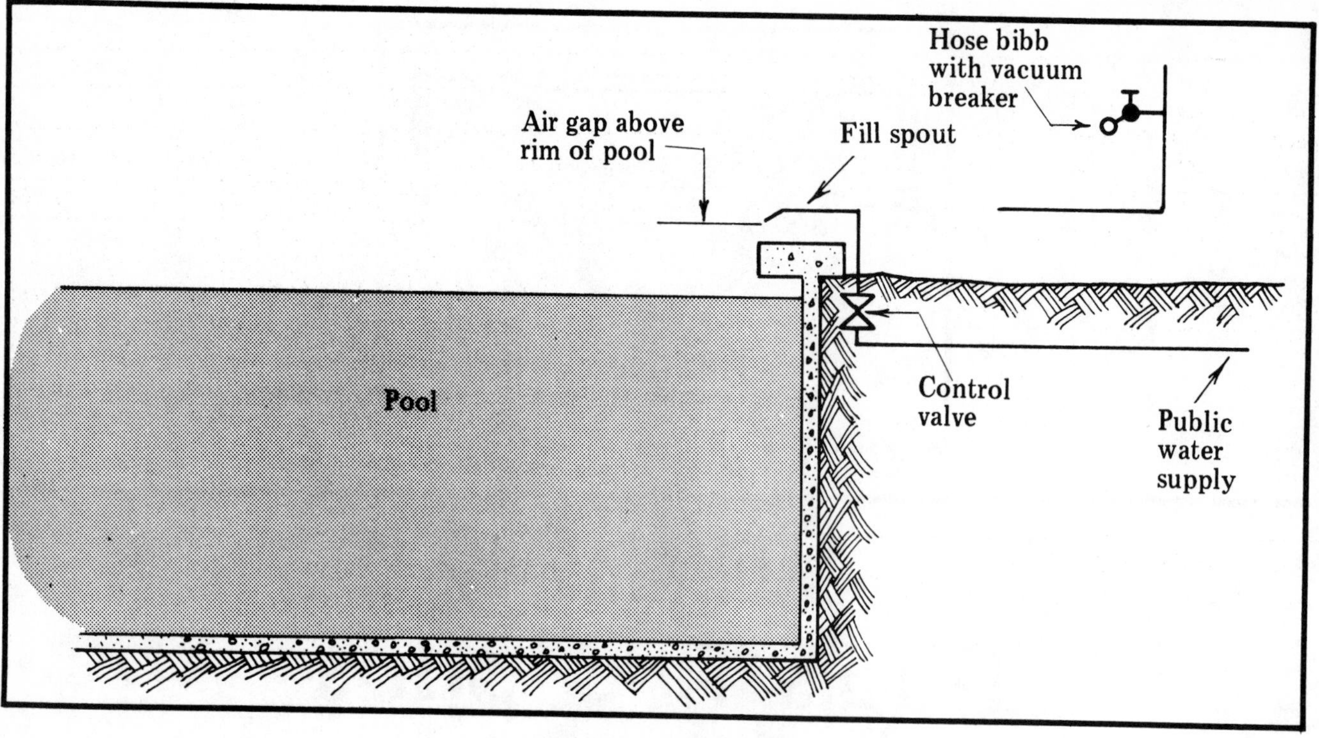

Water supply layout
Figure 12-1

4) *Special Purpose Pool:* a pool operated for water therapy treatments rather than for recreational purposes.

Inground Pool: any pool where the sides rest partially on, or have full contact with, the surrounding earth.

Non-permanent Pool: any pool constructed to be disassembled and re-assembled to its original integrity.

Non-swimming Area: any portion of a pool which is too shallow or which has underwater ledges or walls that prevent normal swimming activity.

Onground Pool: any pool where the bottom and sides rest fully above the surrounding earth.

Permanent Pool: any pool constructed in the ground, on the ground or in a building that cannot be disassembled for storage.

Pool Piping and Equipment

Main Drain (also defined as main outlet or main suction pipe): A drain to empty the pool.

In recent years it has been discovered that children are fascinated by the suction of water through the main drain in swimming pools and spas. Numerous instances have been reported of children sitting on the main drain, being held beneath the water by its strong suction, and drowning. There have been other reports of small children having their intestines literally pulled from their bodies by this powerful drain suction.

Thus, the older, flat main drain previously used in pools and spas is *not* accepted by most codes today. A new dome-type drain is now required on all new construction of swimming pools and spas. It must have either an anti-vortex type cover with a minimum of 6 square inches of open unobstructed area, or an open grate with a minimum open unobstructed area of 85 square inches. The anti-vortex cover must be securely fastened over the main outlet in a manner requiring tools for removal. The main drain must be located at the lowest point of the pool so the pool will drain dry for cleaning, painting and general repairs. See Figure 12-5.

Vacuum Fitting: This fitting is connected to piping which is connected to the suction side of a pump, where it is not provided within the skimmer. Every pool must have a vacuum fitting, located a maximum of 10 inches below the water surface inside the pool, in an accessible location.

Recirculating Piping (also known as *return piping* or *pool inlet piping*): This consists of the piping connected to the discharge side of the pump. It returns water to the pool after filtering.

Backwash Piping: This includes the piping

Pool capacity (gallons)	Diameter S & G filter	Soil percolation rates					Minutes/inch						
		1		2		3		4		5		6	
		Sq. Ft.	Gallons	Sq. Ft.	Gallons	Sq. Ft.	Gallons	Sq. Ft.	Gallons	Sq. Ft.	Gallons	Sq. Ft.	Gallons
17,000	24"	53.5	2,000	96	3,590	130	4,860	158	5,910	182	6,800	202	7,560
17,000 to 26,000	30"	83	3,100	149	5,560	200	7,550	247	9,240	280	10,500	315	11,780
26,000 to 38,000	36"	120	4,490	215	8,050	292	10,910	358	13,400	408	15,290	452	16,900
38,000 to 52,000	42"	163	6,100	293	10,970	400	14,980	485	18,150	555	20,800	618	23,100

Note: Effective depth of soakage pits - 5'-0"
Square feet refers to the area of the bottom of the pit

Minimum area and volume of soakage pits for swimming pools
Table 12-2

connected to the backwash outlet of the filter. It conveys waste water to an approved point of disposal.

Filter Piping: All piping, fittings, and valves necessary to connect the filter system together as a unit.

Water Supply

Water for the pool can come from the public water system. Many owners of private pools use a garden hose attached to a hose bibb to fill the pool. The hose bibb should have a vacuum breaker to prevent cross-connection. Public pools and some private pools have a direct connection to the public water system. If there is a direct water supply, a fill spout with an air gap above the overflow rim of the pool is required. See Figure 12-1.

Where a well is used to obtain the make-up water for a swimming pool, the water supplied must be clean and meet the bacterial requirements for a domestic water supply. If it is not reasonably free of objectionable minerals, the filtration system must correct the deficiency.

Well water color should not exceed the standard of 100 and the iron content should not exceed 0.3 parts per million before filtration. If the raw well water does not meet these specifications, it must be given a preliminary treatment prior to introduction into the pool.

Waste Water Disposal

Swimming pools must have some means of disposing of backwash water, and of being emptied. Pools equipped with pressure diatomite filters require piping which enables backwash waste to be deployed to a settling basin before final disposal. One of the following water disposal methods must be used:

- It may be emptied into a public or privately owned sewage system if approved by the authority having jurisdiction.
- It may be discharged into a disposal well when approved by the authority having jurisdiction.
- It may be discharged into an open waterway, bay or ocean, where this is permitted by the Director of Environmental Resource Management (DERM), or by the local health department, depending upon which agency has jurisdiction.
- It may be expelled into an adequately sized drainfield.
- It may be disposed of through a sprinkler system used for irrigation purposes. The waste must be confined to the property from which it originates. It must not flow on or across any adjoining property, public or private. Backwash water must not be discharged through a sprinkler system.

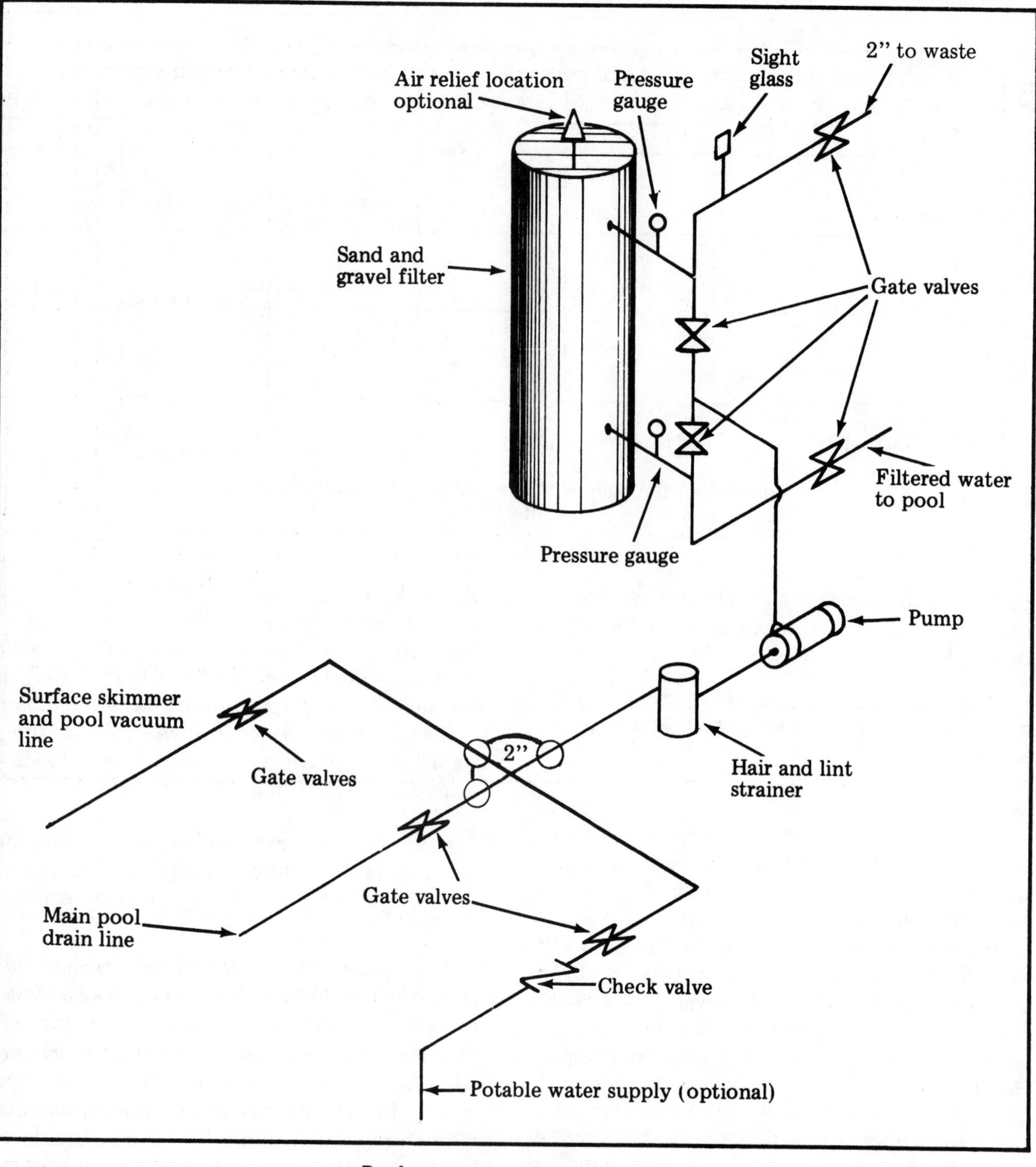

Pool equipment piping diagram
Figure 12-3

- It may be expelled into an adequately sized soakage pit or drainage trench.
- Pool waste and backwash water may be puddled on private property provided that the disposal area is big enough and properly graded to retain the waste water within the confines of the property. The pool of standing water can not remain for more than one hour after discharge. The disposal area must be a minimum of 50 feet from any supply well. Table 12-2 gives percolation rates for soakage pits.

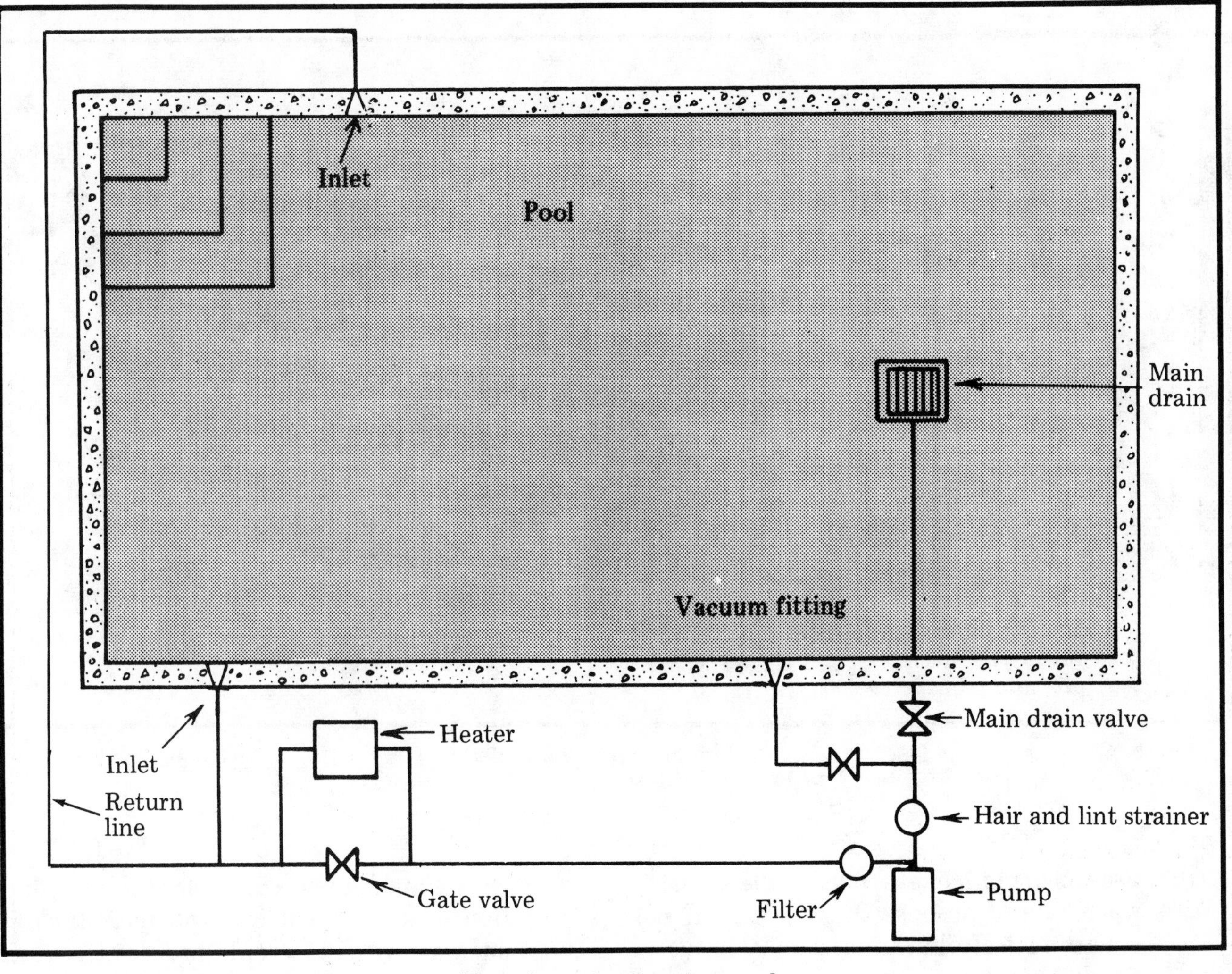

Plan view of swimming pool
Figure 12-4

Minimum Equipment for Swimming Pools

A certain minimum of equipment is required for every pool. See Figures 12-3 and 12-4. Each pool must have a main drain (outlet) located at the lowest point for emptying the pool.

A hair and lint strainer, equipped with an easily removeable screen with a free area 5 times the cross sectional area of the suction pipe must be installed in the suction line ahead of the pump.

The recirculation inlet or inlets must be sized and spaced to produce uniform circulation of the incoming water throughout the pool. One inlet is required for each 350 square feet of pool water surface or fraction thereof, with a minimum of 2 inlets located at least 10 feet apart. The entire recirculation system must be sized so that the velocities do not exceed 10 feet per second (f.p.s) at the design flow. The suction line must not exceed 5 f.p.s. at the design flow during the filtration period. When the main drain is used for a return, it must be considered an inlet but sized as a suction line.

A vacuum fitting sized 1½ inches in diameter must be provided on all pools. This fitting is installed a maximum of 10 inches below the water line and in an accessible location.

A valve must be installed on the main drain (outlet, or suction) line, and must be accessibly located outside the walls of the pool. Residential pools must have a minimum turnover rate of once every 12 hours of operation.

Filtration equipment should be one of the following types. It may be of another type if it is shown by test to be equal in efficiency to a sand and gravel filter.

Sand Filters

- **Pressure sand filters must have a filtration rate not over 5 g.p.m. per square foot of filter area.**

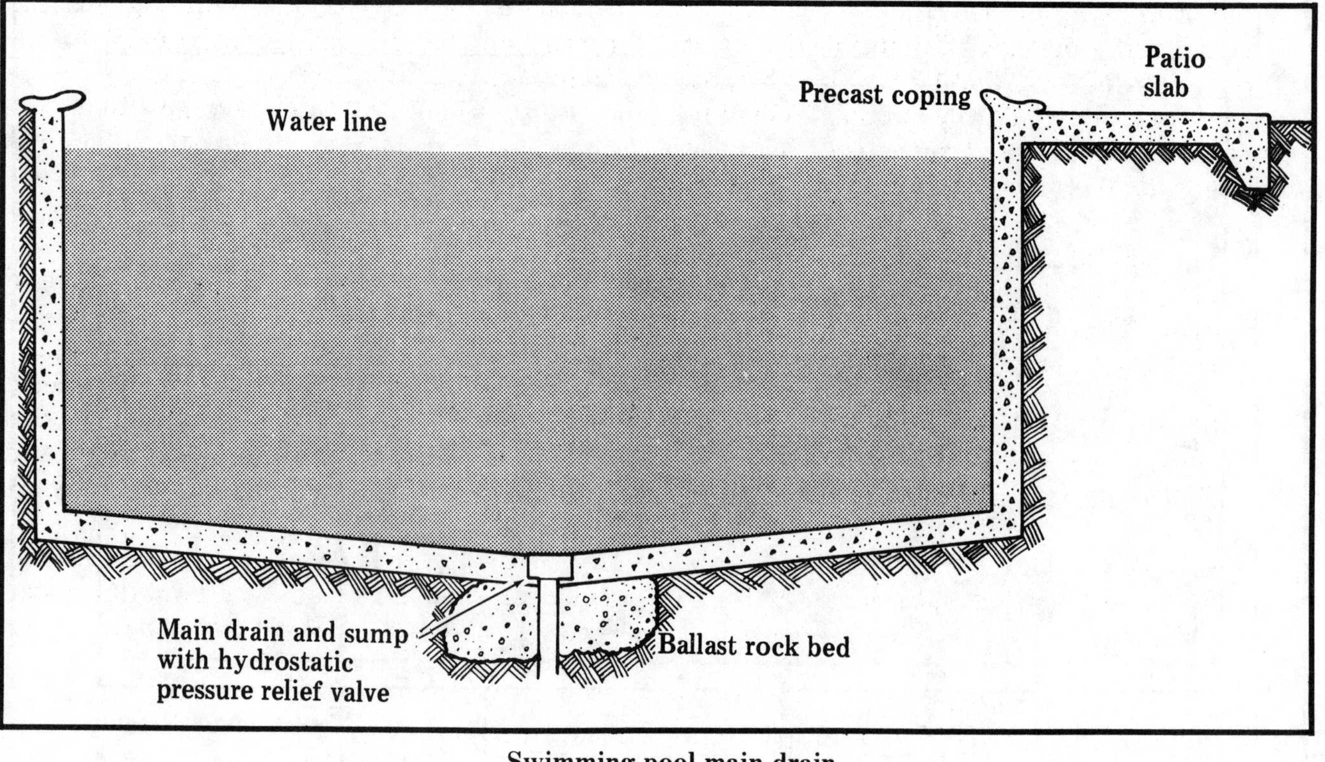

Swimming pool main drain
Figure 12-5

- Pressure sand filters must have a backwash rate of not less than 12 g.p.m. per square foot of filter area.
- Sand filters must contain a minimum of 19 inches of suitable grades of screened, sharp silica sand properly supported on a graded silica gravel bed.
- A sufficient free-board above the surface of the sand and below the overflow troughs or pipes of the filter must be provided to permit 50% expansion of the sand when the backwash cycle is used. There should be no loss of sand.
- The inflow and effluent lines must be provided with pressure gauges.
- The backwash line must have a sight glass installed so backwash water can be visibly checked for clarity.
- Tanks larger than 24 inches must have an access hole measuring a minimum of 11 inches by 15 inches.

Diatomite-type Filters

- Diatomite-type filters may be either the vacuum or pressure type.
- They must have a filtration rate of no more than two g.p.m. per square foot of effective filter area.
- Provisions must be made for the introduction of a filter aid into the filter tank in such a way as to evenly precoat the filter septum or element before it is placed in operation.
- The filter piping must be designed and installed during the precoating operation in such a way that the filter aid recirculates or discharges through the waste pipe and does not return to the pool waters.
- Provisions must be made for removing the caked diatomite either by backwash or disassembly.
- The design and installation of these filters must permit the filter elements to be removed easily.
- Pressure or vacuum gauges must be installed on these filters to determine the differential across the filter and the need for cleaning.
- An air relief device must be installed on each pressure filter tank at its high point.
- If discharge from backwash line is not visible, a sight glass shall be provided in the line to check the clarity of backwash water.

Surface Skimming for Pools

A minimum of one skimming device must be provided for each 600 square feet of pool surface or fraction thereof. Skimmers shall be constructed into the wall of the pool and meet the following requirements:

- The rate of flow through the skimming device should be adjustable up to a maximum 50 percent of the pool filter system.
- The skimmer weirs shall be automatically adjustable to variations in water level over a range of 3 inches. Skimming devices must be a minimum width of 5 inches.
- A basket having a minimum of 75 cubic inches must be located where it can easily be removed for cleaning purposes.

Pumps must be capable of filtering and backwashing the pool water at the proper pressure and rate required for the particular filter and piping system used. For operation, maintenance and inspection, all valves, pumps, filters and other installed equipment must be readily accessible.

Materials and Installation Method

The following types of pipe can be used in pool installations:

Type	Use
Copper, type K or L	All lines
Galvanized steel, standard weight	All lines
Wrought iron, standard weight	All lines
Brass pipe or tubing	All lines
Cast iron, service weight	Gutter lines only
Stainless steel, AISI, type 300 series	All lines
Monel	All lines
Polyethylene pipe	Pressure lines only
ABS and PVC Schedule 40	All lines

Materials and installation method

Thermoplastic pipe must be continuously marked on opposite sides with the size, type, schedule, U.S. Commercial Standard and National Sanitation Foundation seal of approval. All fittings for ABS or PVC plastic pipe must be schedule 40. Polyethylene pipe fittings must be the insert type and stainless steel clamps must be used for each connection. All fittings used in gutter lines (usually public pools) must be the drainage type.

Short radius 90-degree piping elbow fittings should not be installed on pool or spa suction piping below grade.

Suction piping for pools and spas must be a minimum of 2 inches in diameter.

The use of dielectric fittings is required where dissimilar metals are installed in pool and filter piping.

Pipe of like materials must be supported and installed in the same way as water piping. (See Chapter 9.) Where thermoplastic pipe and fittings are used, the trenches and backfill must be free of rock.

The entire pool pressure piping system, including the main drain, must be water-tested at 40 p.s.i. and proved tight before the installation is concealed.

Pool Heaters

Gas-fired swimming pool heaters and swimming pool boilers must comply with A.G.A. and A.S.M.E. standards. Oil-burning equipment must be approved by Underwriters' Laboratory or some other nationally recognized testing agency.

Pool-heating equipment must have a minimum of 70% thermal efficiency across the unit.

Water heaters and boilers must have either a thermostatic or high temperature control with a maximum temperature differential of 15° F or some other acceptable overheat protection device.

The installation must be designed so that the temperature of the heated water can not exceed 110°F.

The installation of pool water heating equipment is the same as for domestic water heaters. If the heater is installed in a pit, drainage must be provided for its protection.

Spas

The growing popularity of spas has forced local authorities to compile and adopt codes to provide adequate protection for the users. Spas and pools have many similar requirements, but those applicable only to spas are listed below.

A residential spa as defined by code is one that is permanent or non-permanent, and used by not more than two families and their guests.

A spa is defined by code by its gallonage

capacity. A spa's capacity cannot exceed 3250 gallons of water.

Spas which have a water depth of more than 3 feet 6 inches must have adequate and suitable handholds around 60 per cent of their perimeter area. Some of the approved handholds are: suitable slip-resistant coping, ledges, flanges or decks located along the immediate top edge of the spa. Ladders, steps, or seat ledges are also acceptable.

Protrusions, extensions, means of entanglement or other obstructions which can cause entrapment or injury to the bather must not be used.

Spa Filter Requirements

Spa filters are similar to swimming pool filters and are almost always designed by the manufacturer to meet the following code requirements:

- Filters shall be designed to maintain spa water under anticipated operating conditions.
  1) so that filtration surfaces can be easily restored to the design capacity.
  2) so that filtration surfaces can be inspected and serviced.
- Any filter and/or separation tank incorporating an automatic internal air release as its principal means of air release must have, as part of its design, a means of providing a slow and safe release of pressure.
- Filters must meet the safety performance standards of the National Sanitation Foundation, or other approved testing agencies covering filters.

Pumps for Spas

A pump must be installed for the circulation of the spa water. Such pump must meet the conditions of the flow required for filtering and cleaning of the water, to wit:

- A hair and lint strainer (equipped with an easily removable screen) to filter out such things as solids, debris, hair, and lint must be installed before the circulation pump.
- The design and construction of the pump and component parts must provide safe operation.
- Pumps must be mounted on a solid formed base, elevating the bottom of the motor at least 4 inches above the surrounding area.

Air Induction Systems

When an air induction system is provided for a spa, it shall totally prevent water backup that could cause electrical shock hazards.

Air intake sources must be positioned so as to minimize introduction of deck water, dirt, etc., into the spa.

Surface Skimming Devices

Spas have the same requirements as pools when it comes to skimming devices, with the following exceptions:

1) Spas must be provided with at least one skimming device.

2) Spa skimmers must be equipped with a vacuum break.

Piping Materials and Installation Methods

Materials and installation methods of like materials are the same as for swimming pools and domestic water. (See Chapter 9.)

Spa Heaters

The maximum temperature of the spa heater is 104 degrees Fahrenheit. A thermostatic control for the water is required, of course.

Maximum spa water temperature limit must be included in the consumer use label provided with each spa. It is a requirement that it be posted in a prominent place on, or in close proximity to, the spa itself.

Harnessing Solar Energy

In the past few years the rapid increase in the cost of electricity and fossil fuels has spurred massive research into alternate sources of energy. Harnessing the sun to heat and cool our homes and businesses will take many years of research and enormous amounts of money. Eventually solar heating and cooling may become the standard for most homes. Today solar heating is competitive for heating water for pools and domestic use where gas is not available. More and more solar energy units will be installed in the 1980's and you should know how professional plumbers fit into the solar energy picture.

When the energy crunch arrived, many energy experts were quick to explain savings that were available from converting existing homes and planning new homes around solar energy.

It is true that approximately eighty five per cent of the energy used in homes goes for heating domestic water and space heating and cooling. It is also true that the technology needed to use solar energy is available now. The problem is that the initial cost and minor inconveniences associated with solar energy are more than most consumers are willing to bear.

Domestic Hot Water Heating

Solar energy is a practical means for heating and this will be the most common home use of solar power for many years. It can be added to existing systems as well as to new construction at a relatively modest cost.

Solar energy for heating domestic water is not new. In Florida and other sun belt areas of the world it has been used successfully since before 1900. In 1955 over 60,000 solar water heaters were in use in Florida. It is estimated that over 100,000 units are in use in Israel. Some older buildings in Florida have used solar energy for many years. A family of four should realize savings of as much as two hundred dollars per year by installing a solar water heater where electricity costs 4 cents per kilowatt hour and gas is not available.

It is likely that within the next few years thousands of new and existing homes will be equipped to use solar energy for water heating. As a professional plumber you should know all you can about solar heating of domestic water.

Heating and Cooling

Economic studies done at the University of Florida Laboratory have shown that solar power is most practical if it can be used for both heating and cooling. The technology required for heating systems - collectors and storage tanks - is readily available now. What is *not* available is an air conditioning system that will operate at the low temperatures produced by efficient solar collection systems. Recent government studies have indicated that up to ten years of research may be required before solar powered cooling technology has advanced to the point where low-cost solar powered air conditioning systems are available.

Research now being done on solar energy is directed toward developing the hardware necessary to make solar energy cost effective. Solar energy will not find a mass market as long as it is the most expensive solution to the energy problem. It is foolish to assume that everyone will install solar systems in their homes because someone has convinced them that solar energy is in style.

Solar Energy Consumer Protection

Plumbing codes have been rewritten to cover solar energy systems so that the consumer and local contractor are protected against untested and inferior products. It is appropriate at this point to review some of the problems solar equipment installers are encountering.

Al Jenkins, a Las Vegas solar collector manufacturer (Enerjex Corporation, the nation's largest producer of solar collectors) and vice president of the Solar Energy Industries Association (SEIA) stated in the March, 1976, *Florida Contractor* magazine, "If there's to be a future for the solar industry, we've got to find ways of integrating to get us all pointed in the same direction. I recognize the man in the middle is the installing contractor. He's the one who gets the flak if his manufacturer makes a faulty product." It should be the manufacturer's responsibility, he feels, to provide support for contractors, to train and educate all the people in the distribution and installation chain who represent his products.

If the manufacturer is not responsible to his dealers and isn't selling a competitive product, "Then he shouldn't be in this business," declares Mr. Jenkins. It is his contention that solar product warranties should be an expression of the manufacturer's confidence in his products. They should reflect confidence about how long the product should last and should state what the manufacturer will do if it fails.

The Solar Energy Industries Association is quite concerned with the "fly-by-nights" in the solar business and even with the claims some of its own members use in advertising. In addition to working with the U.S. Office of Consumer Affairs, the organization is writing a tough code of ethics for its members (again quoting Mr. Jenkins) "to hopefully eliminate fraudulent, exaggerated - and even misleading - claims."

Jenkins states SEIA recommends strongly that until there is a tightening of policy, each manufacturer should have every product tested under the National Bureau of Standards testing procedure. This, he maintains, will provide "a frame of reference on which to base advertising claims. It will solve a lot of confusion in the industry." He points out that the installing contractor must be skeptical of the claims that are made in solar energy literature and must not over-promise to his customer.

Solar Water Heater Components

A solar heater is a simple device. There are at least three components in any heat collecting system:

1. The solar heat collector
2. The circulation system
3. The solar storage tank

A fourth component used in many systems is a back-up heat source to ensure that hot water is available even during periods of increased use or low solar radiation.

There are several obvious features that make solar water heating systems unique. In a conventional fossil fuel water heater the heating element can work continuously to produce hot water. The storage tank can be rather small, 30 or 42 gallons. The tank in a solar powered system must be large enough to carry through cloudy days and at night. The heating element of a solar water heater is known as a solar collector. See Figures 13-1 and 13-2.

For residential use the flat plate solar collector is the most practical and least expensive. It can produce temperatures up to 200° F. The heat deck consists of a metal plate and tubing. The collector plate absorbs heat and transfers it to the liquid in the tubing. Heat deck materials can be copper, aluminum or steel. Thermally it doesn't matter which one is used. Both the tubing and the collector plate should be of the same material so they will expand and contract at the same rates. Codes generally will not permit potable water to flow through aluminum tubing.

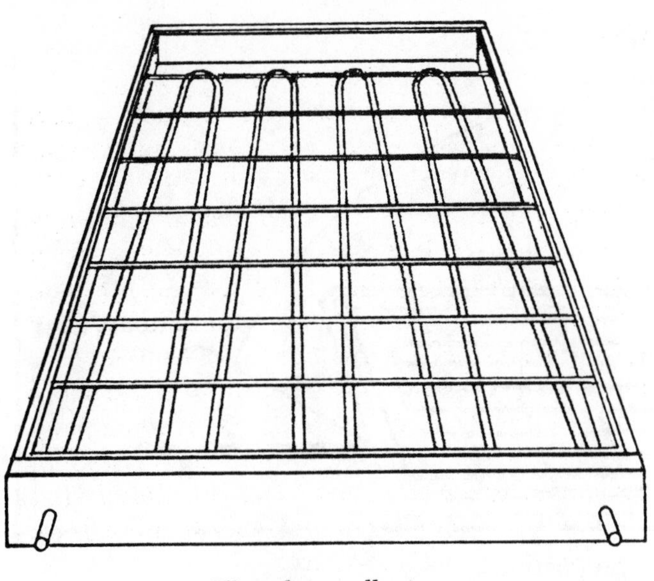

Flat plate collector
Figure 13-1

The heat collector box should be well insulated to shield the heat deck plate from the weather and reduce heat loss. The heat deck plate and tubing are usually painted flat black to maximize absorption of the sun's energy.

A transparent cover on the box permits the sun's rays to strike the metal collector plate and reduces the loss of reradiated heat back to the atmosphere. The glass used should have low iron content so it is as transparent as possible to incoming rays. The glass admits solar radiation but is opaque to the long-wave radiation that is created when the sunlight hits a solid surface. The long-wave energy is trapped inside the box to heat the fluid in the tubing.

A clear plastic sheet is better than no cover at all, but glass is both more efficient and more durable. Plastics tend to transmit both incoming and outgoing energy. Most available plastics deteriorate fairly rapidly where exposed to heat and moisture. If more heat resistant plastics become available they would be more desirable because glass has the disadvantage of breaking easily.

In cold climates a double layer of glass is recommended to prevent heat loss by convection when very cold air strikes the transparent surface.

A 4' x 12' solar heat collector will heat approximately 80 gallons of water per day. This should be sufficient for the average family of four. Allow twelve square feet of heat collector surface per person to meet typical hot water needs of most families at any time of the year in most sun belt areas of the U.S.

As a professional plumber you will not be expected to build solar collectors. But you should have an understanding of how they work and be able to answer your client's questions. Collectors are manufactured by many highly responsible companies. It is your responsibility, however, to check for the approvals required by your local plumbing code before the unit is installed.

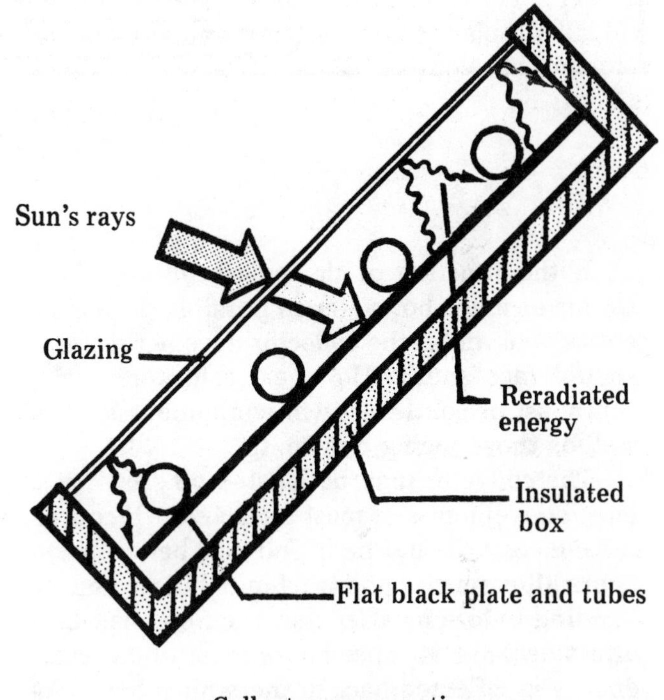

Collector cross section
Figure 13-2

Installation of the Solar Heat Collector

Only about 30 to 65% of the solar energy which strikes the glass surface of a collector during the day actually heats water circulating through the tubing. The rest is lost back to the atmosphere through the glass plate.

The most efficient collector would be perpendicular to the sun's rays at all times during the day at all seasons of the year. However, this is possible only if the collector turns and tilts to follow the sun's path during the seasons of the year and during the day. The best compromise is a flat collector tilted in the general direction of the sun's path across the sky and at an angle equal to the latitude plus ten degrees.

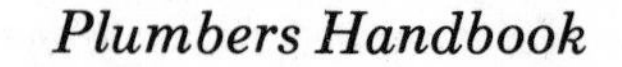

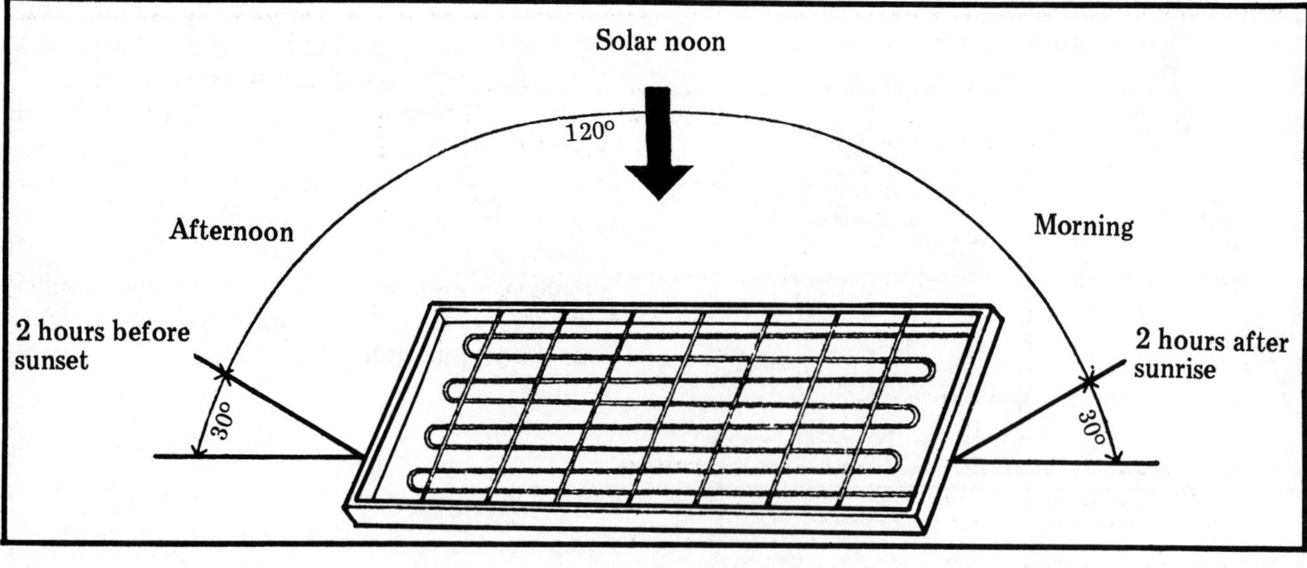

The optimum heat collection period
Figure 13-3

If the function of the solar collector is to supply as much hot water as possible during the entire year and if the collector does not move, it should face south. However, collectors facing southeast or southwest will work about 75% as well as those facing due south.

The collector may be located wherever it is most convenient and most attractive. Of course, the collector should be in full sun between two hours after sunrise and two hours before sunset. Shading before or after these periods has little effect because at those hours most of the sun's energy is reflected back to the atmosphere. See Figure 13-3.

The collector should be mounted as close as possible to the storage tank to reduce heat losses and friction in the pipes. Pipes should be well insulated.

In northern areas the transparent cover should be shielded to protect it from winds that would otherwise cool the surface. The collector should face south, as the greatest efficiency of the collector is needed in cold winter temperatures.

Collectors that are to be installed as awnings or as fixed overhangs on a residence should be approved by your local authority before work commences. This type of installation is generally not acceptable.

Collectors can be mounted on the ground almost anywhere provided they are securely anchored. But the glass box cover is more subject to accidental breakage or vandalism when the collector is mounted at ground level.

In climates subject to freezing temperatures it is important that all tubes drain dry. Draining the system is the best and cheapest way to protect it from freezing and bursting the pipes.

Mounting the Heat Collector

In many areas, plumbers are responsible for mounting collectors for solar systems. This is especially true in new construction where collector systems are built into the roof as an integral part of the home. These collectors look better because they can be designed to blend with the exterior, resist wind loads better and can be coordinated with other roof surfaces so there are fewer leaks.

Roofing contractors have made several recommendations for mounting collectors. They suggest that the collector have a structural frame just as is required for any other mechanical equipment. The frames should be welded or bolted to the main roof structure. Flashing and a rain collar are put around the pipes and the collector. The collector is then fitted and anchored securely to the frame.

Roof leaks can be a serious problem with roof mounted collectors. It is important to seal every hole that is drilled through the roof membrane. If a hole is too large or not properly sealed, a leak will occur. To prevent leaks, bore a small hole and caulk around it carefully. If you use a pitch pan, seal up as much as possible before securing the pan and filling it with pitch.

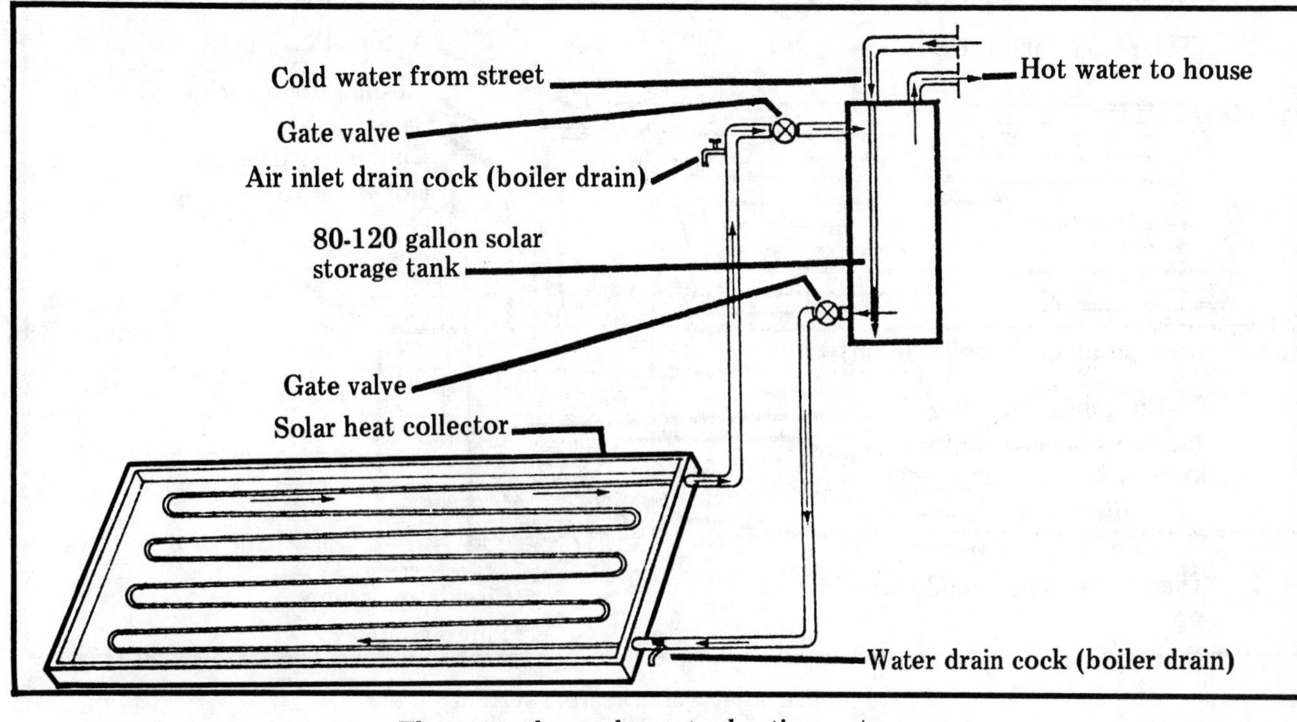

Thermosyphon solar water heating system
Figure 13-4

Otherwise the pitch will heat up under summer sun and create a pocket. Water can then get into the pocket. This moisture will eventually rust the top of the bolt. The pitch pan is a practical solution on built-up roofs. The owner should be advised to check his roof annually because expansion, contraction and wind effects can open fissures at the joints.

A bearing plate should be installed under the cliff angle to prevent rocking by the collector. Screw the bolt all the way through the sheeting and well into the main roof structure. Lag bolts fastened only to the sheeting will pull out under the load of vibration of the collector.

On roofs with asphalt shingles a layer of plastic cement should be placed on top of the shingles and beneath the cliff angle. Another layer of plastic cement should be placed on top of the cliff angle. The cliff angle should be securely fastened to the main roof structure with lag bolts.

Water Circulation

Water circulates through the collector to move hot water from where it is generated to where it is needed. A pump can do this, but is not necessarily required. Natural thermosyphon circulation requires no external energy source and no pumps, controls or other moving parts. The rate of movement of hot water from the collector to the storage tank, and of cold water from the tank to the collector, is controlled by the intensity of the sunshine. A thermosyphon results when hot water which is lighter rises to the storage tank and replaces heavier cold water which is drawn into the collector. See Figure 13-4.

The thermosyphon will not work properly unless the storage tank bottom is at least two feet above the top of the solar collector. Locating a large, heavy storage tank higher than the solar collector may mean that the tank must be on a roof or in an attic. This presents problems of weight, construction and appearance. A leak in an attic-mounted tank can cause considerable water damage inside the home. For this reason, solar energy codes usually require that an adequate sized drain pan with a safe waste pipe to the exterior be placed beneath all hot water storage tanks located above the ground floor.

In a thermosyphon system at least ¾ inch inside diameter piping must be used both in the collector and the circulation system. This reduces flow resistance. Be sure that the connecting pipe or tubing has a continuous fall with no sections that would permit the formation of an air pocket. An air pocket will stop the circulation.

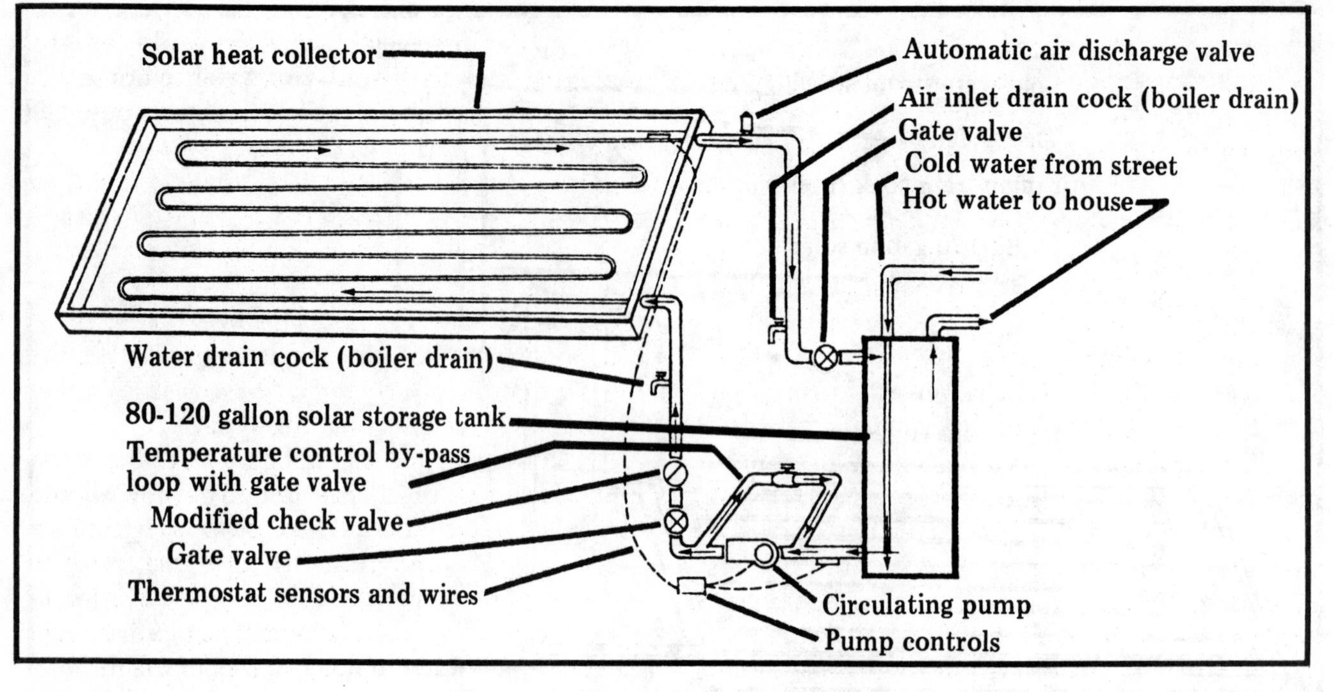

Pumped solar water heating system
Figure 13-5

A pumped system uses the same basic parts as the thermosyphon system, but adds a pump to force hot water from the collector to the large storage tank. See Figure 13-5.

In a pumped system the heavy solar tank can be located in any convenient place. This avoids the problems which go with roof or attic-mounted tanks. One-half inch copper tubing can be used in a pumped system. The pump must be controlled so that it circulates water through the heat collector only when water in the tank is cooler than the water leaving the collector. The pump and controls will add to the total cost, but this may be offset by the lower cost of installing the heavy storage tank at ground level.

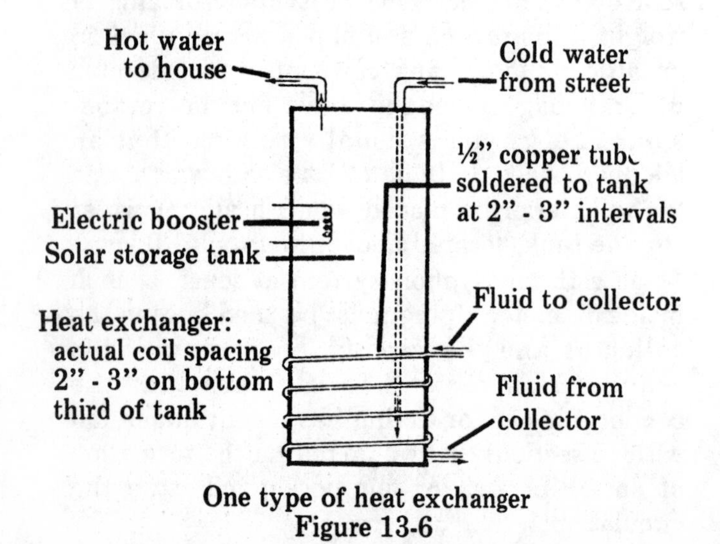

One type of heat exchanger
Figure 13-6

In a closed solar energy collection system, a fluid such as antifreeze is heated in the collector. See Figure 13-6. The fluid circulates through the solar collector and transfers its heat to water in the storage tank through a heat exchanger. This system, although more complicated and expensive, eliminates the possibility of freezing. A hard freeze can destroy a heat collector if it is not properly protected. It is important that a closed system be used or that the entire system be installed so that it can be drained dry if freezing temperatures are expected.

Materials and Installation

The rapid development of interest in solar energy has created a problem for many building code administrative authorities. Building inspectors are faced with new installation problems. Existing code regulations are not adequate for regulating installations of solar systems. However, until more adequate codes are developed, existing codes are being used to control installation methods and materials.

It is likely that a fairly standard solar energy code will be developed and adopted by most jurisdictions within the next few years. In the interim, you need guidance for solar installations. The following material will help you understand how existing codes and the "Uniform Solar Energy Code" are being applied to solar energy installations.

Quality and weights of materials for use in the installation of a solar energy system are the same as for all other plumbing. Refer back to Table 4-2. Pipe and fittings used for conveying fluids within a solar system must comply with the requirements for a potable water system. Only cast iron pipe, cast iron threaded pipe, wrought iron pipe, brass or copper type K or L pipe and steel pipe and fittings can be used. Cast iron, wrought iron, steel pipe and their fittings up to and including two inches in size should be galvanized. Plastic pipe must not be used for conveying heated fluids within a solar system since standards restrict the use of plastic pipe where temperatures could exceed 180° F. Aluminum tubing can not be used in potable water systems. Its use within a solar system (if first approved by jurisdictional authorities) would be limited to the closed type system using a heat exchanger.

Copper tube heat exchangers must not be less than type L, without seams, joints, fittings or valves. They must be constructed of double wall material designed to prevent leaks that could result in a cross connection.

Fittings used in a solar piping system should be of the same materials as the pipe. An exception is made for valves and similar devices. Where the use of dissimilar piping and fitting materials cannot be avoided, these materials should be electrically isolated.

Valves installed in a solar piping system up to and including two inches must be of brass or other approved materials. The fully opened valve must have 80 percent of the cross-sectional area of the nominal size of the pipe to which the valve is connected. Control valves must be installed so that they can isolate the solar system from the potable water supply. All control valves must be readily accessible.

Any solar energy system with control valves which isolate heat generating or transfer equipment and pressure storage tanks must have an adequate pressure relief valve. The rate of discharge of the valve should be sufficient to limit the pressure rise within the system to below the recognized maximum working pressure of the pipe or storage tank.

Where excessive water pressure is possible, an approved type pressure regulator must be installed. Since a solar energy system can be an integral part of the building's water supply system, the pressure regulator could be the regulator installed in the water service pipe. Therefore a second pressure regulator would not be necessary in the solar energy system.

Temperature relief valves must be installed on all equipment used for the heating or storage of domestic hot water. A combination pressure and temperature relief valve is acceptable under most codes. The temperature sensing element must be installed in the hottest water within the top six inches of the tank. Temperature relief valves must be the reseating type.

Relief valves located inside a building must have a full size discharge line. The line should discharge to the outside, turn down to within six inches of grade and drain dry after use. Be sure the line is securely strapped to the exterior of the building. The end of the discharge line must not be threaded. A discharge line can terminate at other locations if first approved by the controlling authority.

Automatic air discharge vents must be installed at all high points of a solar piping system. See Figure 13-5.

Storage Tanks

All storage tanks used for domestic hot water must meet the applicable A.S.M.E. requirement and the requirements listed in Table 4-2. Some codes require that any hot water storage tank located above the ground floor have an adequately sized drain pan with a minimum three-quarter inch indirect waste pipe which discharges to the outside of the building.

All storage tanks must be equipped with an adequate and accessible drain cock. The tank and any devices attached to it must be accessible for repair or replacement.

Storage tanks must be permanently labeled with the maximum allowable working pressure and the hydrostatic test pressure which the tank is designed to withstand. These markings must be accessible to the inspector. The minimum hydrostatic test pressure is 300 pounds per square inch. The working pressure can never be more than 42½ percent of the hydrostatic test pressure.

Non-pressure type tanks must be tested by filling the tank with water for a period of 24 hours prior to inspection. The tank must not leak during the test.

With special approval from the administrative authority, tanks designed and constructed to resist trench loads and corrosive soil effects can be buried underground. But no part of the tank can be covered or concealed prior to inspection and approval.

Heat Collectors

Administrative authorities, when they deem it necessary, can require that solar collectors be tested by an approved laboratory to determine the thermal performance. Some codes require the testing of solar collectors in accordance with A.S.T.M. procedures to check their integrity under wind loads.

Frames for securing heat collectors to a building must be built of materials durable enough for exterior use. The heat collectors must be anchored securely so they can withstand the expected dead, live and wind loads. Anchoring devices and pipes that penetrate the roof surface can result in leaks if not sealed adequately. Joints around pipes, ducts, bolts, or anything which penetrates the roof must be made water tight. Use a pitch pan, lead, copper, galvanized sheet steel or other approved flashings. Voids around piping, ducts and other appurtenances passing through walls, ceilings and floors must also be sealed. Collector panels that are not an integral part of a roof surface must be mounted a minimum of 3 inches above the roof surface.

Collector panels and related piping must be installed so that it will drain dry. Collector boxes should have drainage holes located at the low point for draining rain or other liquids that might collect in the box.

Glass used in the collectors should have a low iron content. Only tempered glass should be used.

All piping in the system must be installed to allow for expansion, contraction and normal movement of the structure. Piping which might be subject to undue corrosion, erosion or mechanical damage must be protected. Refer back to Chapter Nine.

Insulation

All piping carrying heated water, fluids or gases from a collector or heat exchanger to storage tank must be insulated to minimize heat loss. The insulation thickness must limit maximum heat loss to 50 B.t.u.'s per hour per linear foot of pipe up to two inches in diameter. Tanks must be insulated to limit heat loss to no more than two percent of the stored energy in a 12 hour period.

Fittings

Threaded joints, soldered joints and flare joints must be made as though for water piping. Refer back to Chapter Nine.

Cross Connection

Piping in the system must be isolated so that it is impossible for gases, fluids or other substances to enter any portion of the potable water system. The solar piping, tank, receptacle and all equipment must remain isolated in the event of back siphonage, suction or any other irregularity. An approved backflow prevention device will be adequate to isolate the solar energy system.

Two sets of plans are generally required by the administrative authority. Plans in most cases must be prepared by a professional engineer and must reflect structural calculations, mounting frames and anchorage detail. Plumbing drawings should show the entire solar system and specify only approved plumbing items and UL approved electrical components.

A building permit is usually required to install, repair, or alter any solar energy system.

When you have completed installation of the solar energy system, it must be tested and proved tight under a water, fluid, or air pressure test. The system must be able to withstand 125 psi without leaking for a period of 15 minutes.

Gas

Every substance known to man is either a liquid, a solid, or a gas. Although gas is lighter than the other two forms of matter, it does have weight. It can be forced through very small spaces, a valuable characteristic. Gas has neither a fixed shape nor a fixed volume. It is made of constantly moving atoms. When these atoms are forced into a container they will take on the container's shape but occupy only about one thousandth of the container's interior space. The spaces between these particles are empty.

Gas particles are liquefied when they are cooled below their boiling point. When this temperature is reached, the gas particles are pulled together to form a liquid. This principle is used in making liquid oxygen.

The first recorded discovery of natural gas was made by a Greek shepherd. He noticed his sheep acting strangely at a certain grazing location. He investigated and discovered that a substance coming from the ground made him lightheaded and talkative. Ancient Greeks thought these vapors were the breath of the god Apollo. A temple was erected at this site and was called Delphi. It soon became the religious center of Greece.

Nearly 3,000 years ago the ancient Chinese discovered natural gas and learned it would burn. They are credited with being the first to use this fuel for industrial purposes. They used hollow bamboo to pipe natural gas to where it could be used to evaporate brine to make salt.

Belgian physician Jan Baptista Van Helmont invented the word gas in 1652 to describe this amazing substance. He is also credited with producing the first manufactured gas from coal.

In 1775 natural gas was first discovered in America in the state of West Virginia. In 1806 a man who lived in Newport, Rhode Island (David Melville, by name), lighted his home with gas manufactured from coal. Baltimore was the first city to light its streets with manufactured gas. This was in 1817. The first commercial use of natural gas was from a well drilled to a depth of twenty-seven feet at Fredonia, New York, in 1821. Three years later the town of Fredonia was using gas from this well for lighting its streets and homes. By 1850 many cities and towns were using gas for lighting purposes. Gas piped to these lighting outlets was yellow and gave off a poor light.

A German, Robert Wilhelm Von Bunsen, developed the Bunsen burner. People in the plumbing and gas fitting trades are familiar with the Bunsen burner because it is still in use today. A Bunsen burner mixes air with the gas before combustion to get the best flame. The flame can be adjusted for a particular use and is nearly smokeless and free of soot. Gas was used almost exclusively for lighting streets and homes until Thomas A. Edison invented the electric light.

Gas is used today in cooking, refrigeration, heating and many industrial applications.

Texas and Louisiana produce about 70 percent of the natural gas found in the United States. Most of the rest comes from the Middle West and Rocky and Appalachian Mountain areas. Transporting this gas to all areas of the United States is a major industry. A network of natural gas pipelines totaling approximately 700,000 miles stretches to all parts of the country.

Since natural gas is clean, dry and has no odor, a gas leak in a pipeline or in pipes within a building might go undetected until an explosion occurred. Therefore a chemical odorant is added to gas before it enters the pipelines. This odor warns anyone in the area of escaping gas before the concentration can reach a danger level.

KINDS OF GAS

Natural gas contains chemical impurities which are valuable for uses other than as a fuel. These impurities are removed before being piped to the consumer. The natural gas which millions of consumers use as fuel in homes and industries is known as dry or sweet gas. Natural gas (methane) is not poisonous but can cause suffocation in a closed space. It is also explosive under certain conditions.

Manufactured gas is produced chiefly from coal. It burns with a blue flame and is generally added to other fuels to increase its heating capacity. Manufactured gas is also used by consumers as fuel in homes and industries. It can be poisonous since it contains carbon monoxide. It is explosive under certain conditions.

Liquified petroleum gas is also known as LP or bottled gas. It is produced in plants that process natural gas. LPG consists primarily of butane or propane, or a mixture of both. LP gas under moderate pressure becomes a liquid, thus making it easy to transport and store in special tanks. When an LPG tank supplies a building's gas piping system, the liquid is allowed to drop to normal atmospheric pressure and temperature and returns to its original gaseous state.

LP gas is heavier than air, colorless, and nonpoisonous. Since it is easily containerized and transported, it is convenient to use as fuel for homes and businesses in remote areas.

GAS PIPING AND THE PLUMBER

The plumber's responsibility for sizing and installing a building's gas supply system is different from his responsibility for sizing and installing a building's water-supply system. Water service pipe from the meter (located at property line) to a building is sized and installed by a plumber. Gas service pipe from its connection with the gas supplier's distribution pipes (located on public property) to the gas meter (generally outside the building) is sized and installed under the direct control of the gas supplier. The plumber has responsibility for sizing and installing the gas supply piping only within the building.

Plumbing codes seldom, if ever, govern the sizing or installation of gas supply systems. The plumbing code usually refers the plumber to the local gas code. The gas code is a separate book but is similar to the plumbing code in complexity. Trying to learn and comply with these codes can be frustrating and discouraging to a plumber no matter what his experience level.

Anyone in the specialty trade of gas fitter and installer must be familiar with all the key requirements of the gas code. A plumber has to know only part of the gas code. Unfortunately information relevant to the plumber is scattered throughout the code. This chapter and the next are intended to give plumbers a working knowledge of the essential requirements. Information is grouped according to use and is sufficiently simplified for easy application. But the information here does not replace the gas code. The code is always your final authority.

SIZING GAS SYSTEMS

The gas main and branch lines of a building can be sized once you know the maximum gas demand at each appliance outlet and the length of piping required to reach the most remote outlet. Other sizing factors such as pressure loss, specific gravity and diversity are already accounted for in the tables in the gas code.

Length in feet	Nominal Iron Pipe Size, Inches							
	½	¾	1	1¼	1½	2	2½	
10	176	361	681	1,401	2,101	3,951	6,301	
20	121	251	466	951	1,461	2,751	4,351	
30	98	201	376	771	1,181	2,201	3,521	
40	83	171	321	661	991	1,901	3,001	
50	74	152	286	581	901	1,681	2,651	
60	67	139	261	531	811	1,521	2,401	Residential
70	62	126	241	491	751	1,401	2,251	
80	58	119	221	461	691	1,301	2,051	
90	54	111	206	431	651	1,221	1,951	
100	51	104	196	401	621	1,151	1,851	
125	45	94	176	361	551	1,021	1,651	
150	41	85	161	326	501	951	1,501	Commercial
175	38	78	146	301	461	851	1,371	
200	36	73	136	281	431	801	1,281	
Column One	two	three	four	five	six	seven	eight	

More complete sizes will be found in your code.

Maximum capacity of pipe in cubic feet of gas per hour
natural gas 1,000 B.t.u./cubic foot
Table 14-1

Gas appliance manufacturers always attach a metal plate in a visible location on each appliance. This data plate shows the B.t.u. input rate which is the maximum gas demand. B.t.u. is the abbreviation for British thermal unit, the quantity of heat required to raise the temperature of one pound of water one degree Fahrenheit. The manufacturer gives you the maximum input rate in B.t.u.'s. The tables in the code prescribe sizing of gas piping in cubic feet of gas rather than B.t.u.'s. Thus each B.t.u. input rating has to be converted to cubic feet of gas before sizing the distribution piping.

You can assume that each cubic foot of natural gas releases 1,000 B.t.u.'s per hour. Some gas has B.t.u. ratings that vary from this figure, but using 1,000 B.t.u. per cubic foot is a safe assumption. Assume you are sizing pipe for a range with a maximum demand of 68,000 B.t.u.'s per hour. Divide the value in B.t.u.'s by 1,000 to find the demand in cubic feet per hour. Thus, 68,000 B.t.u.'s ÷ 1,000 = 68 cubic feet per hour.

In cases where used appliances are to be installed, the B.t.u. rating may not be legible or may be missing. In such a case, the following is a safe practice: make the appliance inlet pipe no smaller than the supply pipe serving the appliance. A larger size supply pipe could be used without violating the code, but will not cause the appliance to function any better. The supply pipe should never be smaller than the appliance's inlet pipe, and under no circumstances can it be smaller than ½ inch.

Table 14-1 gives you a sizing method that is quick and easy. Use it only to help you understand how to use the table in your local gas code. The low pressure gas table is the one most commonly used by plumbers. Low pressure gas is used in millions of home and business installations.

Figure 14-2 shows a gas piping arrangement for a restaurant. This is a typical commercial gas piping system similar to what you might find on the journeyman or master plumber examination.

Figure 14-3 shows a simple gas piping system similar to what you might find in most single family residences. Note that each section of piping must be sized to serve the B.t.u. input rating of the appropriate appliance outlet. Study this illustration and the explanation in the rest of this chapter until you can size each section of pipe correctly. When you master these illustrations you can size the pipes in any type of low pressure gas system.

To size the gas piping system in Figure 14-2,

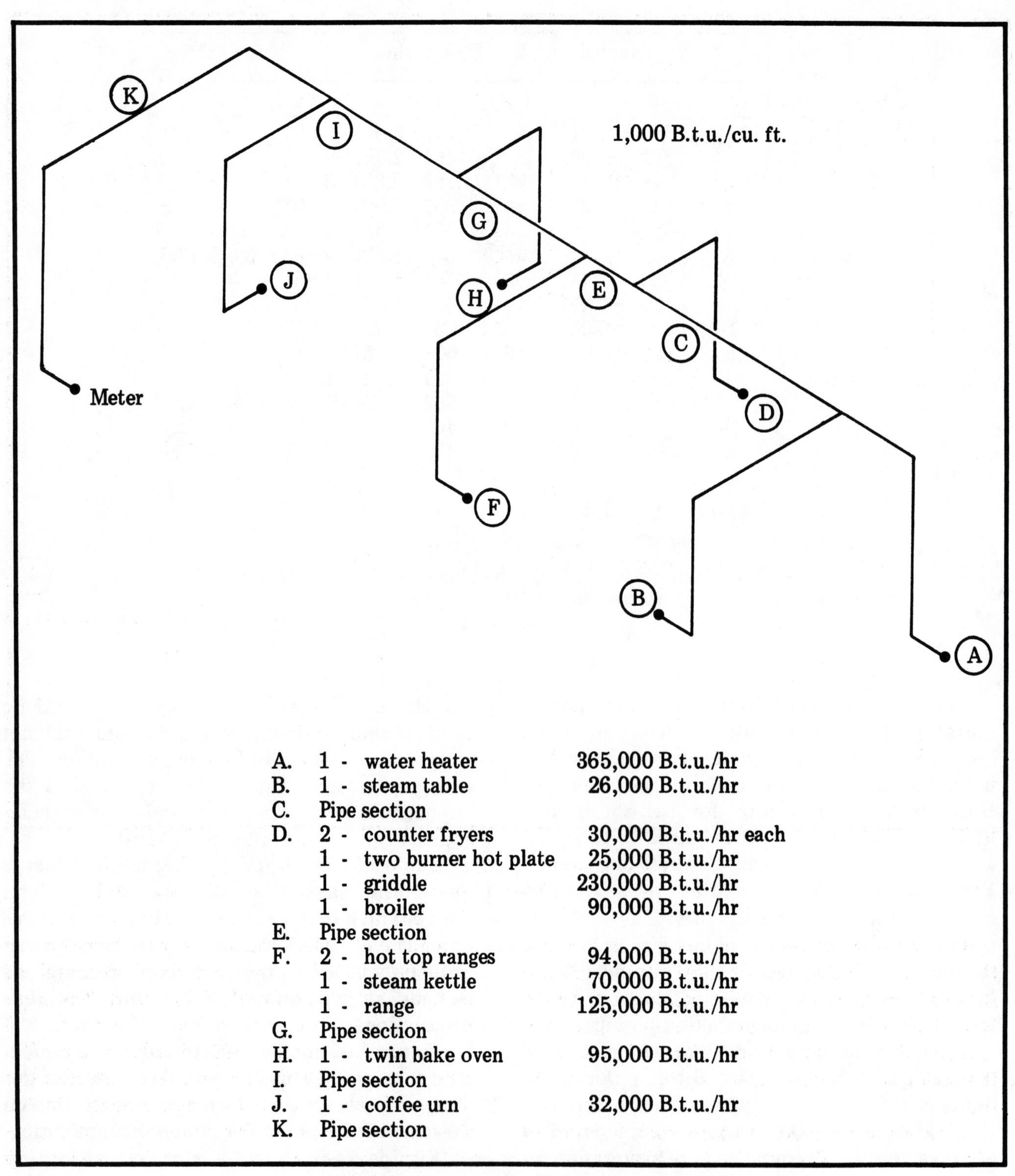

Natural gas - commercial kitchen installation
Figure 14-2

use Table 14-1 and follow the procedure outlined below. Assume that the total developed length of gas piping in Figure 14-2 is 147 feet from the meter to outlet A. In Table 14-1 find that distance. You have to use the next longer distance if the exact length is not given in the first vertical column (length in feet). *This is most important. The developed length is the only distance used to determine the size of any section of the gas piping.*

In Table 14-1 (first column), the length of piping (147 feet) exceeds 125 feet but not 150

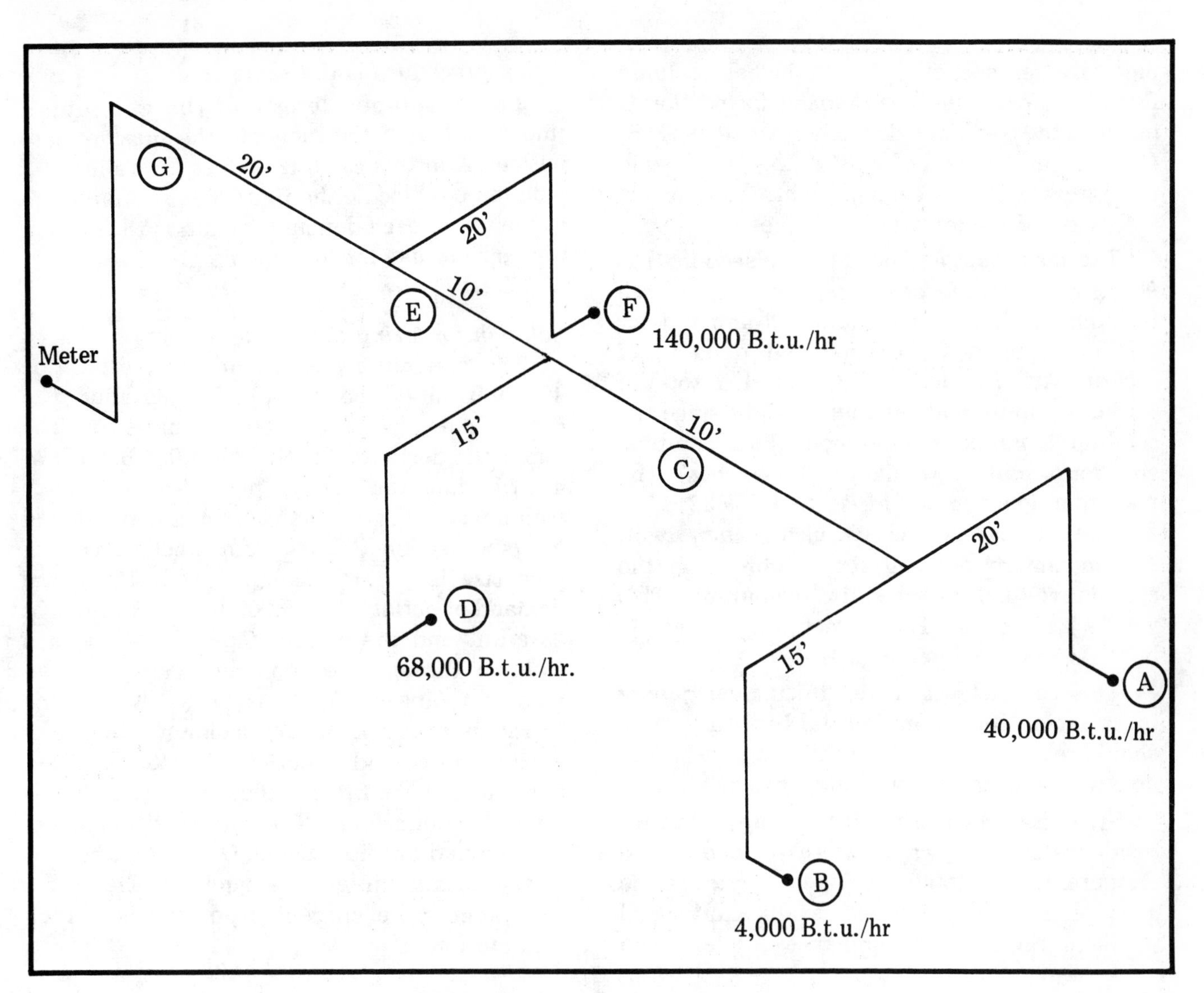

Natural gas - residential installation
Figure 14-3

feet. The figure 150 must be used to determine the size of each section of pipe. Underline all figures in each vertical demand column opposite 150. (This row is shaded in Table 14-1.) The pipe size listed at the top of each demand column is the size that will carry the volume listed in the table.

You are now ready to size each section of pipe shown in Figure 14-2. Maximum gas demand of outlet A is 365,000 B.t.u.'s per hour, or 365 cubic feet per hour assuming 1,000 B.t.u. per cubic foot. Looking across the 150 foot row, the first figure exceeding 365 is found in column six (501 c.f.h.). At the top of the column where 501 appears is the pipe size to use (1½ inches). The correct pipe size for section A is 1½ inches.

Use the same procedure to size each section. The maximum gas demand for section B is 26 c.f.h. and the correct pipe size is ½ inch (column two). The maximum gas demand for section C includes the combined demand of sections A and B, at total of 391 c.f.h. The pipe size would be 1½ inches as the total exceeds 326 in column five but not 501 in column six. The maximum gas demand for section D includes the combined demand of five appliances: 30,000 plus 30,000, plus 25,000, plus 230,000, plus 90,000. The conversion to cubic feet yields 405 c.f.h. Again, using column 6, the pipe size would be 1½ inches. The maximum gas demand for section E includes the combined c.f.h. of sections A, B, C and D, for a total of 796. The total calculated demand of 796 exceeds 501 in column six but not 951 in column seven. This section of pipe must be 2 inches. The maximum gas demand for section F is 289 c.f.h. Note carefully that the two hot top burners have a *combined* B.t.u. rating of 94,000. The correct pipe size for section F is 1¼ inches (column five). The maximum gas demand for section G includes the combined demand of sections A, B, C, D, E and F, a total of 1,085 c.f.h. We have to use figure 1,501 in column eight. The correct pipe size for section G is 2½ inches.

The maximum gas demand for section H is 95 c.f.h. and the correct pipe size is 1 inch (column four). The maximum gas demand for section I includes the combined demand of sections A, B, C, D, E, F, G and H, a total of 1,180. Again we will use column eight since the combined total does not exceed 1,501. The pipe size for section I would be 2½ inches. The maximum gas demand for section J is 32 c.f.h. and the correct pipe size is ½ inch (column two). The maximum demand for section K is the cumulative total for the entire restaurant. This will be 1,212 c.f.h. The correct pipe size would be 2½ inches (column eight).

This completes a more difficult gas piping installation for a commercial building. You should review this section until you understand how to size each section in the system.

Size the gas piping for a single family residence the same way as was explained for the restaurant. The total developed length of the gas piping for a residence is usually shorter and the B.t.u. ratings of the appliances are less. The sizing procedure is the same.

The developed length of the gas piping (measured from the meter to the most remote outlet - A in this case) is 60 feet. In Table 14-1 (column one) locate the figure 60. The pipe size at the top of each demand column is the correct pipe size to use for the volume given.

Solution for sizing the piping in Figure 14-3

The maximum gas demand of outlet A is 40,000 B.t.u. per hour or 40 c.f.h. per hour. The correct pipe size is ½ inch (column two). The maximum demand of outlet B is 4,000 B.t.u.'s or 4 c.f.h. and the correct pipe size is ½ inch (column two). The combined maximum demand for pipe section C is 44 c.f.h. and the correct pipe size is ½ inch (column two). Maximum demand of outlet D is 68,000 B.t.u. per hour or 68 c.f.h. and the correct pipe size is ¾ inch (column three). The combined maximum demand for pipe section E is 112 c.f.h. and the correct pipe size is ¾ inch (column three). The maximum demand of outlet F is 140,000 B.t.u. per hour or 140 c.f.h. and the correct pipe size is 1 inch (column four). The combined maximum gas demand for pipe section G is 252 c.f.h. This is the maximum gas demand for the entire residence. The correct pipe size is 1 inch (column four).

Material And Installation Methods For Gas

Your local gas code regulates the materials and installation methods for gas systems. You will note that certain materials are limited in use and others are prohibited.

When selecting the materials for gas supply pipes, tubing, or fittings, consider the characteristics of your particular gas supply and its effect on the pipes' interior. For example, gases in certain areas are classed as corrosive. These gases contain an average of 0.3 grains of hydrogen sulfide per 100 cubic feet. If your community gas supply is corrosive, certain types of materials in common use for gas piping will not be acceptable to your local gas supplier. If the character of the gas to be used is not known, call your gas supplier before the gas system is installed.

MATERIALS

A wide variety of gas piping materials are acceptable. Some are acceptable for underground installations, some for above ground only. Some materials can be used for both.

Some piping materials are very common because of their versatility. These materials can be used outside, above ground and underground (except under a concrete slab): galvanized steel pipe, black steel pipe, galvanized wrought iron pipe. Pipe up to two inches is generally threaded. Fittings should be of the same material as the pipe. Threaded joints must be sealed tight with an approved pipe compound. Larger size pipes must have fittings of like materials and can be welded or flanged. Galvanized steel pipe, black steel pipe and galvanized wrought iron pipe are also acceptable for use with corrosive gas.

Brass, copper and aluminum pipe of a thickness that can receive a standard iron pipe thread can be used for interior gas piping. Brass and copper pipe can be installed outside underground but not under a concrete slab. Aluminum pipe is never used outside or underground. Brass, copper and aluminum pipe can not be used if the gas is corrosive. Fittings must be of the same materials as the pipe.

Copper seamless tubing, type K or L, aluminum or steel tubing can be used for interior gas piping *only* if first approved by the local code authority or the area gas supplier. These tubings can not be used if the gas is corrosive. Fittings used must be of the same material as the tubing selected. Joints between fittings and tubing should be soldered with a hard solder. This generally means a silver solder. Joints between fittings and tubing can also be brazed. This usually means that a filler of brass is used for joining the metals. This type of gas piping generally requires a special installation and is seldom used in construction work. Approved gas flare fittings can also be

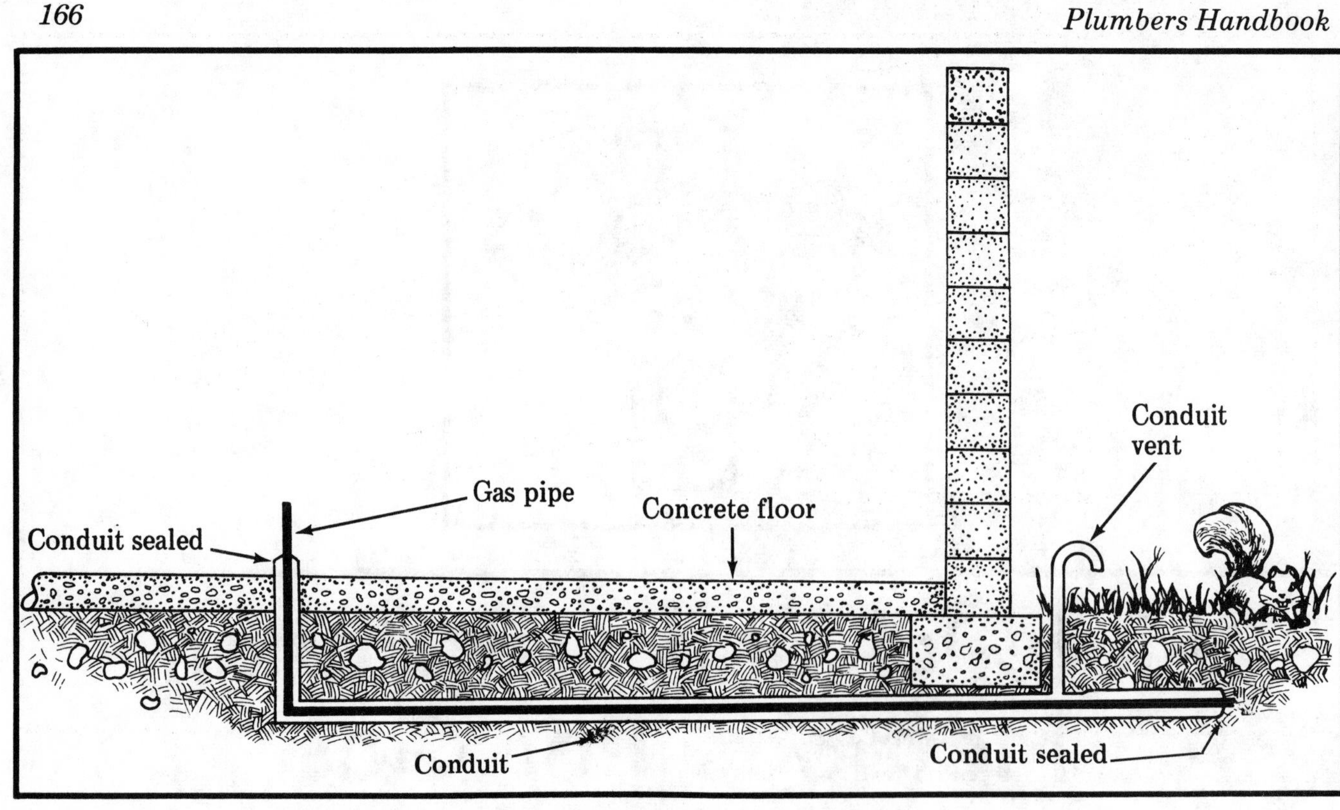

Installation beneath concrete floor
Figure 15-1

used for joining two like metals if the joint is not concealed.

Plastic pipe and fittings, when approved for use by the local authority, must conform with the ASTM (American Society for Testing and Materials) specifications and may be used for outside natural gas piping underground only.

- Plastic pipe, tubing and fittings shall be jointed by either the solvent cement method, heat fusion method, or by compression couplings or flanges.
- Solvent cement joints and heat fusion joints shall be made to produce gas-tight joints at least as strong as the pipe or tubing being joined.
- Solvent cement or heat fusion joints shall not be made between different kinds of plastics.
- Heat fusion or mechanical joints shall be used when joining polyethylene pipe, tubing or fittings.
- When compression type mechanical joints are used, the gasket material in the fitting shall be compatible with the plastic piping and be approved by the gas company.
- Connections between metallic and plastic piping shall be made only outside the building and underground.
- Plastic pipe or tubing shall not be threaded.

INSTALLATION OF GAS PIPING

Gas piping installed underground should be placed deep enough to protect the pipe from damage from sharp tools. A six inch depth is a good minimum, but this is not governed by the code. When installed in corrosive soils, the pipe should be protected with an approved wrapping or one or two coats of asphaltum paint. In areas where freezing temperatures occur, the trench bottom should be below the frost line to prevent freezing and rupturing of the pipe. Insulate the pipe where it enters a building above ground, in crawl spaces, or anywhere it is not protected from the cold. Many gases contain moisture which can freeze and block the pipe.

Underground gas piping should be laid in open trenches on a firm bed of earth. The pipe should be securely supported to prevent sagging and excessive stress during backfill. Only fine materials should be used for backfilling.

Occasionally gas piping must be installed underground under a building slab. When this is unavoidable, the code will permit this type of

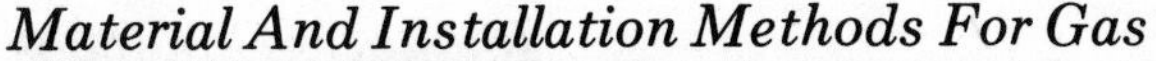

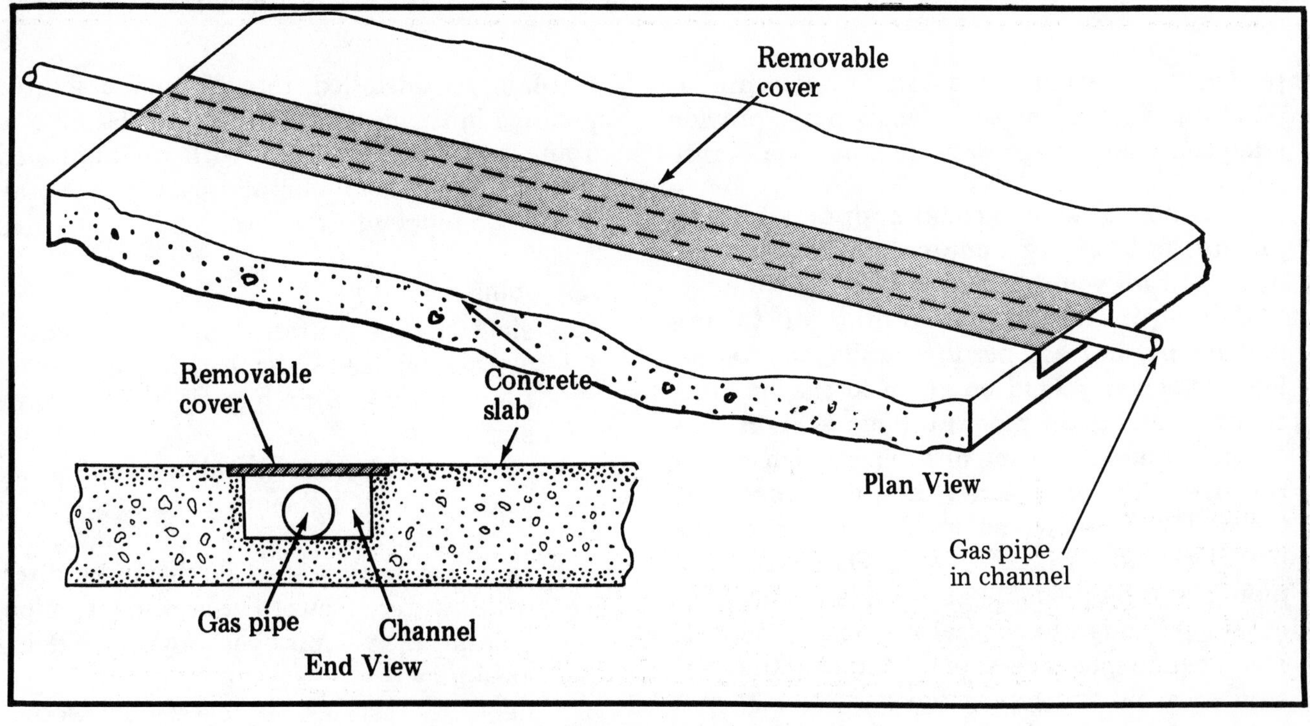

Concrete floor open-channel installation
Figure 15-2

installation only if the following three conditions are met: 1) The entire length of gas piping up and through the floor must be encased in conduit. 2) The termination of the conduit above the floor must be sealed to prevent the entrance of any gas into the building in case of a leak. 3) The termination of the conduit outside the building must be tightly sealed to prevent water from entering the conduit. A vent must be extended above the grade and secured to the conduit. This vent conveys any leaking gas to the outside of the building. See Figure 15-1.

Gas appliances that are located in the center of a room away from adjoining partitions can be a problem because walls are not available to conceal the gas piping. In this case piping must be installed in an open channel in the concrete floor. The channel must have a removable grill or cover so there is access to the piping. See Figure 15-2.

Gas equipment or appliances subject to vibration or requiring mobility can be connected with an approved flexible gas hose connector. The gas hose should be no longer than necessary and not more than six feet in any case. Only approved gas hose connectors can be used to connect the hose to the gas outlet pipe.

In rare cases where it is unavoidable, you can get approval from your local authority and gas supplier to embed the pipe in concrete. But

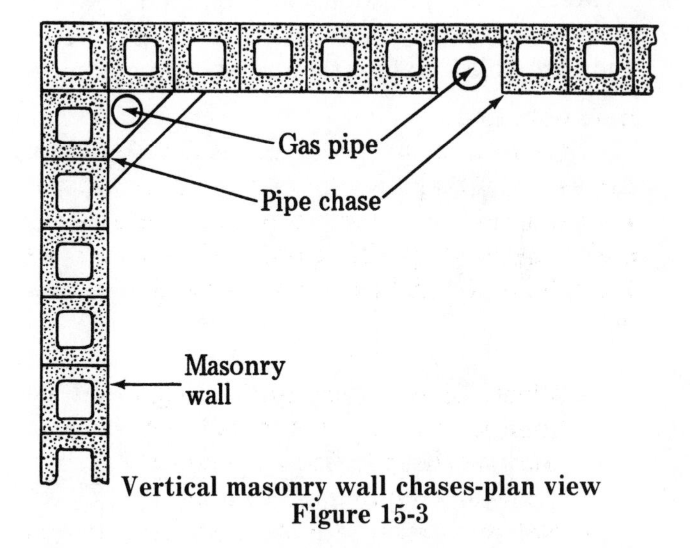

Vertical masonry wall chases-plan view
Figure 15-3

you must meet the following four conditions: 1) The concrete must not contain cinder aggregates or additives designed to set concrete quicker than normal. 2) The pipe must be embedded directly in a Portland cement concrete slab with a minimum of 1½ inches of concrete on all sides. 3) The piping must not be in contact with any metallic materials. 4) Gas piping at the point of entry to and exit from a concrete slab must be adequately protected from the corrosive effect of the concrete. An approved coating or sleeve is required.

When gas piping must pass through masonry walls, the pipe must be properly protected against corrosion by sleeving or painting. Vertical masonry walls must provide adequate chases to protect the pipe. See Figure 15-3.

Horizontal and vertical supports for gas piping and tubing should be the same as described for water piping in Chapter Nine.

When installing gas piping or tubing horizontally in *wood partition* walls, the following protection should be given to the building structure and to the pipe or tubing: 1) Short runs of horizontal gas piping or tubing which do not require additional joints should be installed through a hole drilled in the center of the partition stud. 2) Longer runs of horizontal gas piping or tubing should be installed in notches cut deep enough to conceal the pipe or tubing. Don't cut deeper than ⅓ of the total width of the stud to avoid weakening the partition. 3) Soft tubing in a notched partition should be protected by a metal stud guard to avoid penetration by lath nails.

Metal stud partitions are replacing wood partitions in many new buildings. The metal studs are hollow rather than solid. The gas pipe or tubing is installed through manufactured openings in the center of the stud. The pipe or tubing must be wrapped with an approved material to prevent contact with the metal. Secure the pipe with tie wire.

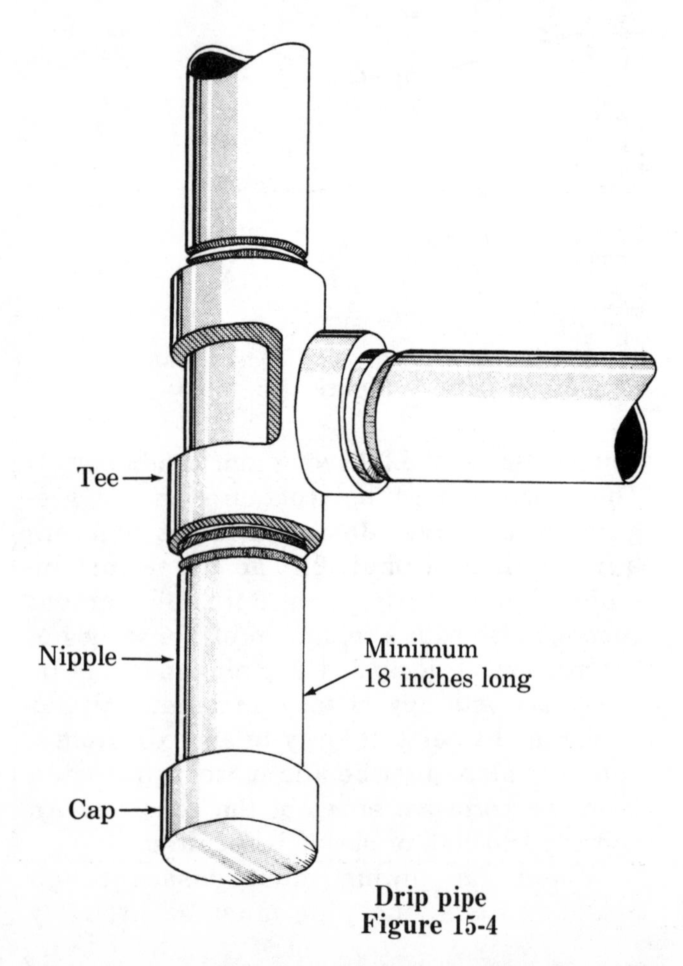

Drip pipe
Figure 15-4

Gas piping:

- shall not be installed in any air duct, clothes chute, chimney or vent, ventilating duct, dumbwaiter, or elevator shaft.
- shall not be installed underground closer than eight (8) inches to a water pipe or a sewer line.
- shall not be installed in the same ditch with water, sewer or drainage pipe, unless first approved by the local authority.
- may be installed in accessible above-ceiling spaces used as an air plenum, but the valves shall not be located in such space.

Drip Pipe

Gas mains must be installed so they can drain dry. The pitch or grade has to be toward the gas meter. Appropriate drip pipes must be provided to receive condensation that may form within the pipe. The drip pipe must be accessible for emptying. It is usually assembled from a tee, nipple and cap as shown in Figure 15-4 and should not be smaller than the pipe or pipes it serves or less than 18 inches long. The drip pipe should be protected from freezing in colder climate zones.

Gas branch pipes connecting to other horizontal pipes must connect at the top or the side of the feeder pipe and never from the bottom. This will keep condensate from filling and obstructing the branch lines. See Figure 15-5.

Shutoff Valves

Buildings with multiple tenant occupancy and a master meter must have a gas shutoff valve for each apartment. Each shutoff valve should be located on the outside of the building or at some equally accessible location. To avoid accidental or malicious tampering with the valve, it usually has a square nut head which can be turned only with a special tool.

Each gas appliance within a building must have an accessible, manually operated shutoff valve. This valve must have a lever handle that

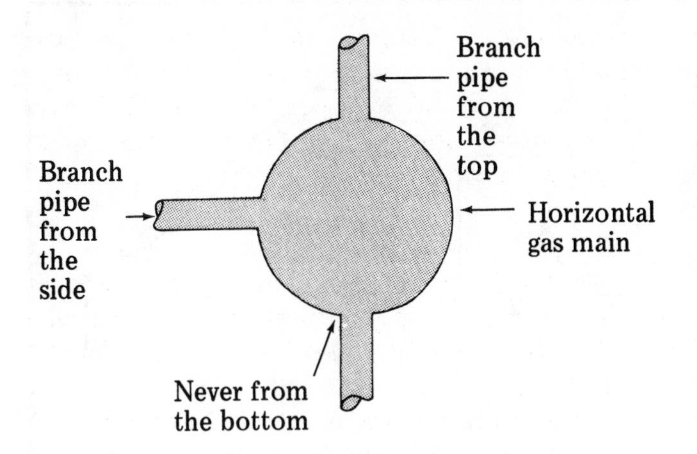

Branch pipe connection to horizontal pipe
Figure 15-5

does not require a tool. This valve should be located as close as possible to the gas outlet pipe serving each appliance.

Shutoff valves are manufactured in two types: one a straight pattern, the second an angle pattern for convenient use. The shutoff valve should not be installed more than six feet from the appliance it serves.

Gas piping or tubing installed in a concealed location must not have unions, right or left couplings, tubing fittings, running threads, bushings or swing joints.

To make a new connection to an existing line in a concealed threaded gas piping system, use a ground joint union. The center nut is punched to prevent it from working loose under vibration. No new connection is permitted on a concealed gas *tubing* installation, regardless of the tubing material used.

Threads for gas pipe must conform with the standards adopted by the American Standards Association. See Table 9-4 for the number and length of standard pipe threads. Pipe with chipped or torn threads should not be used. The procedure for cutting, threading and reaming of gas pipe is the same as described for water pipe in Chapter Nine.

Don't conceal the completed gas installation until it is pressure tested. Cap each outlet and install a pressure gauge on one of the outlets. Pressure test at 10 to 20 pounds. The system must remain air tight (no loss of pressure) until after inspection. Leave the caps in place until the appliances are ready to be connected. If a leak should occur during the test of the rough piping, test for leaks by running a small brush dipped in liquid soap around each joint. The leaking joint will blow bubbles.

It is the plumber's job to connect the gas from the wall outlet to the appliance. This connection may be made with either a rigid pipe

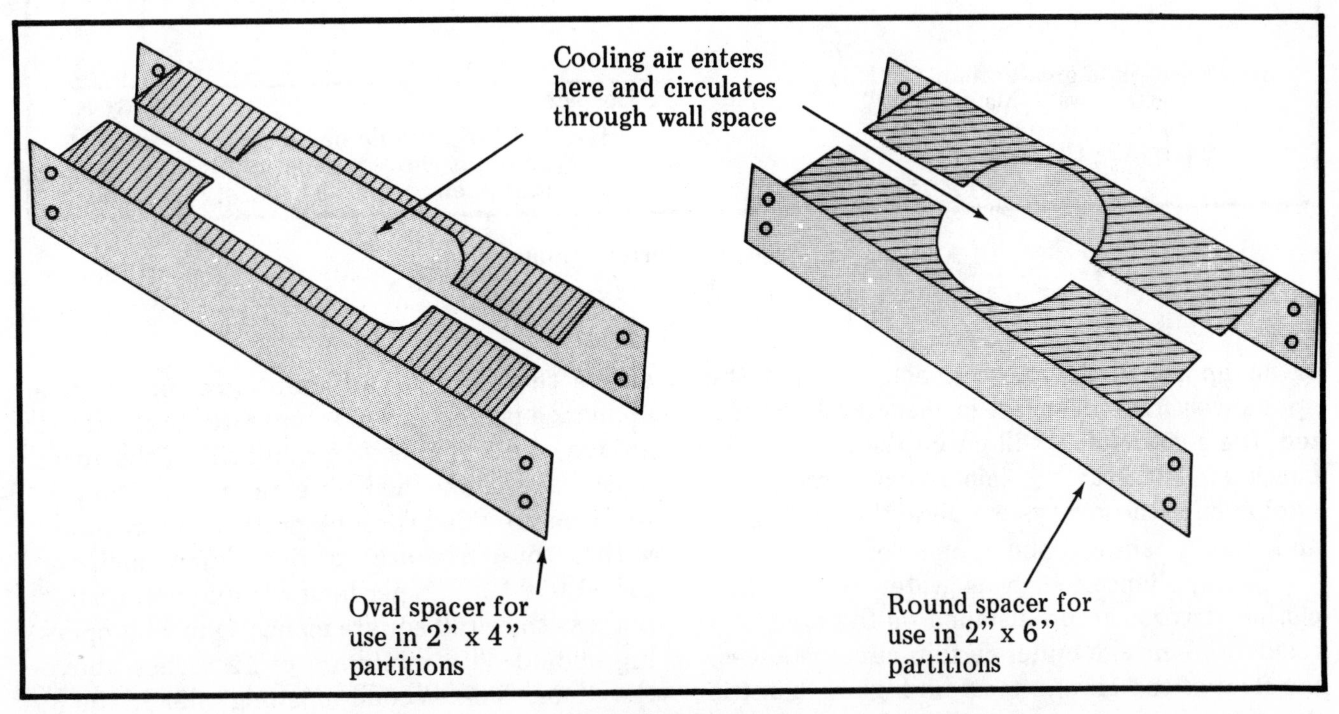

Vent pipe spacers
Figure 15-6

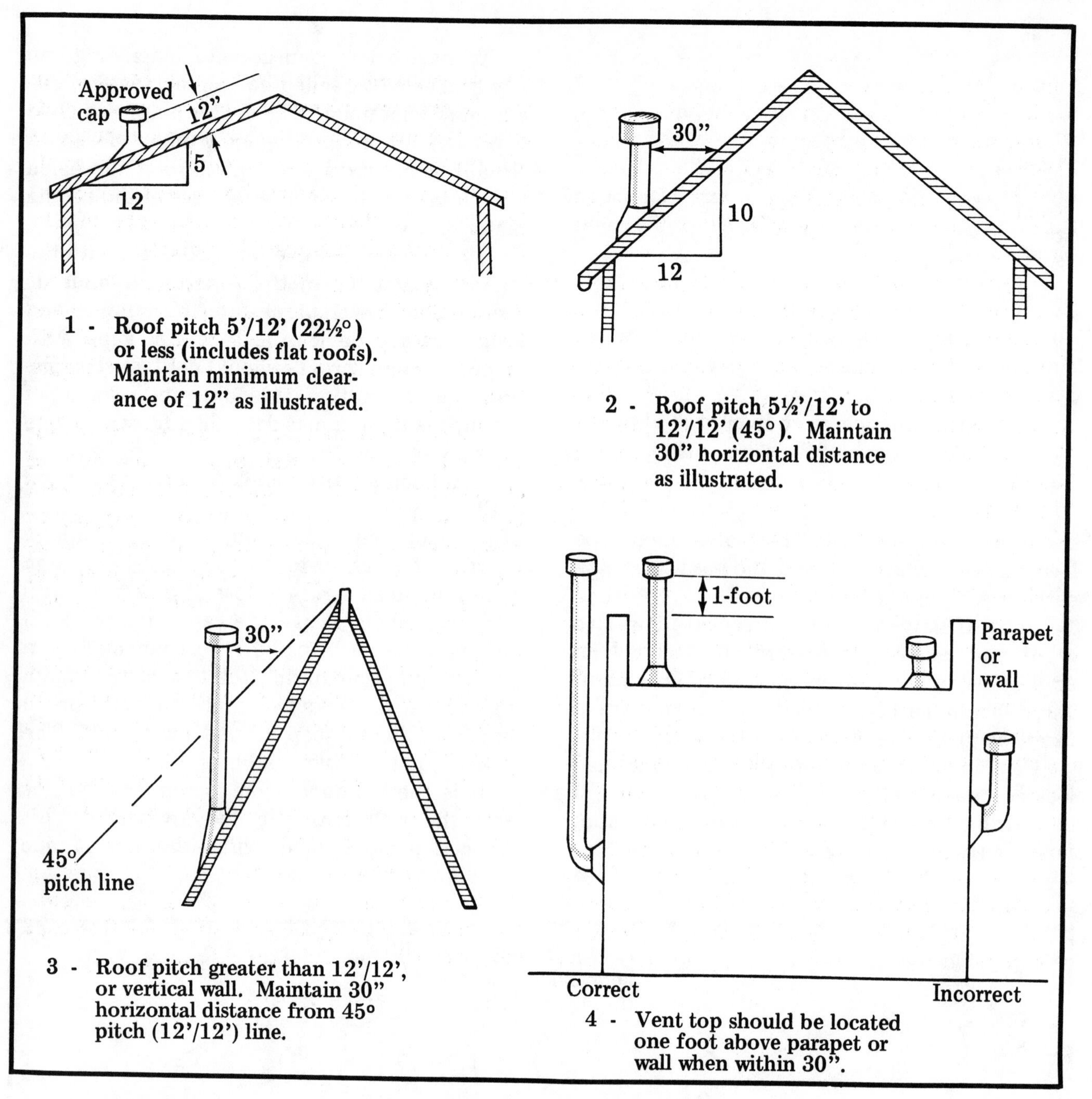

Vent pipe terminations
Figure 15-7

or an approved flexible connector. When the appliances have been set in place and connected, the gas supplier will purge the lines of air, check for leaks at the joints that connect the appliance to the gas system, light the pilot lights and, finally, adjust each appliance.

Gas appliances such as water heaters and clothes dryers can be installed on the floor of a residential garage under certain circumstances: the floor of the garage must be higher than the driveway or adjacent ground and the combustion chamber must be a minimum of 18 inches above the floor or adjacent ground. If the appliance is installed in a separate room off the garage, the walls, ceiling and door of the room must have a one hour fire rating. Ventilation must be provided through permanent openings with a total free area of one square inch for each 1,000 B.t.u.'s per hour of input rating, but not less than 100 square inches. One vent opening should be a minimum of 12 inches above the floor. The second opening should be a minimum of 12 inches below the ceiling. This permits air circulation for combustion and dilution of the flue gases.

Gas appliances should be installed so there is access to the appliance for repairs and cleaning as well as the intended use. Water heaters must not be installed in any living area that may be closed such as bedrooms or bathrooms.

Appliances with low B.t.u. input ratings do not have to be vented if the input of the appliance does not exceed 30 B.t.u. per hour per cubic foot of room space. For example, a water heater with an input rating not over 5,000 B.t.u.'s may be installed in a room (not normally closed) 6 feet long x 4 feet wide with a ceiling height of 7 feet. No vent would be required. Water heaters with an input rating of 5,000 B.t.u.'s or less are rarely used.

Gas water heaters with an insulated jacket must not be installed closer than two inches to any combustible material. Water heaters should not be installed closer than one inch to enclosures constructed of one hour fire rated materials.

Appliances that must be vented should be installed as close to the vent pipe as possible. If a draft hood is required, the vent pipe should never be smaller than the opening of the draft hood. Gas appliances and their vents should be installed so there is sufficient clearance from combustible materials to avoid a fire hazard.

There are two acceptable types of concealed vent pipe material. The most common is the double wall metal pipe and fittings. Stamped in the metal is the clearance distance recommended by the manufacturer. One inch clearance is recommended from any combustible material. The second vent material is asbestos cement flue pipe. It should have a clearance of 1½ inches from any combustible material.

Vent pipes installed in partitions built of combustible material must have an approved metal spacing device. This spacer keeps the surface temperature below 160° F. See Figure 15-6.

Single wall vent pipe can be used for exposed vent pipe installed in a room built of non-combustible material. All horizontal vent pipes must be supported to prevent sagging or misalignment. Straps or hangers should be at least 20 gauge sheet metal. The horizontal vent section length must not exceed 75% of the vertical vent length.

The exit terminals of vent pipes shall not be located closer than twelve inches from any opening through which combustion products could enter the buildings, or closer than two feet from an adjacent building, and not less than seven feet above grade when located adjacent to public walkways.

All vent pipes above the roof of a building must terminate in a UL approved cap. Figure 15-7 shows several examples of acceptable vent pipe terminations outside a building.

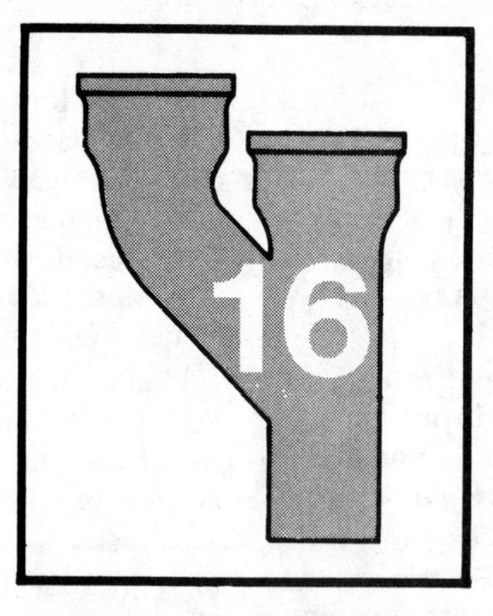

Plumbing Fixtures

Quality must be built into plumbing fixtures. As a plumbing professional it is your responsibility to install only quality fixtures that conform in design to one of the standards in Table 4-2. Fixtures must be free from defects and concealed fouling surfaces. Fixtures constructed of pervious materials (such as Roman baths or showers constructed of tile or marble) must not have a waste outlet that can retain water. A quality plumbing job requires quality fixtures. Don't let substandard fixtures detract from the quality of the work you do.

All fixtures should be set level and in proper alignment with the adjacent walls. They should be located in adequately lighted and ventilated rooms. Where natural ventilation such as a window is not available, a fan and duct are required. The lack of adequate lighting or ventilation promotes unsanitary conditions. Most codes prohibit locating fixtures in such locations.

Important and well recognized standards have been developed for the various types of plumbing fixtures over the past 100 years. These standards control the quality and design for all plumbing fixtures in use today.

Fixture Overflow

Bathtubs and lavatories are two of the more common fixtures provided with overflows. The code does not require that certain plumbing fixtures have overflows. Thus there is a current trend toward omitting overflows on lavatories. Integral overflow passageways do provide secondary protection against self-siphonage of a fixture trap seal as well as an escape for excess water at a level below the flood-level rim of the fixture.

Where overflows are provided on plumbing fixtures, the waste pipe must be arranged to both prevent water from rising into the overflow when the stopper is closed and prevent water from remaining in the overflow when the drain is open for emptying.

The overflow pipe or passageway from a fixture must be connected on the inlet side of the fixture trap. This prevents sewer gases and odors from entering the room through the overflow. In other words, the code prohibits connecting the overflow of a fixture to any other part of a drainage system. See Figure 16-1.

Fixtures must have durable strainers or stoppers. An exception is made for fixtures with integral traps. The strainer or stopper must not prevent rapid drainage of the fixture. The strainer should not be smaller than the fixture waste outlet it serves and (except for fixed strainers) should be easy to remove for cleaning.

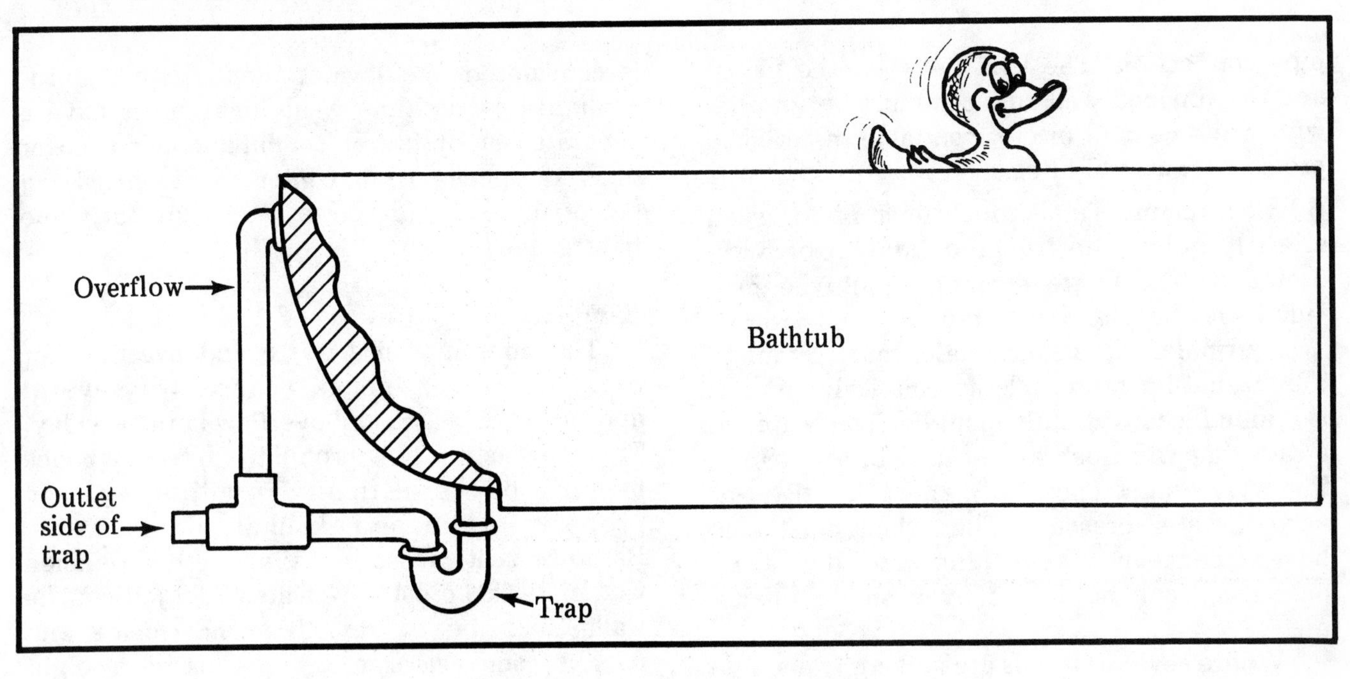

Prohibited overflow connection
Figure 16-1

Water Closets

Both wall-hung and floor mounted water closets must be grouted with white cement or another suitable material to provide a water-tight seal at the joint with the wall or floor. This prevents the accumulation of odor-causing materials, avoids other unsanitary conditions and keeps roaches and other insects away from these "ideal" areas.

Water closets installed for public use (this is any place other than a single family residence or apartment building) must have an elongated bowl equipped with an open front seat.

Wall-hung water closets should be rigidly supported with brass bolts on a concealed metal carrier. Load should not be transmitted to the fixture pipe connection.

Seats for water closets must be constructed of smooth nonabsorbent materials and must fit the water closet bowl. For example, do not install a round front seat on an elongated bowl.

Water closets with tanks designed to use ballcocks should refill after each flushing and then close tight when the tank is full. The tank must have a refill tube reaching and turning down into the overflow tube. Water from this tube automatically restores the closet bowl water seal. An anti-siphon valve must be built into the unit to prevent contamination of the potable water supply. The flush valve is to be operated manually but the flushing operation must be automatic after manual activation. Each tank has to have an overflow tube adequate to prevent tank overflow and remove excess water at the rate it enters the tank. Consider what would happen if the flush ball is securely in place on the flush valve seat and the ballcock should become locked in an open position. The flush valve seat must be a minimum of one inch above the rim of the bowl.

The flushing device and the connection between the tank and the bowl should have enough flow capacity to allow the water to flush all surfaces of the bowl.

A water closet using a flushometer rather than a tank must have a vacuum breaker located a minimum of 6 inches above the rim of the bowl. It must complete the normal flushing cycle automatically after being manually activated. It must deliver water at a rate which is adequate to flush all surfaces of the bowl. It should open fully and close tight under the normal water working pressure. Flushometers must be installed so they are readily accessible for repair. Each flushometer can serve only one water closet. The valve must have some way to regulate the water that flows with each flush.

Urinals

Wall-hung urinals must be rigidly supported by a concealed metal carrier or other approved backing so that no strain is transmitted to the pipe connection. The joint between the urinal and the finished wall surfaces must be grouted with white cement or other suitable material to provide a water-tight seal.

Floor mounted stall urinals must be recessed slightly below the finished floor to provide drainage. The waste opening should be provided with beehive type strainers.

A urinal using a flushometer must complete the normal flushing cycle automatically after it is manually activated. It should deliver water at a rate that will flush all surfaces of the urinal. The valve must open fully and close tight at normal water pressure. The urinal must also have some means of regulating water flow. Only one urinal can be served by a single flushometer.

Where several urinals are served by a single flushing tank, the flushing device must operate automatically once it is manually activated. The tank must have enough capacity to cleanse all urinals simultaneously.

Lavatories

Wall-hung lavatories for commercial use are generally supported by a concealed metal carrier. The carrier provides the lavatory with the necessary support so that no strain is transmitted to the fixture pipe connection or to the finished wall. There is little possibility that a lavatory so secured can pull away from the wall. Wall-hung lavatories for residential use are generally supported by a metal bracket screwed securely to wooden backing material fastened to the bathroom partition studs.

The point of lavatory contact to finished wall surfaces must be sealed with white cement or other suitable material.

Cabinet-mounted lavatories are securely fastened to the counter top by special rim clips and are made water-tight with a caulking compound or other adhesive. The weight is transferred to the cabinet top and places no strain on the fixture piping.

Other wall-mounted fixtures must be adequately supported and grouted for sanitary purposes. Even shower rods must have a suitable backing so they don't work loose from the wall.

Waste outlets for lavatories must be a minimum of 1¼ inches o.d.

Where circular type multiple wash sinks are used, each 18 inches of wash sink circumference is considered one lavatory (one fixture unit). Straight-line multiple wash sinks must have a separate set of faucet combinations no closer than 18 inches from center to center. Each faucet set is considered to be one lavatory (one fixture unit).

Bathtubs

The minimum size waste and overflow for bathtubs is 1½ inches. There are several approved tub waste and overflows in use today. The trip waste is now prohibited by some codes because of the difficulty in adjusting it to properly retain and discharge tub water. Bathtubs that recess into tile or other finished wall materials must have waterproof joints. The walls must be of smooth, noncorrosive and nonabsorbent waterproof materials to a height of 4 feet above the rim of the tub.

Showers

Waste outlets for shower compartment floors must be a minimum of 2 inches in diameter. Water should drain from the shower floor without puddling. The free area of the shower strainer must be a minimum of three and one-half square inches. The strainer has to be removable so the trap can be cleaned. Shower traps must not be smaller than the waste outlet pipe used in the shower compartment.

Shower compartments need a minimum floor area of 1,024 square inches. This requires a minimum span between walls of 32 inches and is considered adequate for use by adults. Floors of shower compartments must be smooth and sound. Floors of institutional or gang showers used by more than one bather at a time should be designed so that waste water from one bather does not pass over areas occupied by other bathers.

Where shower pans are required, pans of lead, copper or other approved materials should be used. Lead pans should weigh not less than 4 pounds per square foot. Copper pans must weigh not less than 12 ounces per square foot. Shower pans of lead or copper must be painted with asphaltum paint inside and outside to protect the pan from corrosion where it joins concrete or mortar. Put either a layer of 30 pound asphalt saturated felt or a ½ inch thick layer of sand under the pan. This protects the pan against rough surfaces and should avoid accidental puncturing of the pan before it is

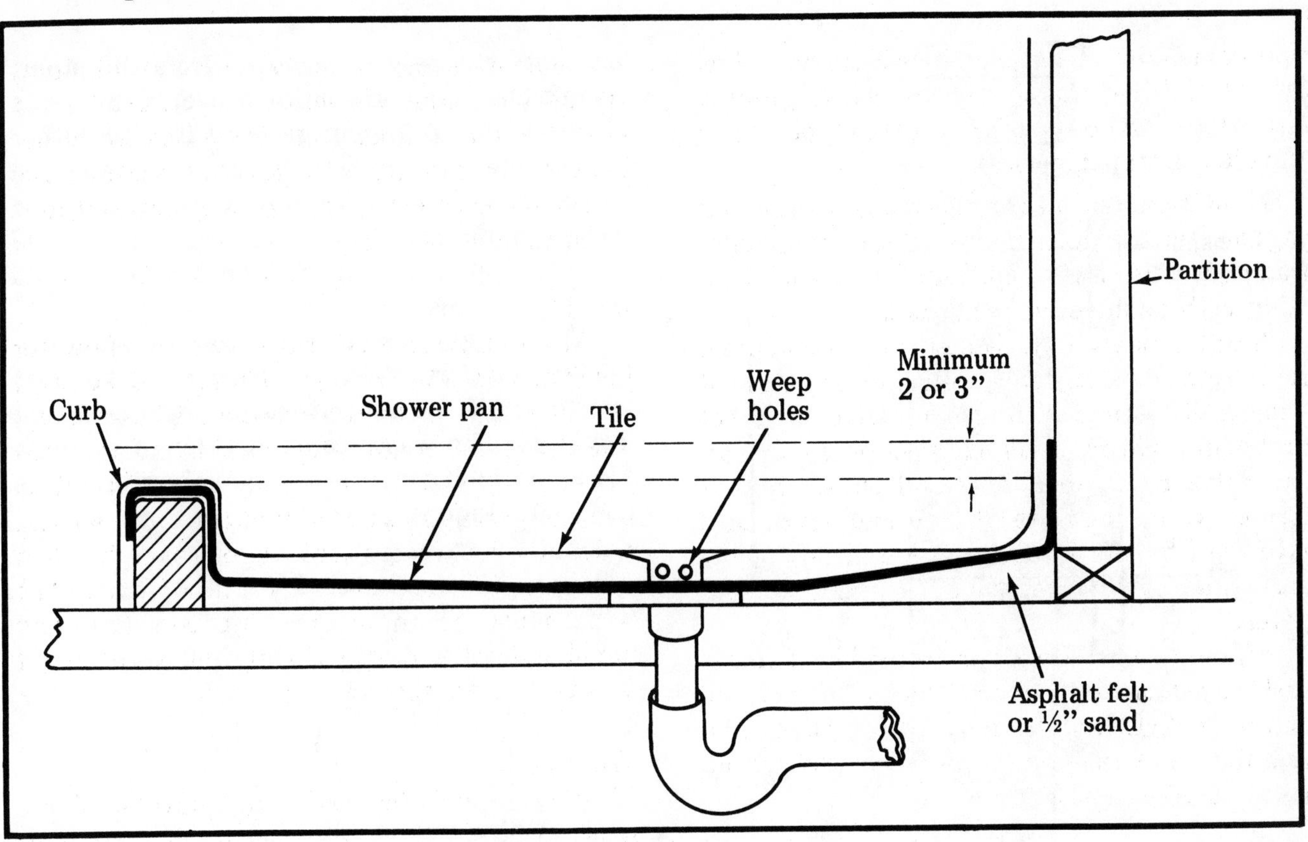

Shower pan installation detail
Figure 16-2

given the protection of the finished floor material.

The shower pan must be cut large enough so that, when properly folded to fit the shower compartment, it has a turned up edge on all sides at least 2 inches above the finished curb or 3½ inches above the rough curb. Shower pans must be securely fastened to the shower strainer base at the invert of the weep holes. Use a clamping ring to make a water-tight joint between the shower waste outlet stub and the pan.

Shower pan material must be securely supported by an adequate backing secured to the partition studs. This should prevent sagging of the pan sides until the interior of the shower compartment is in place to hold the pan rigid. If the shower pan material has to be punctured to secure it in place, the penetration must not be lower than one inch from the top of the pan's turn-up. Figure 16-2 shows an installed pan.

You must test each shower pan for inspection. Remove the shower strainer plate and plug waste outlet. Fill the pan with water. The pan must be full and ready for inspection during the tub and water pipe inspection. Otherwise, the contractor may have to pay for a reinspection.

Shower pans can be omitted in shower compartments built on a concrete slab on the ground floor. The bottom, sides and curbs of the shower compartment must be poured at the same time the floor slab is poured. A curb one inch higher than the existing slab must be poured around all sides of the shower compartment. This will usually keep the water level below the height of any surrounding wood plates or studs and will help keep the compartment water-tight. See Figure 16-3.

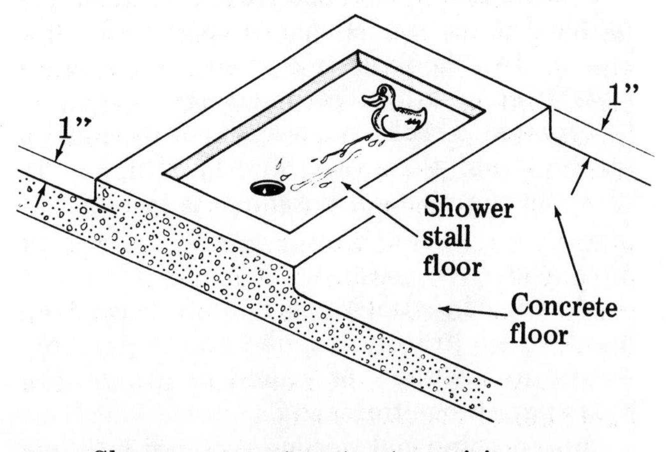

Shower compartment not requiring pan
Figure 16-3

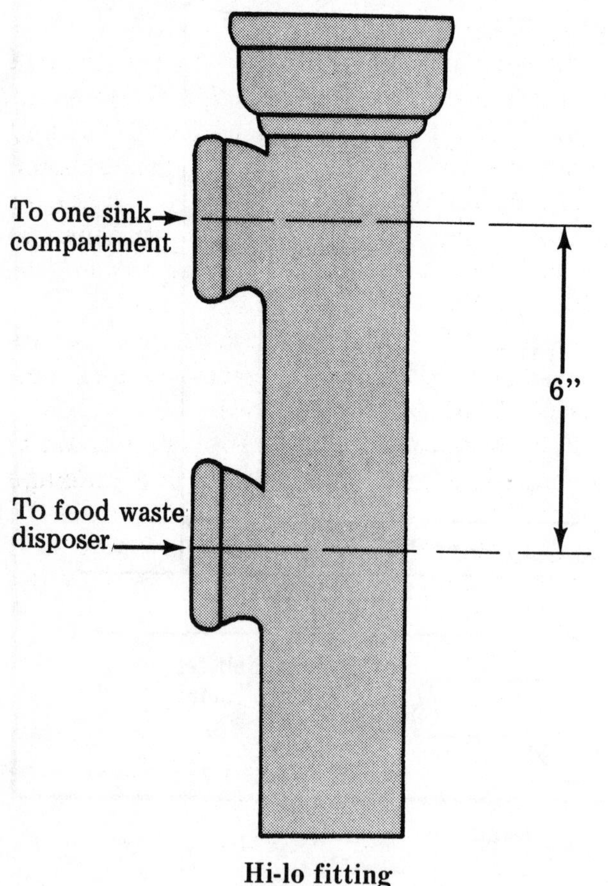

Hi-lo fitting
Figure 16-4

Shower pans are not required in prefabricated shower stalls. But these stalls require individual approval for water-tightness by the plumbing inspector. Walls of shower compartments must be waterproof, smooth, non-corrosive and non-absorbent to six feet above the floor.

Sinks and Laundry Tubs

Waste for sinks and laundry tubs are a minimum of 1½ inches in diameter. Tail pieces and continuous waste pipe must be a minimum of 1½ inches o.d. Each compartment in a laundry tub should have a waste outlet with a suitable stopper for retaining water.

Domestic sinks need a waste opening a minimum of three and one-half inches in diameter if a waste disposal unit is installed.

Food Waste Disposers

A two-compartment sink in a residence often must be separately trapped and separately wasted to the stack or vented branch. A few codes permit one trap and one waste line. This double trapping and wasting required by some codes permits use of one compartment of the sink until stoppage is removed from the other. Double trapping and venting prevents the waste disposal from pumping garbage into the other compartment of the sink. Where two traps and waste lines are required, use a hi-lo fitting two inches in diameter with two 1½ inch double vertical tappings not more than 6 inches apart. See Figure 16-4.

You can still install a food waste disposer in the two compartment sink in an existing home or apartment if the second waste opening is not available. The waste can exit through a single 1½ inch trap if a special directional tee or wye is used. See Figure 16-5.

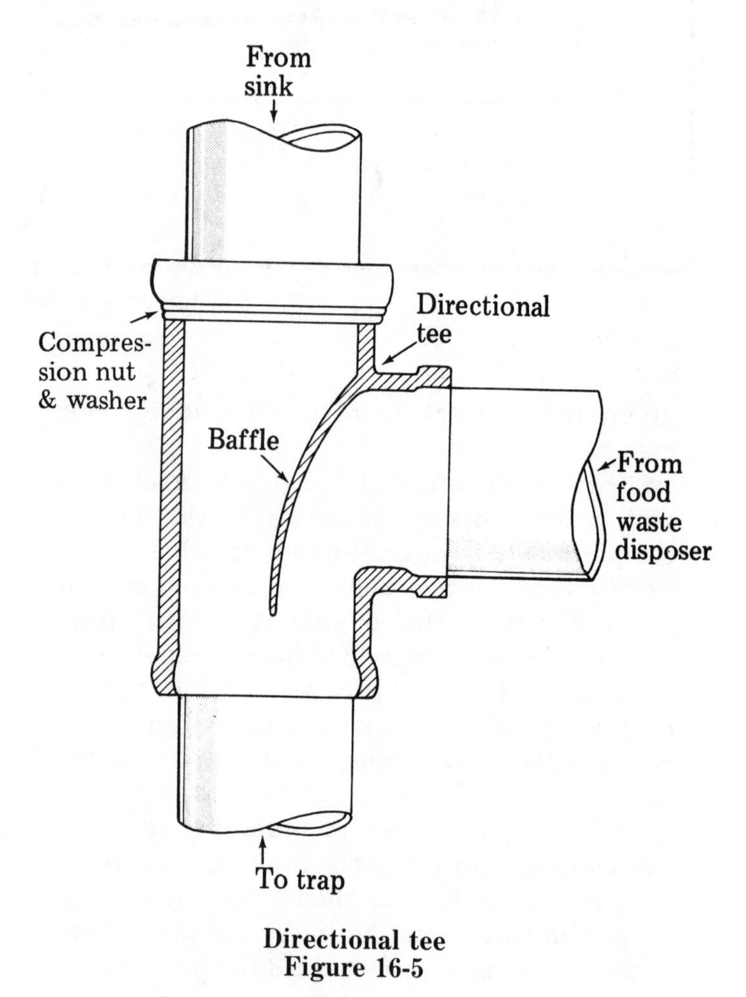

Directional tee
Figure 16-5

Commercial food waste grinders must waste directly into the sanitary drainage system, never through a grease interceptor. The waste pipe out of commercial food waste grinders can not be smaller than two inches and must be equal in size to the discharge opening of the machine. The grinder has to be individually trapped and vented just like any other fixture.

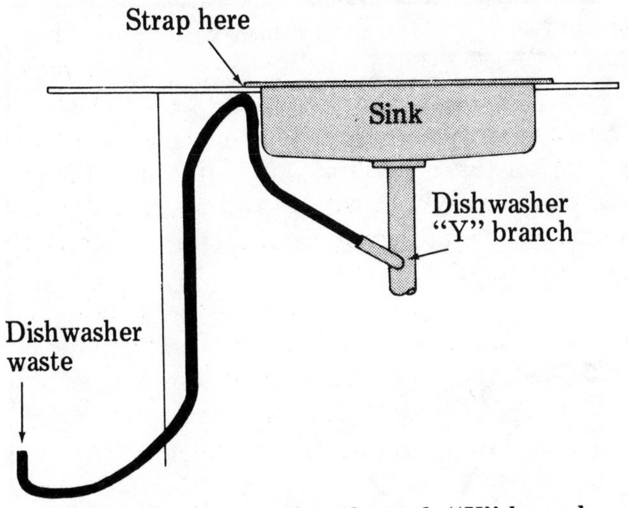

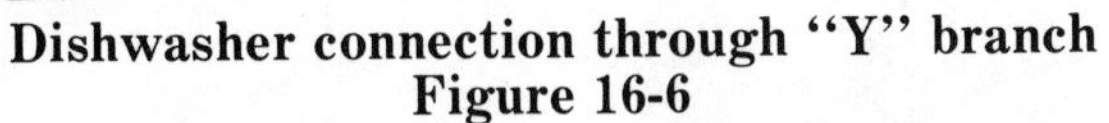

Dishwasher connection through "Y" branch
Figure 16-6

Dishwashers

Waste from a domestic dishwasher with a pump discharge must rise to a height equal to the height of the underside of the dishwasher top. It should connect to the sink waste with a dishwasher "Y" branch as shown in Figure 16-6.

Some codes require that this connection be made with a directional tee installed in the sink tail piece (see Figure 16-7). Other codes require a waste air gap fitting, either deck-mounted to the sink or cabinet top or wall-mounted. See Figure 16-7 and 16-8. Dishwashers should not be located more than five feet from the sink waste connection.

If a food disposal unit is installed in a sink, the waste from the dishwasher must connect to the tap provided in the body of food disposer. See Figure 16-8.

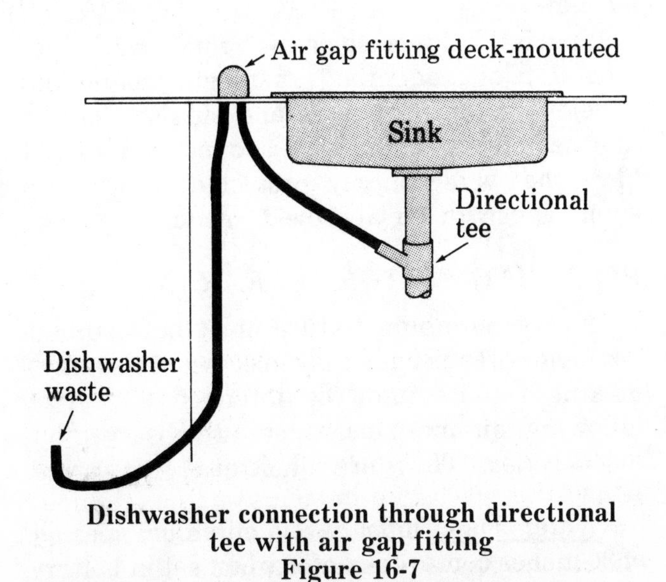

Dishwasher connection through directional tee with air gap fitting
Figure 16-7

Floor Flanges

All floor fixtures should be securely fastened to an approved floor flange with brass bolts or screws. Plumbing fixtures with a flanged connection between the fixture and the drainage pipe must have an approved gasket, washer, or be set in setting compound. Approved gasket materials are graphite-impregnated asbestos and felt.

Flanges for floor fixtures have to be set on top of the finished floor, not recessed flush with finished floor. See Figure 16-9.

Floor flanges must be of the same material or compatible with the materials in a drainage system. Where lead stubs are used to secure the fixture to the drainage system, a brass or hard lead flange must be soldered securely to the stub. In a copper drainage system where copper stubs are used to secure the floor flange, a brass flange should be soldered securely to the copper stub. In a cast iron drainage system, cast iron stubs are acceptable under some codes. A cast iron flange with a lead and oakum joint is used to secure the floor flange to the cast iron stub. In a plastic drainage system a plastic flange with a cement welded connection is used to secure the floor flange to the plastic stub.

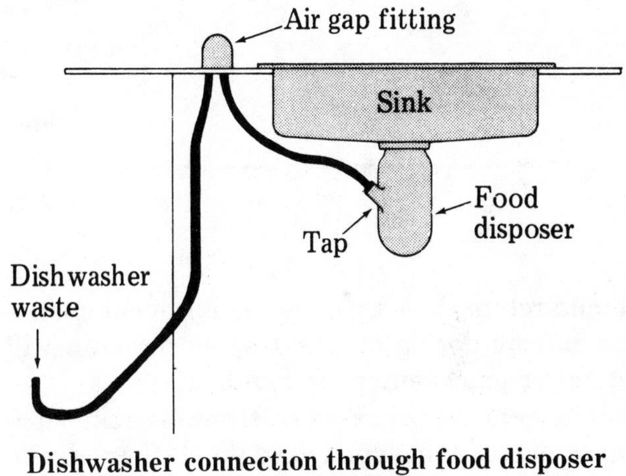

Dishwasher connection through food disposer with air gap fitting
Figure 16-8

Floor Drains

Floor drains are considered fixtures. Traps of floor drains must have a permanent water seal fed from an approved source of water, or an automatic priming device designed and installed for that purpose. This permanent water supply

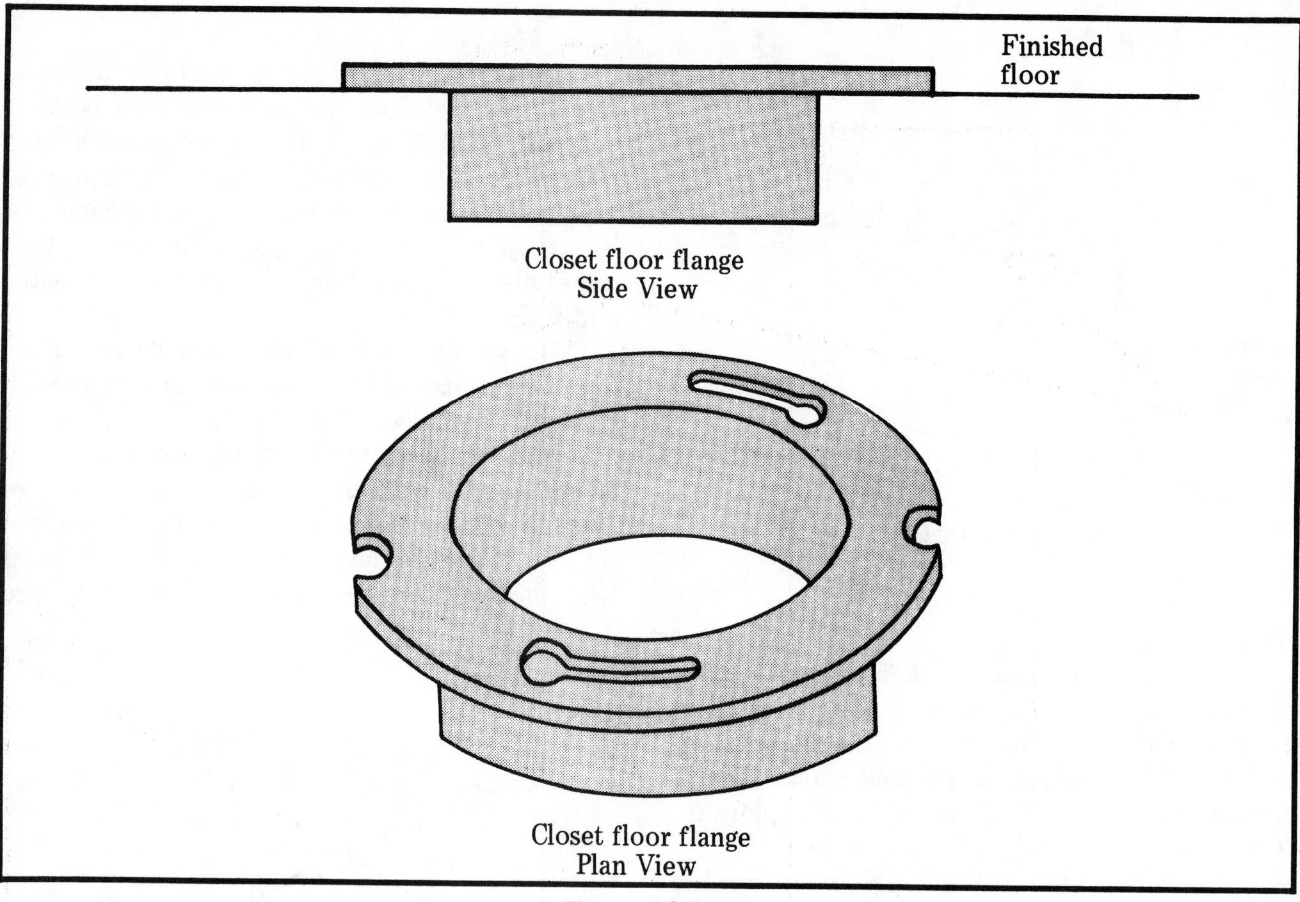

Figure 16-9

should retain the trap's seal and prevent evaporation from drying out the trap. A dry trap will let sewer gases enter the building.

Condensate drain waste from air conditioning units can be connected to a floor drain. However, this is not considered enough waste to supply a permanent water seal. A drinking fountain waste can discharge to one floor drain if the drain is not in a restroom. This is usually considered an adequate water supply to protect the trap seal. Waste from only one drinking fountain can be connected to a floor drain.

The discharge from a garbage can washer should not discharge through a trap serving any other device or fixture. The waste pipe should connect directly into the greasy waste line and discharge through a grease interceptor. The receptacle (floor drain) receiving the waste from the garbage can washer must be equipped with a basket to prevent passage into the drainage system of solids 1/2 inch or larger in size. It is essential that the basket be easily removable for cleaning purposes.

The hot and cold water connection must be properly valved and have an approved vacuum fitting to prevent cross-connection. See Figure 16-10.

Floor drains serving indirect waste pipes from food or drink storage rooms or appliances must not be located in toilet rooms or in any inaccessible or unventilated closet or store room. No type of plumbing fixture can ever be installed in a room containing air handling machinery.

Special fixtures such as baptistries, ornamental pools, aquaria, ornamental fountains, developing tanks or sinks and similar fixtures that have a waste and water connection should have the water supply protected from backsiphonage with an approved vacuum breaker.

PLUMBING FIXTURE CLEARANCE

Every plumbing fixture must be installed and spaced to permit easy access for cleaning, making repairs and the intended use. The following minimum clearances are observed by most codes. They are illustrated in Figure 16-11.

Water closets must have a minimum spacing of 30 inches center-to-center when set in battery

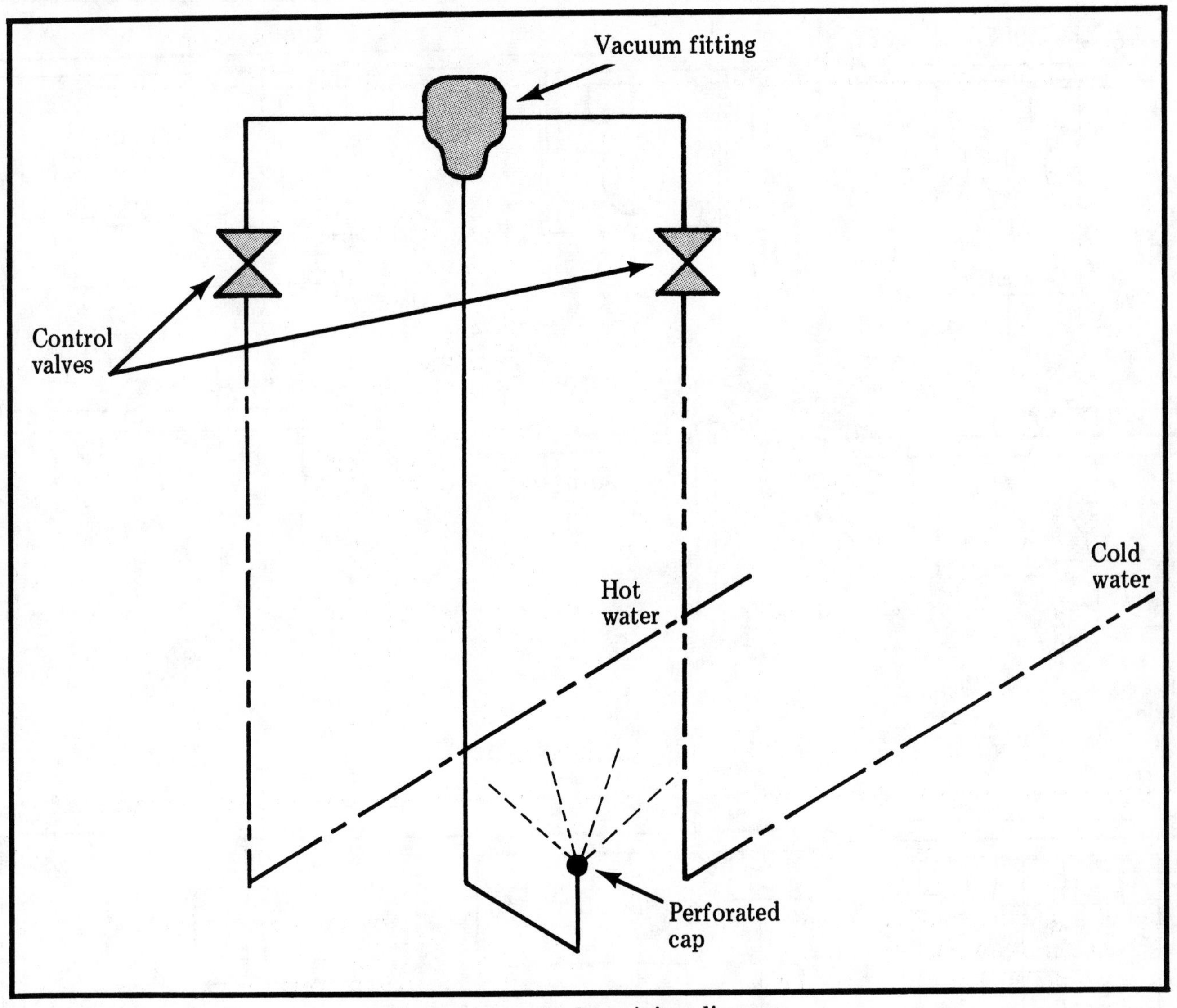

Garbage can washer piping diagram
Figure 16-10

installations. They must be set a minimum of 15 inches from the center of the bowl to any finished wall or partition and must be set a minimum of 12 inches from the center of the bowl to the outside edge of a tub apron. Finally, they must have a minimum clearance of 21 inches from the front of the bowl to any finished wall, door or other plumbing fixture.

Pedestal, stall or wall hung urinals must have a minimum 30 inch center-to-center spacing when set in a battery installation. They must have a minimum 15 inch clearance from the center of the urinal to any finished wall. Pedestal urinals must have a minimum clearance of 18 inches from the front of the urinal to any finished wall, door or other plumbing fixture.

Stall and wall hung urinals must have a minimum clearance of 21 inches from the front of the urinal to any finished wall, door or other plumbing fixture.

Lavatories are manufactured in various designs and widths, so center-to-center measurements are not applicable. The minimum clearance is measured from the edge of the lavatory to the nearest obstruction. Lavatories must have a minimum clearance of four inches from its edge to any finished wall. They must have a minimum clearance of two inches from the edge to the edge of a tub. Lavatories in battery installation must have a minimum clearance of four inches from lavatory edge to lavatory edge. Finally, they must have a clearance of 21 inches from the front of the lavatory to any finished wall, door or other plumbing fixture.

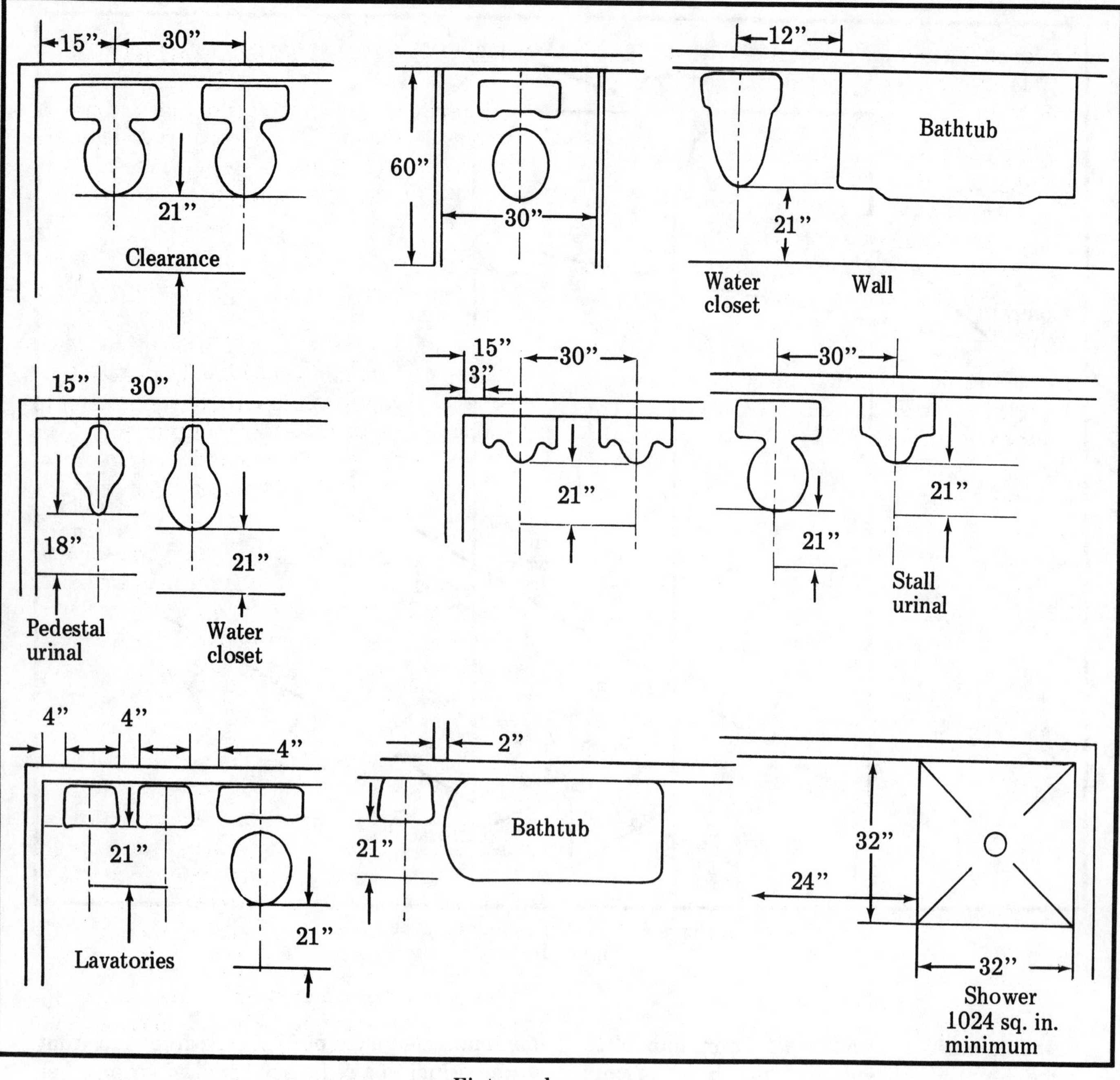

Fixture clearances
Figure 16-11

The opening to a shower compartment or stall must have a minimum clearance of 24 inches for easy entry and exit from any finished wall, door or other plumbing fixture.

FIXTURES FOR THE HANDICAPPED

Federal and state laws now require that public and private buildings have toilet facilities available for the physically handicapped. Single family residences and buildings considered hazardous are excepted from this requirement. The law states that all *required* restrooms in a building must provide this service for the handicapped. In large multi-story buildings facilities should be provided on each floor for each sex. Halls and restroom doors are to be marked with readily visible signs denoting the location of such restrooms. In each *required* restroom having one or more water closets, one toilet must comply with standards created by the President's Committee on Employment of the Physically Handicapped and by the American

National Standards Institute (ANSI). These standards are seldom included in the plumbing code or in other reference books.

1. Toilet rooms must have at least one toilet stall that:

- is 3 feet wide
- is a minimum of 4 feet 8 inches deep
- has a door (if one is used) that is a minimum of 32 inches wide and that swings out
- has handrails, one on each side, 33 inches high and parallel to the floor. Handrails should be 1½ inches in outside diameter and have a clearance of 1½ inches between the rail and wall. Handrails must be securely fastened at each end and at the center.
- has a water closet with the seat 17 to 19 inches from the floor.

2. Toilet rooms must have at least one lavatory that:

- has a narrow apron which, when mounted at standard height, is usable by individuals in wheelchairs
- may be mounted higher, when the particular design demands, so as to be usable by individuals in wheelchairs
- has the drain pipes and hot water pipes insulated so that a wheelchair individual without sensation will not burn himself.

3. Toilet rooms for men must have at least one urinal that:

- is floor-mounted and that is level with the main floor of the toilet room
- is wall-mounted with the opening of the basin no higher than 19 inches from the toilet room floor.

4. An appropriate number of water coolers or fountains must be accessible to and usable by the physically handicapped. These are to be hand operated or hand and foot operated and are to have up front spouts and controls.

Wall-mounted, hand operated coolers can serve the able bodied and the physically handicapped equally well when mounted 36 inches from the floor.

Water fountains that are fully recessed are not acceptable for use by the handicapped. Water fountains should not be recessed but can be set in an alcove if the alcove is 32 inches wide.

MINIMUM FIXTURE REQUIREMENTS

Wherever the installation of plumbing fixtures is required, the code regulates the minimum number and type of fixtures. This includes all premises intended for human occupancy or use.

The minimum number and type of fixtures set by most plumbing codes is generally based on the type of occupancy and the anticipated number of people who will use these toilet facilities. Codes vary considerably in how they arrive at the number of fixtures needed. Refer to your local code for the exact requirements. This book will help you understand and interpret your particular code's method of computing the required toilet facilities.

Single family residences minimum requirements for this type dwelling are: one kitchen sink, one water closet, one lavatory and one bathtub or shower unit. Provision must also be made for a clothes washing machine. Hot water is optional in some codes and mandatory in others.

Duplex residential units must have: one kitchen sink, one water closet, one lavatory, one bathtub or shower unit. Provision must also be made for a clothes washing machine for each unit. One machine is adequate for both units if it is available to all residents. Hot water is optional in some codes and mandatory in others.

Apartment units must be equipped with one kitchen sink, one water closet, one lavatory, one bathtub or shower unit. Provision may be made for a clothes washing machine for each individual dwelling unit unless a centrally located laundry room has the following equipment ratios: One machine for the first five dwelling units and one machine for each additional 15 units or portion thereof. For example, an apartment building with eight units would require two machines. An apartment building with twenty units would still require two machines. An apartment building with twenty one units would require three machines.

When a central washing facility is used by residents in a complex of several buildings, a maximum distance to the most remote unit is established in most plumbing codes. The distance from the entrance of the most distant apartment building should not exceed 400 feet.

MALES				FEMALES		
No. of males	Water closets	Urinals	Lavatories	No. of females	Water closets	Lavatories
1- 9	1	--	1	1- 25	1	1
10- 25	1	1	1	26- 50	2	2
26- 60	2	1	2	51- 75	3	3
61-100	3	2	2	76-100	4	3
101-140	4	2	3	101-125	5	3
141-180	5	3	3	126-150	6	4
181-220	6	3	4	151-175	7	4
				176-225	8	5

For each group of 40 males over 220, add one water closet, one urinal, and one lavatory.

Note: In men's toilet rooms open to the public, one urinal is required for each 3 required water closets. The number of required water closets cannot be reduced to less than 2/3 of the minimum specified for men.

For each group of 50 females over 225, add one water closet and one lavatory.

Note: Female urinals may be substituted for water closets but the number cannot exceed ¾ of the number of water closets required.

Places of employment
Table 16-12

Hot water generating facilities are generally installed in all buildings. Again, hot water is optional in some codes and mandatory in others.

Places of Employment

The number of toilet fixtures in manufacturing plants, heavy industry, warehouses and similar establishments is based on the number of employees. The percentage ratio and type of fixtures required for males and females may be changed by the administrative authority. The authority in your area will alter the requirements in Table 16-12 if you can provide data which shows that some other fixture ratio is more appropriate.

Let's consider one such example. Assume that toilet facilities are needed for a medium sized manufacturing plant employing 100 persons. Some building codes require a ratio of 50% male and 50% female facilities. Other codes for the same type occupancy use a percentage ratio of 75% male and 25% female. Obviously, if these ratios were used rigidly, an imbalance of toilet fixtures would result in many installations. The administrative authority (plumbing plans examiner) may request a notarized letter from the owner giving the maximum number of probable male and female employees in his particular plant. Then the correct number and type of plumbing fixtures could be determined from Table 16-12.

A drinking fountain must be provided for each 75 persons or portion thereof. It must be accessibly located within 50 feet of all operational processes. Drinking fountains must not be located in any restroom or vestibule to a restroom.

Wash up sinks may be substituted for lavatories where the type of employment warrants their use.

Establishments that have 10 or more offices or rooms and employ 25 persons or more must provide a service sink on each floor.

Manufacturing plants that may subject its employees to excessive heat, infection or irritating materials must provide a shower for bathing for each 15 persons. Establishments employing 9 persons or less which do not cater to the public (such as storage warehouses and light manufacturing buildings) have less rigid requirements. In these applications *some codes* would consider as adequate 1 water closet and 1 lavatory for both sexes. But observe the following conditions:

MALES				FEMALES		
No. of males	Water closets	Urinals	Lava-tories	No. of females	Water closets	Lava-tories
1 - 100	1	1	1	1 - 50	1	1
101 - 250	2	1	1	51 - 140	2	1
251 - 360	2	2	1	141 - 250	3	2
361 - 470	2	3	2	251 - 360	4	2
471 - 580	3	3	2	361 - 470	5	3
581 - 700	3	4	3	471 - 690	6	3
701 - 820	3	5	3	691 - 960	7	4
821 - 975	4	5	4	961 - 1300	8	4
976 - 1150	4	6	4	1301 - 1640	9	5
1151 - 1325	4	7	4	1641 - 2000	10	6
1326 - 1490	5	7	5	2001 - 2350	11	7
1491 - 1675	5	8	5	2351 - 2700	12	8
1676 - 1875	5	9	5	---	---	---
1876 - 2075	6	9	6	---	---	---
2076 - 2250	6	10	6	---	---	---
2251 - 2475	6	11	6	---	---	---
2476 - 2700	6	12	7	---	---	---

For each group of 500 males over 2700, add one water closet and one lavatory. For each additional 300 males, add one urinal.

For each group of 350 females over 2700, add one water closet. For each additional group of 500 females, add one lavatory.

Drinking fountains must be provided at a ratio of one for each 200 persons.

Public Assembly
Table 16-13

- If the minority sex exceeds 3 persons, separate toilet facilities are required. For example, where 4 males and 5 females (or vice versa) are employed, separate toilet facilities must be provided.

- If the number of males employed exceeds 5, a urinal must be provided.

Establishments which are frequented by the public must provide toilet facilities for the number of employees and the public reasonably anticipated, unless they have special permission or special invitation to do otherwise. There are two classifications of public use which determine the number and type of plumbing fixtures required:

1) Establishments that provide countable seating capacity such as churches, theaters, stadiums, restaurants are in the first class.

2) Establishments such as in retail stores, office buildings, and similar establishments that have no countable seating capacity. This is the second class. The facilities required are determined by the square foot area. See Table 16-16.

Public places having seating capacities

Public assembly facilities such as churches, theaters, stadiums and similar establishments use the percentage ratio for determining the number and type of plumbing fixtures required. Many building codes use the percentage ratio of 50% male and 50% female.

Some building codes use a different ratio for theaters than for churches. For example, the Standard Plumbing Code uses a percentage ratio of 50% male, 50% female for theaters but a ratio of 40% male, 60% female for churches. To bring into clearer focus the difference in fixture requirements for similar occupancies, consider the minimum fixture requirements from two model codes for a church that seats 400 persons.

MODEL PLUMBING CODE

Toilet facilities would have to be provided for

MALES				FEMALES		
No. of males	Water closets	Urinals	Lavatories	No. of females	Water closets	Lavatories
1- 40	1	1	1	1- 40	1	1
41- 90	2	1	2	41- 90	2	2
91-150	2	2	2	91-150	3	2
151-225	3	2	2	151-225	4	2
226-300	3	3	3	226-300	5	3
301-400	4	3	3	301-400	6	3

For each group of 125 males over 400, add one water closet, one urinal and one lavatory.
For each group of 125 females over 400, add one water closet and one lavatory.
Note: See Table 16-12 for urinal substitution in lieu of required water closets for both male and female toilet rooms.

Food and/or drink establishments
Table 16-14

200 males and 200 females. Table 16-13, "Public Assembly," would require the following:

MALES			
No. of males	Water closets	Urinals	Lavatories
101-250	2	1	1

FEMALES		
No. of females	Water closets	Lavatories
141-250	3	2

STANDARD PLUMBING CODE

Toilet facilities would have to be provided for 160 males and 240 females. Their reference table would require the following:

MALES			
No. of males	Water closets	Urinals	Lavatories
101-200	3	2	1

FEMALES		
No. of females	Water closets	Lavatories
201-400	4	2

Water closets for public use must be separated from the rest of the room and from each other by stalls made of some impervious material.

Toilet rooms connected to public rooms or passageways must have a vestibule or must otherwise be screened or arranged to insure decency and privacy. Such a vestibule must not be common to the toilet rooms of both sexes.

Food and Drink Establishments

Table 16-14 is used to determine the minimum toilet facilities for establishments where food and drink are served and consumed on the premises. This includes cafeterias, barbeque stands, private clubs and similar establishments having countable seating capacities. The percentage ratio of 50% male to 50% female is usually followed.

Establishments catering to drive-in service must provide toilet facilities at the ratio of one person for each 100 square feet of parking area.

Public food service establishments that offer only a take-out service need not provide guest toilet facilities. Only toilet facilities for employees are required.

The floors and walls of public toilet rooms must have tile or other impervious materials to a height of five feet.

In a case where the seating capacity is unknown, the following method is used to determine the number of persons who can be expected to occupy the premises. In establishments serving food and drink but no alcoholic beverages, each 30 inches of counter space or 40 square feet of dining room area are considered

Type of occupancy	Square feet NET = one person
Schools	Classrooms
Nurseries, day care	20 square feet
Elementary	20 square feet
Secondary	20 square feet
Office buildings	100 square feet
Retail stores	200 square feet
Dining rooms, clubs, lounges, etc.	40 square feet
Laundries	50 square feet
Barber and beauty shops	50 square feet
Assembly areas, standing or waiting spaces	70 square feet
Theaters, auditoriums, churches, etc.	70 square feet

Square feet per occupant based on net floor area
Table 16-15

to be equal to 1 person. In establishments serving food and drink and alcoholic beverages, each 18 inches of counter or bar space and each 40 square feet of serving area are considered to be equal to 1 person.

Toilet rooms serving such establishments must have easy and convenient access for both patrons and employees. The restrooms must be located within 100 feet along a line of travel from the nearest exit to the dining room, bar or food service area. Toilet rooms must be located on the same floor as the area they serve.

In food or drink establishments where dishes, glasses, or cutlery are to be reused, a dishwashing machine or suitable 3-compartment sink must be used.

In establishments where food or drink are prepared or served, a hand sink must be installed for employees' use. Lavatories in adjoining toilet rooms may not serve this purpose.

Public places without seating capacities

In computing restroom facilities for public places such as shopping centers, retail stores, office buildings and similar establishments, you will need to determine the amount of habitable floor space (area).

Uninhabitable space such as corridors, stairways, vertical shafts, and equipment rooms may be deducted from the gross floor area.

The net square footage will then provide the occupant load factor for individual establishments. Table 16-15 shows the net square footage per expected occupant. Table 16-16 gives the division of facilities based on sex specifications. The minimum number of facilities for each type of establishment is given in Table 16-16A.

Type of occupancy	Males %	Females %
Schools	50	50
Office buildings	40	60
Retail stores	30	70
Eating establishments	50	50
Clubs and lounges	65	35
Laundries	20	80
Barber shops	90	10
Beauty shops	10	90
Waiting rooms	50	50
Dormitories	---	--- (whether men or women)
Theaters	50	50
Churches and auditoriums	40	60

Division of toilet facilities based on sex
Table 16-16

ROUGHING-IN

In a typical single family residence there are approximately 300 feet of concealed piping beneath the floor and in the partitions. The piping includes the drainage, waste, vents and hot and cold water lines. These pipe systems were discussed in previous chapters. Roughing-in the fixtures is possibly the single most critical point where a plumber displays his skill or lack of skill. Knowledge and good workmanship are required for proper roughing-in of the waste and water outlets. The plumbing fixtures will be an important part of the building all during the life of the building.

It is important that you have a thorough knowledge of roughing-in measurements for various types of plumbing fixtures. You must know the height, distance and location for waste and water outlets for wall-hung and floor-mounted fixtures. Memorize these measurements for *common* fixtures or jot them down in a

Type of Occupancy	MALES No. of males	Water closets	Urinals	Lava-tories	FEMALES No. of females	Water closets	Lava-tories
Schools, nurseries, day care	15 or fraction thereof	1	--	1	15 or fraction thereof	1	1
Elementary	60	1	--	1	35	1	1
Secondary	30	1	1	1	45	1	1
Office buildings	1- 15	1	1	1	1- 15	1	1
	16- 35	2	1	2	16- 35	2	2
	36- 55	3	1	3	36- 55	3	3
	56- 80	4	1	4	56- 80	4	4
	81-100	5	2	5	81-100	5	5
Retail stores	1- 15	1	--	1	1- 15	1	1
	16- 35	2	--	2	16- 35	2	2
	36- 55	3	1	3	36- 55	3	3
	56- 80	4	1	4	56- 80	4	4
	81-100	5	2	5	81-100	5	5
Dining rooms, clubs, lounges, etc.	See Table 16-14				See Table 16-14		
Barber and beauty shops	1-35	1	--	1	1-35	1	1
	36-75	2	--	1	36-75	2	1
Assembly areas standing/ waiting	1-100	2	1	2	1-100	2	1
	101-200	3	2	2	101-200	3	2
	201-400	4	3	3	201-400	4	3
Theaters, auditoriums, churches, etc.	1-100	2	2	1	1-100	2	1
	101-200	3	3	2	101-200	3	2
	201-400	4	4	3	201-400	4	2

Note: More complete tables and miscellaneous fixtures may be found in your Code book.

Occupant content and minimum toilet facilities
Table 16-16A

notebook that can be kept in your tool box. Complete roughing-in information is generally available from the manufacturer and distributor for *special* fixtures you use.

Roughing-in dimensions and installation information for the more commonly used fixtures are illustrated here. Measurements are given for bathtubs, water closets, lavatories, showers, kitchen sinks, service sinks, urinals, bidets and a drinking fountain. The roughing-in measurements are for American Standard fixtures but are similar or identical to the measurements for like plumbing fixtures of other manufacturers.

Note that the standard roughing-in measurement for all water closets is 12 inches from the finished wall. If the water closet outlet is roughed too close or too far from the finished wall, special water closets that can be set at 10 or 14 inches from the wall are available.

PLUMBING FIXTURE CARRIERS

Toilet rooms are the most susceptible of all rooms to unsanitary conditions. The bases of on-the-floor fixtures are natural areas for accumulated filth that is nearly impossible to remove. Bathroom floors made of wood tend to deteriorate next to and beneath toilet fixtures. Off-the-floor water closets can solve much of this problem and have been gaining popularity for both commercial and residential use.

Some apprentices and journeymen may never have an opportunity to work with off-the-floor plumbing fixtures because of the type of construction their company contracts for. The following pages present some illustrations of the more commonly used carriers so that you may become familiar with their appearances and installation methods. (Continued on page 189).

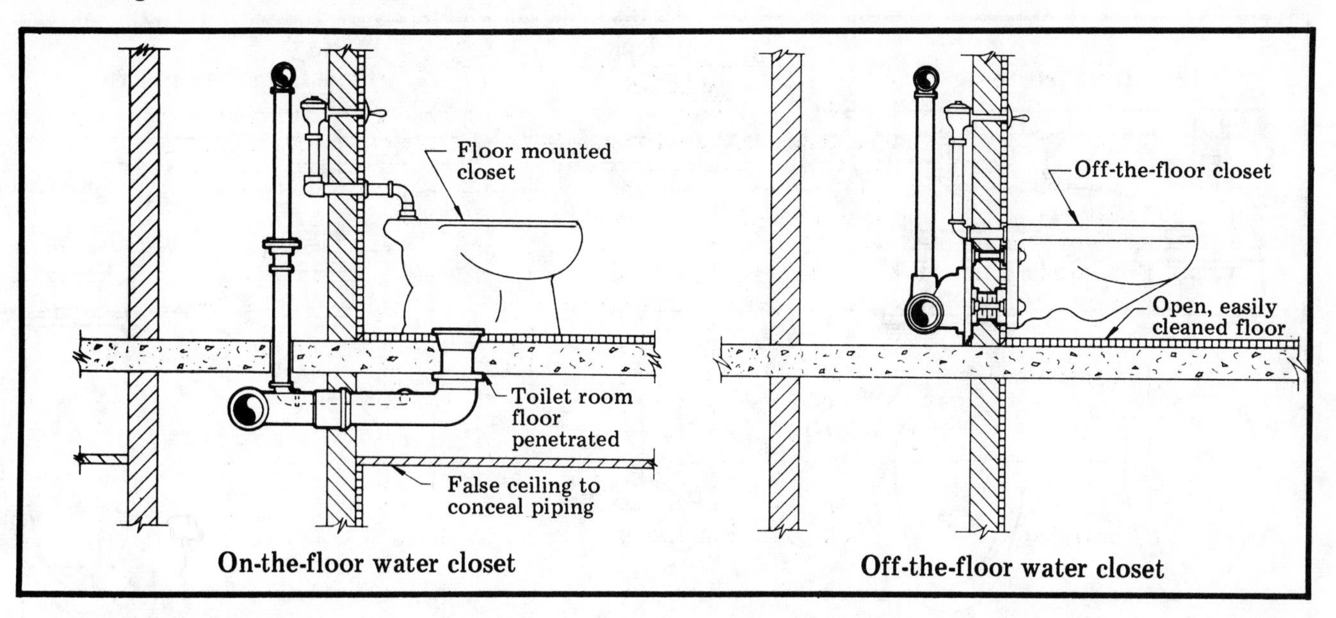

Installation
Many factors, including cost, material, labor required, space, handling and others must be thoroughly evaluated when planning an installation. With on-the-floor water closets, slabs must be penetrated and sleeved at each fixture to accommodate waste piping which, in many instances, has to be suspended below the slab. Such suspended piping requires concealment in multi-story installations, thereby necessitating drop or furred ceilings in the rooms below.

Commercial
Figure 16-17

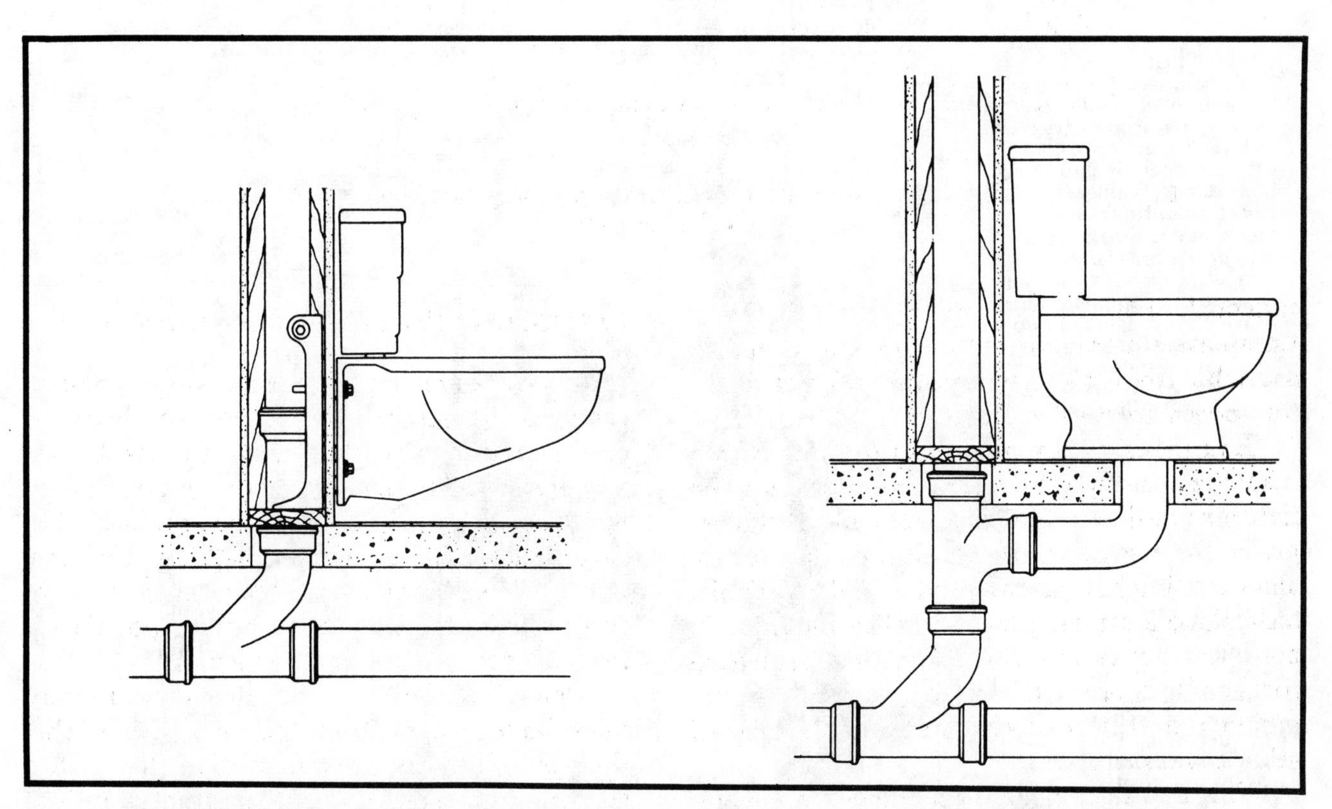

The above illustrations show some of the advantages of off-the-floor closet installation. With on-the-floor water closets, slabs or floor must be penetrated at each fixture to accommodate waste piping. Conversely, with off-the-floor fixtures, there is no slab penetration in toilet room areas. A clear, unobstructed floor is available for cleaning, and deterioration isn't a problem.

Residential
Figure 16-18

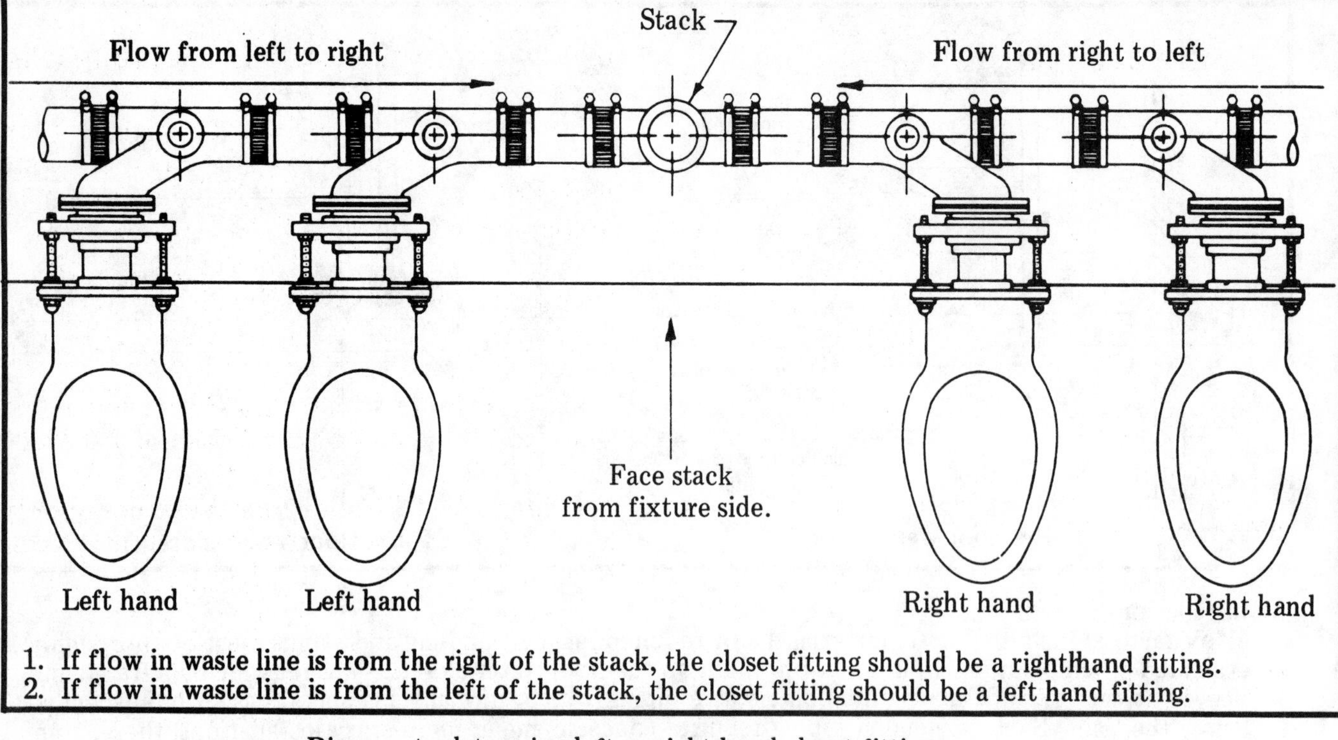

Diagram to determine left or right hand closet fittings
Figure 16-19

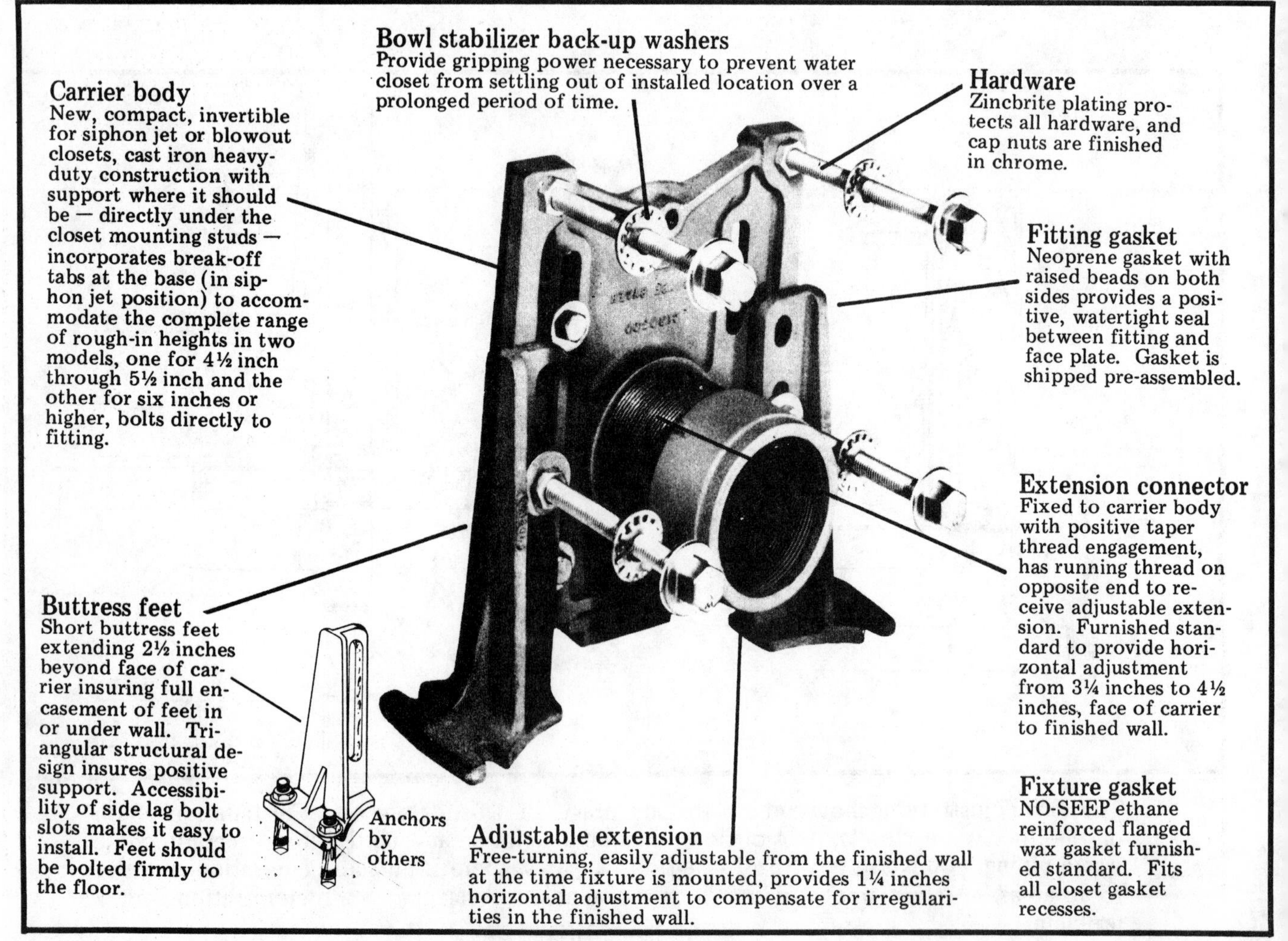

Figure 16-20A

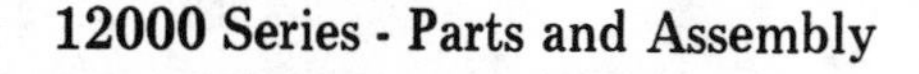

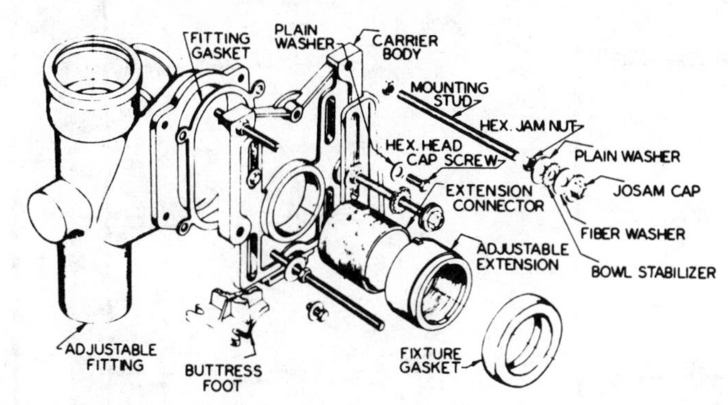

Carrier individually packaged
All hardware, feet, extension components, trim and installation instructions are furnished in individually marked cartons.

Figure 16-20B

We gratefully acknowledge the courtesy of the Josam Manufacturing Company in providing these illustrations.

Figure 16-17 illustrates a flush valve on-the-floor and off-the-floor water closet commonly used in *commercial* buildings. Figure 16-18 illustrates a tank type on-the-floor and off-the-floor water closet commonly used in *residential* buildings. From these illustrations you can readily see the differences in the two installations.

In Figure 16-19 is a typical floor plan showing a battery of water closets. This illustration is important in that it shows you how to identify left and right hand closet carriers.

Figure 16-20 illustrates a Josam closet carrier of the 12000 series. Note that the illustration includes parts and assembly for this particular unit.

Figure 16-21 illustrates the seven steps required for the proper installation of the 12000 closet carrier.

Figure 16-22 illustrates three commonly used installation methods for off-the-floor water closets.

Residential carriers have fewer parts and are therefore easier to assemble. They are designed to receive the waste from a single water closet or at most two water closets (when back to back installations are required). A single family residence never has battery installations as do commercial buildings.

Residential carriers have been designed to be compatible with the newer piping materials that have come to the forefront in recent years. Figure 16-23 shows an off-the-floor residential

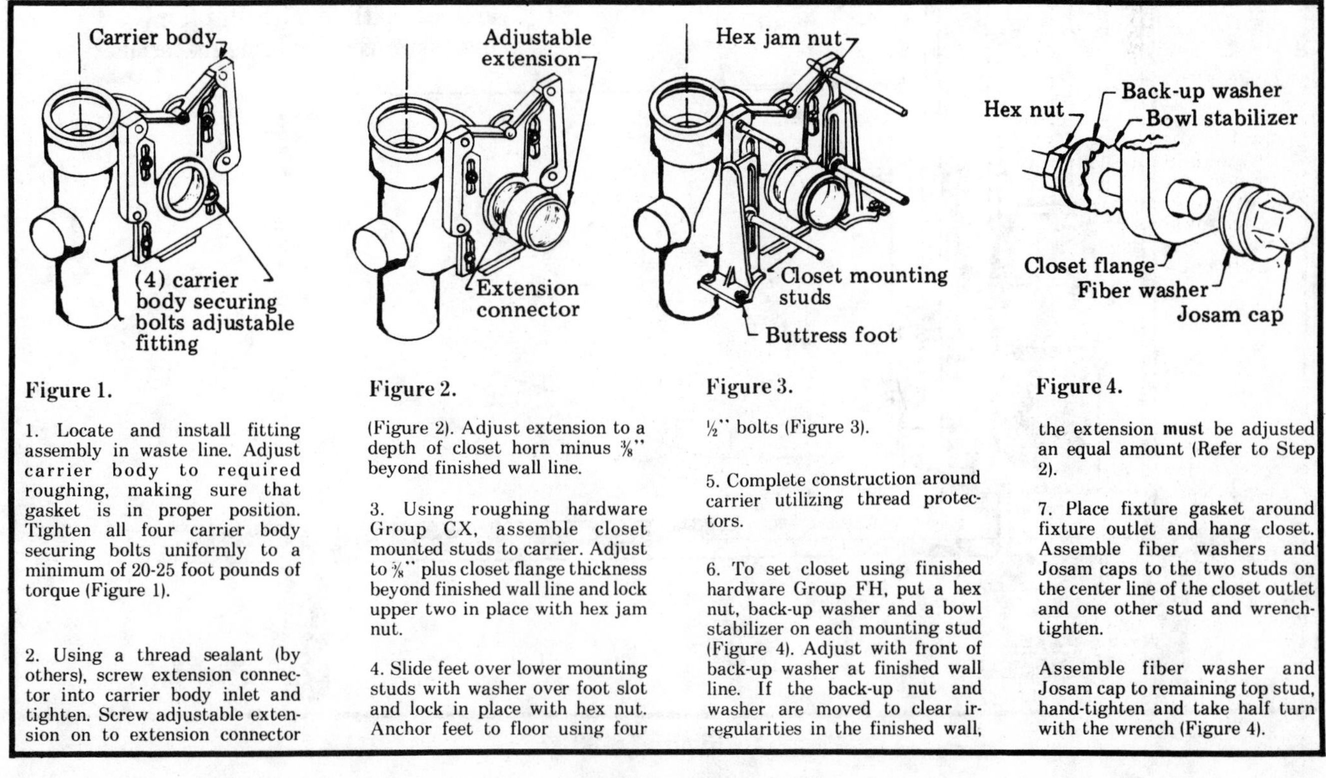

Figure 1.

1. Locate and install fitting assembly in waste line. Adjust carrier body to required roughing, making sure that gasket is in proper position. Tighten all four carrier body securing bolts uniformly to a minimum of 20-25 foot pounds of torque (Figure 1).

2. Using a thread sealant (by others), screw extension connector into carrier body inlet and tighten. Screw adjustable extension on to extension connector (Figure 2). Adjust extension to a depth of closet horn minus ⅜" beyond finished wall line.

Figure 2.

3. Using roughing hardware Group CX, assemble closet mounted studs to carrier. Adjust to ⅜" plus closet flange thickness beyond finished wall line and lock upper two in place with hex jam nut.

4. Slide feet over lower mounting studs with washer over foot slot and lock in place with hex nut. Anchor feet to floor using four ½" bolts (Figure 3).

Figure 3.

5. Complete construction around carrier utilizing thread protectors.

6. To set closet using finished hardware Group FH, put a hex nut, back-up washer and a bowl stabilizer on each mounting stud (Figure 4). Adjust with front of back-up washer at finished wall line. If the back-up nut and washer are moved to clear irregularities in the finished wall, the extension must be adjusted an equal amount (Refer to Step 2).

Figure 4.

7. Place fixture gasket around fixture outlet and hang closet. Assemble fiber washers and Josam caps to the two studs on the center line of the closet outlet and one other stud and wrench-tighten.

Assemble fiber washer and Josam cap to remaining top stud, hand-tighten and take half turn with the wrench (Figure 4).

12000 Chase-Saver® II
Figure 16-21

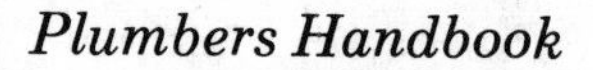

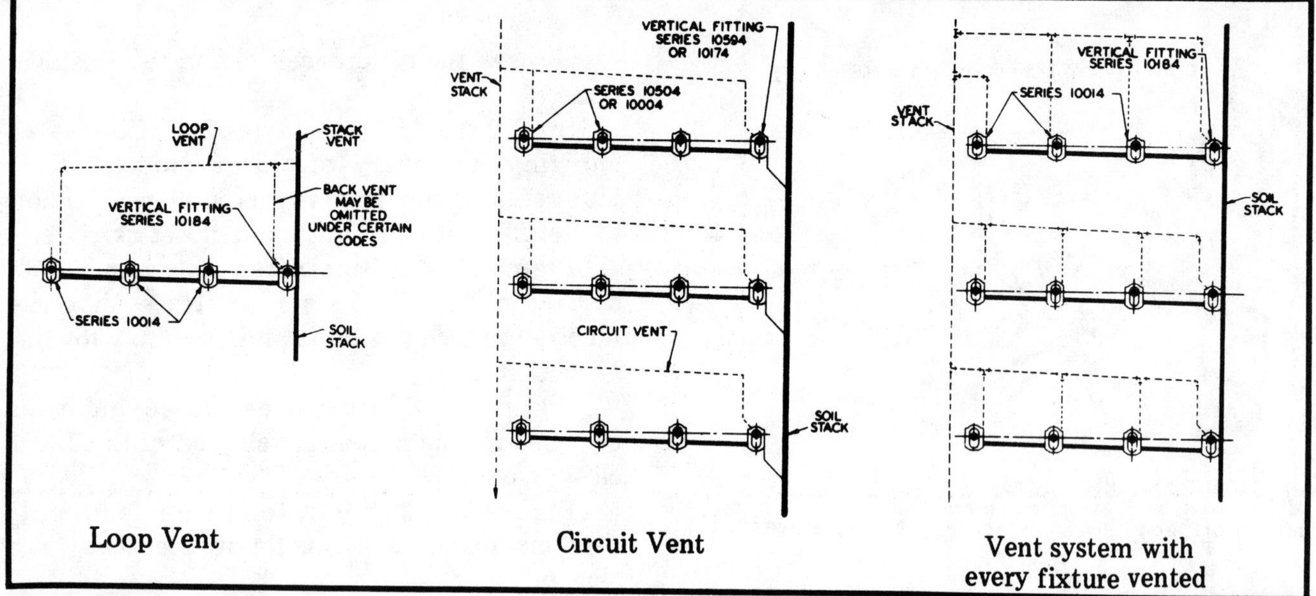

Josam Carriers Venting
Some examples of venting a horizontal run of fixtures are detailed above. Local code requirements should be consulted for specific requirements for the area of your projects.

Figure 16-22

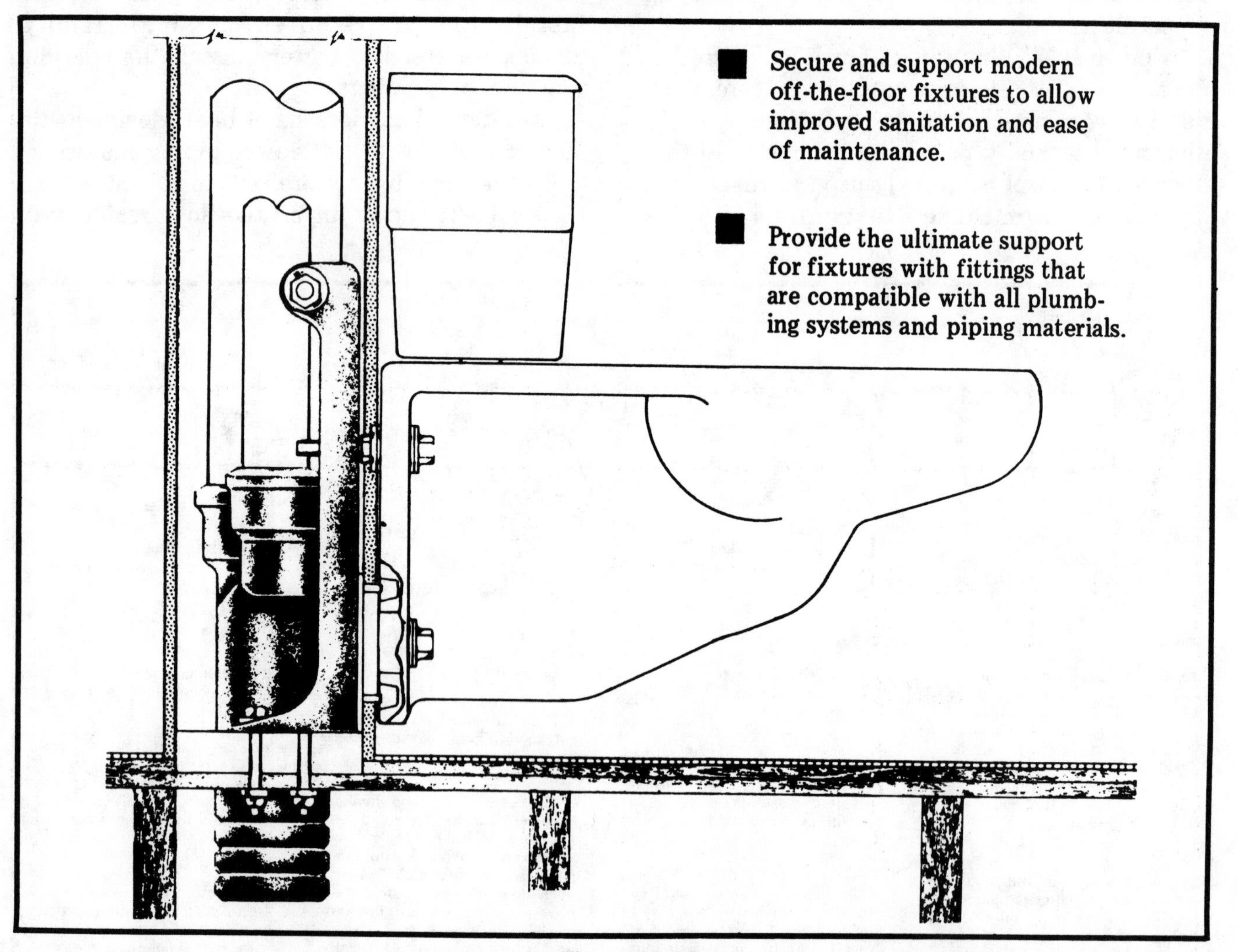

Figure 16-23

water closet. Figure 16-24 shows the same closet carrier with its parts identified.

Figures 16-25 and 16-26 show units generally used in commercial installations.

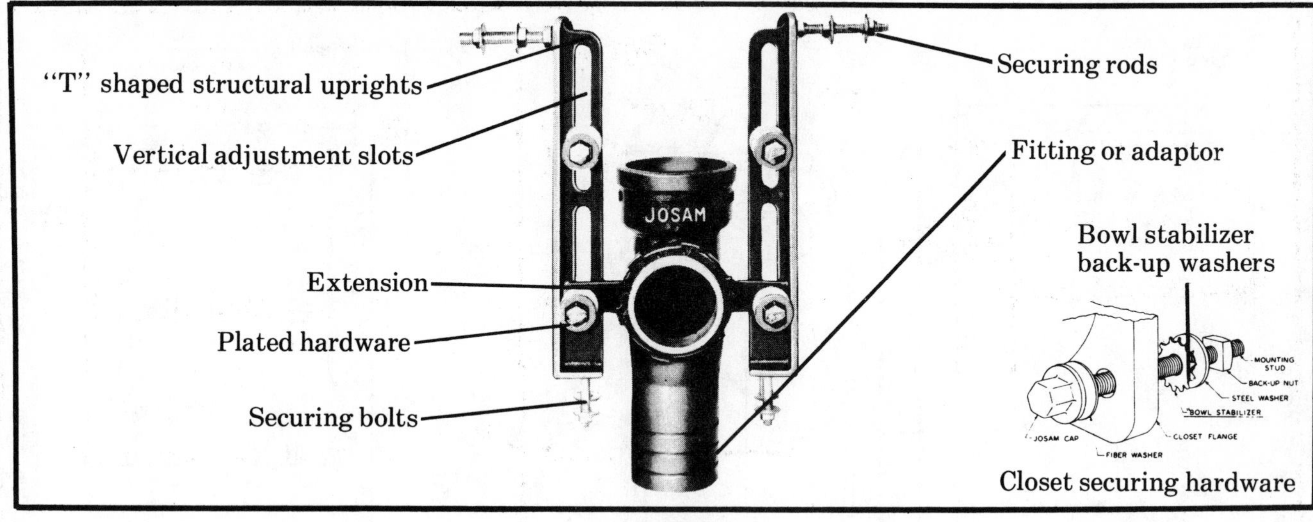

Figure 16-24

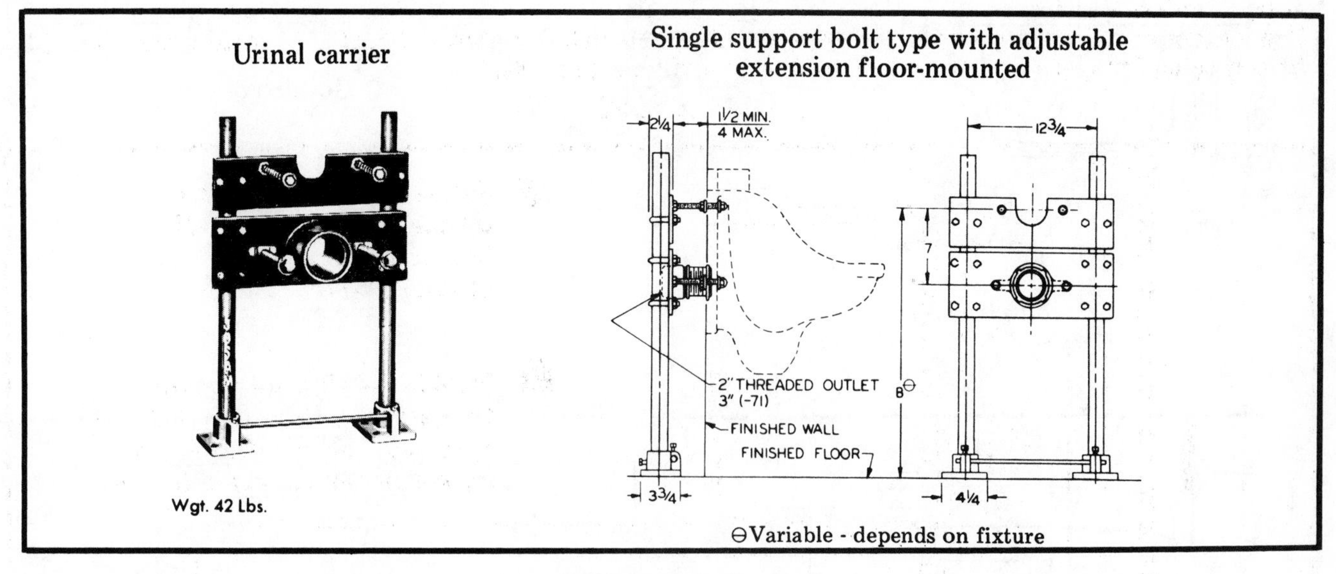

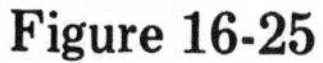

Figure 16-25

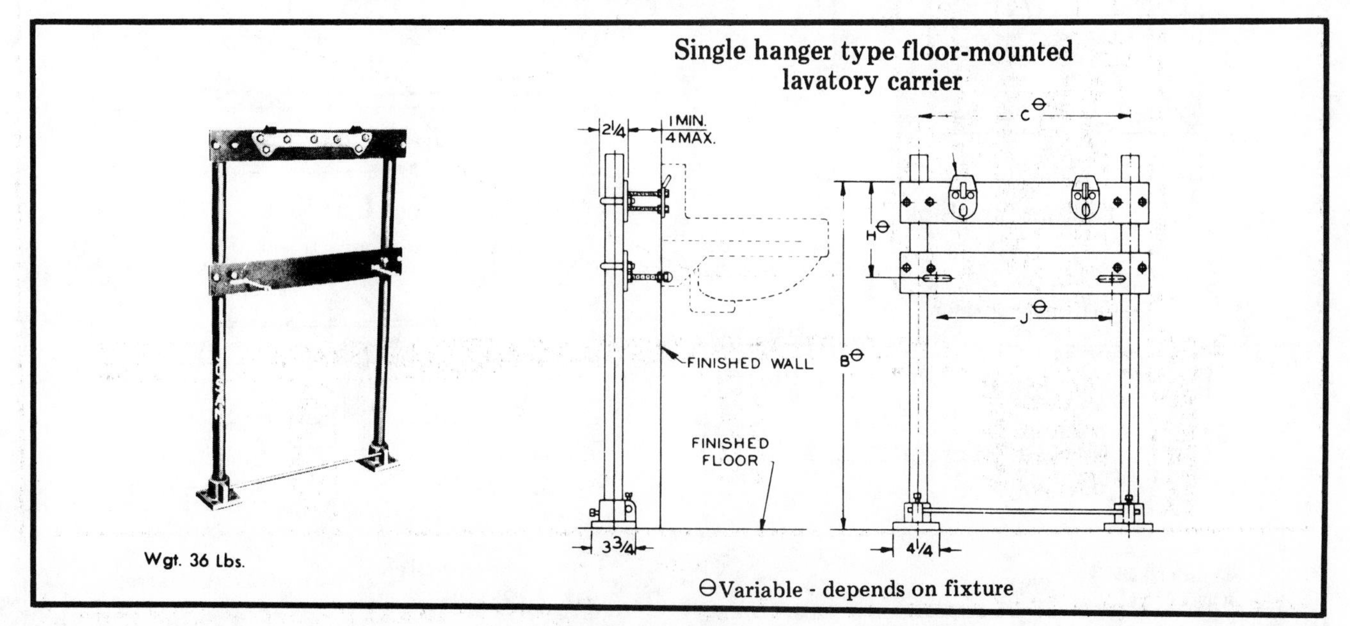

Figure 16-26

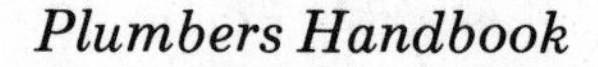

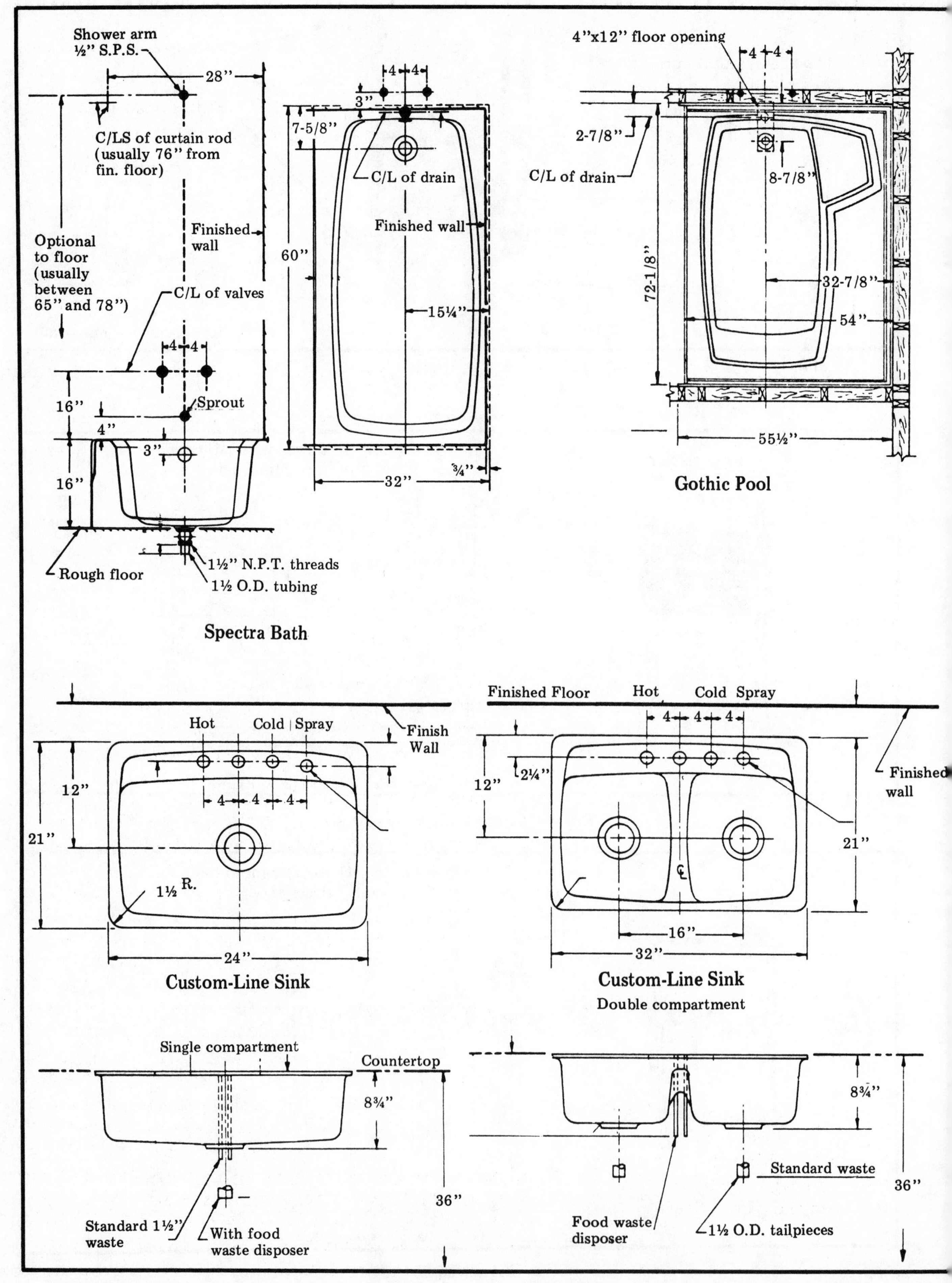
Shower arm
½" S.P.S.
28"
C/LS of curtain rod
(usually 76" from
fin. floor)
Finished
wall
Optional
to floor
(usually
between
65" and 78")
C/L of valves
4
4
16"
Sprout
4"
3"
16"
Rough floor
1½" N.P.T. threads
1½ O.D. tubing
3"
7-5/8"
C/L of drain
Finished wall
60"
15¼"
32"
¾"
4"x12" floor opening
2-7/8"
C/L of drain
8-7/8"
72-1/8"
32-7/8"
54"
55½"
Gothic Pool
Spectra Bath
Hot
Cold
Spray
Finish
Wall
12"
21"
1½ R.
24"
Custom-Line Sink
Finished Floor
2¼"
Finished
wall
16"
32"
Custom-Line Sink
Double compartment
Single compartment
Countertop
8¾"
36"
Standard 1½"
waste
With food
waste disposer
Food waste
disposer
Standard waste
1½ O.D. tailpieces

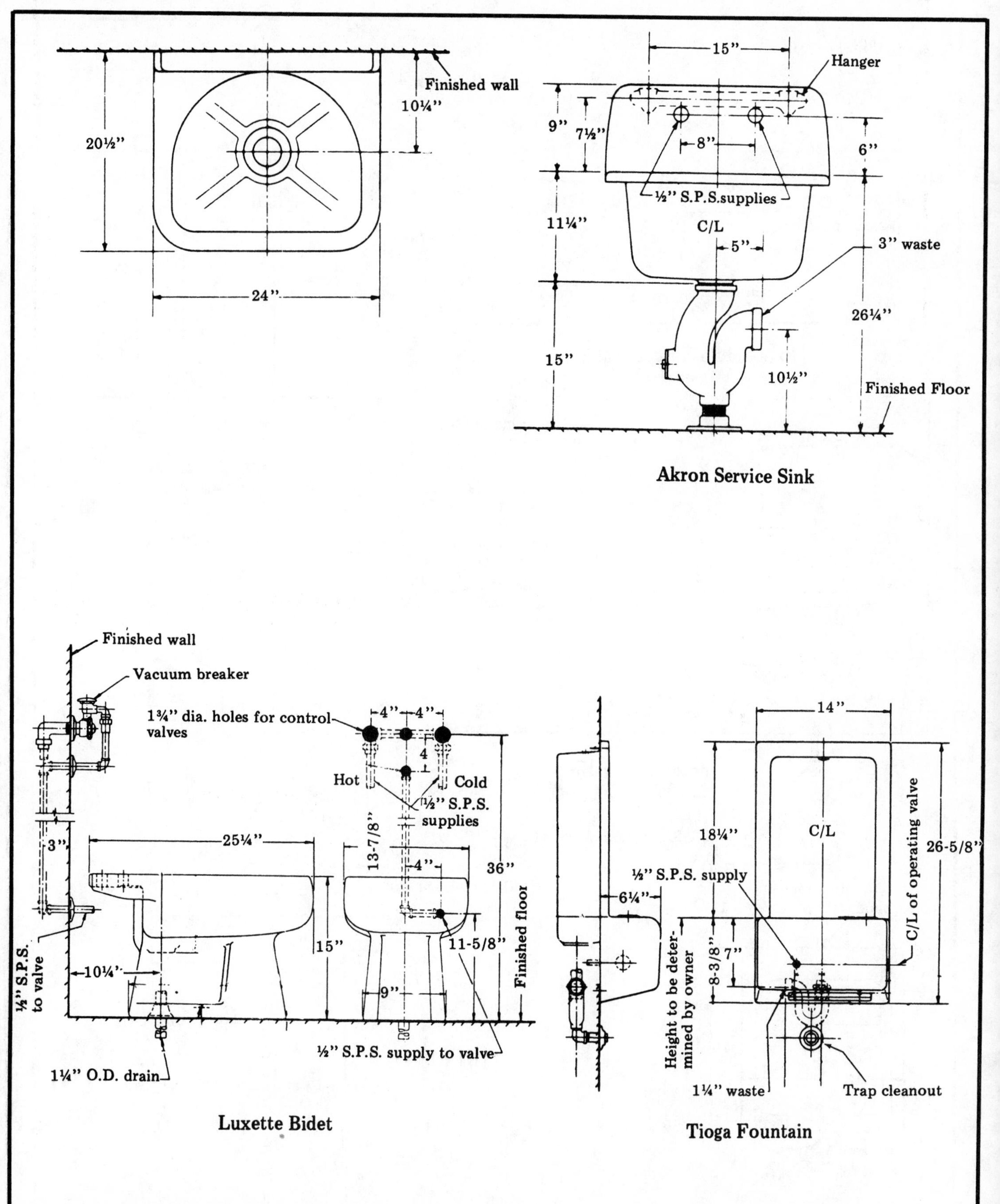

Akron Service Sink

Luxette Bidet

Tioga Fountain

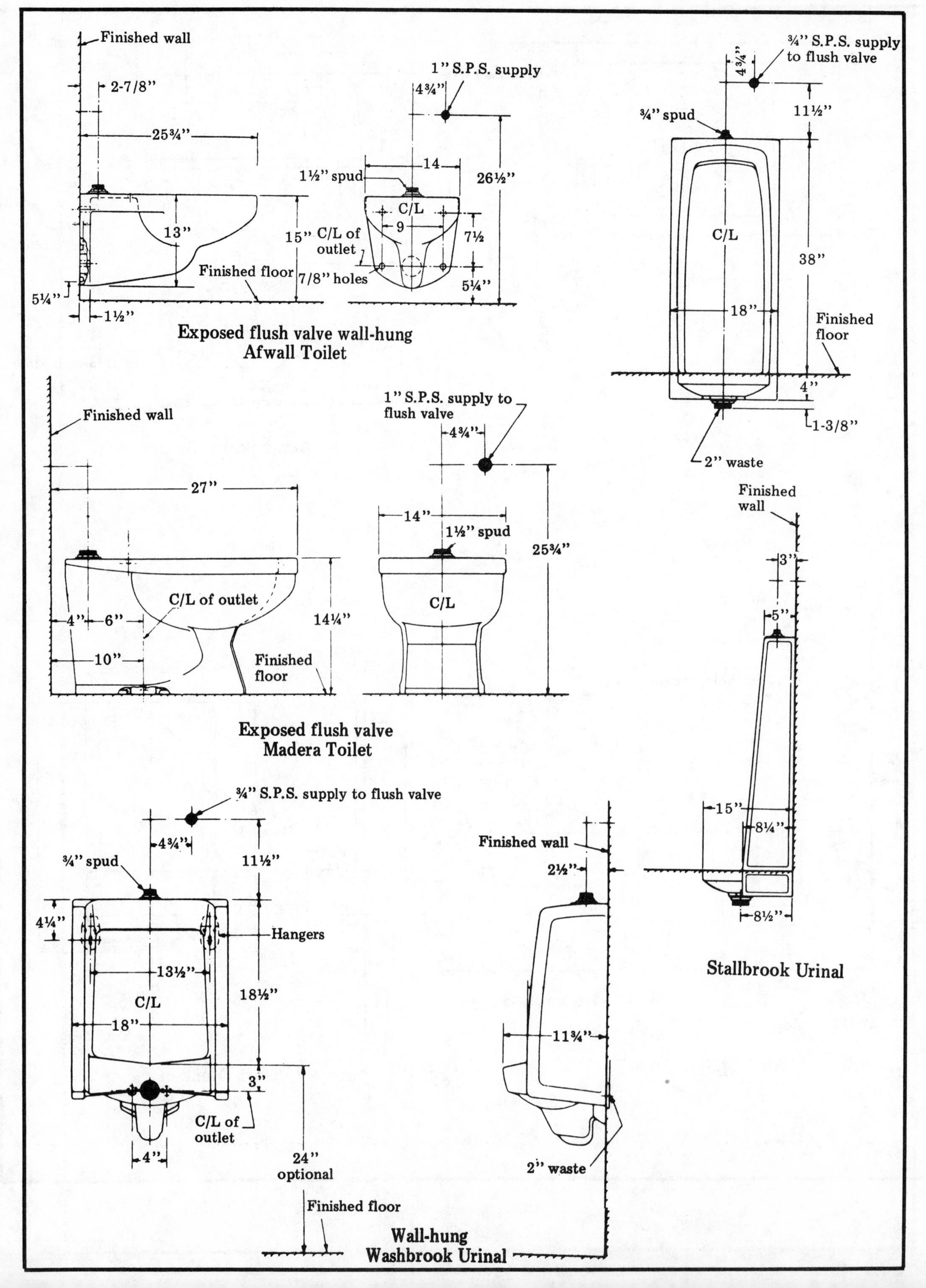

Exposed flush valve wall-hung Afwall Toilet

Exposed flush valve Madera Toilet

Wall-hung Washbrook Urinal

Stallbrook Urinal

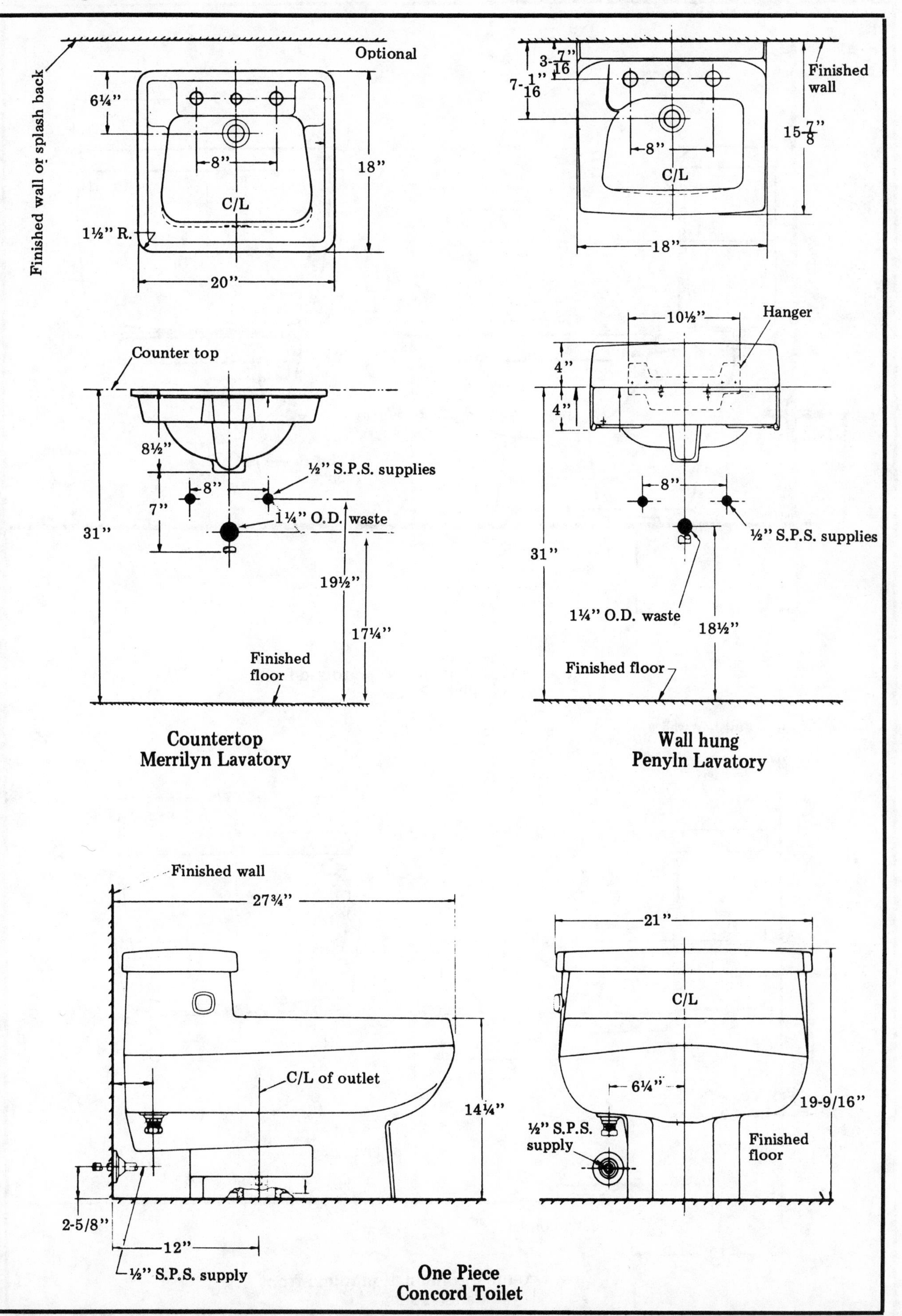

Countertop
Merrilyn Lavatory

Wall hung
Penyln Lavatory

One Piece
Concord Toilet

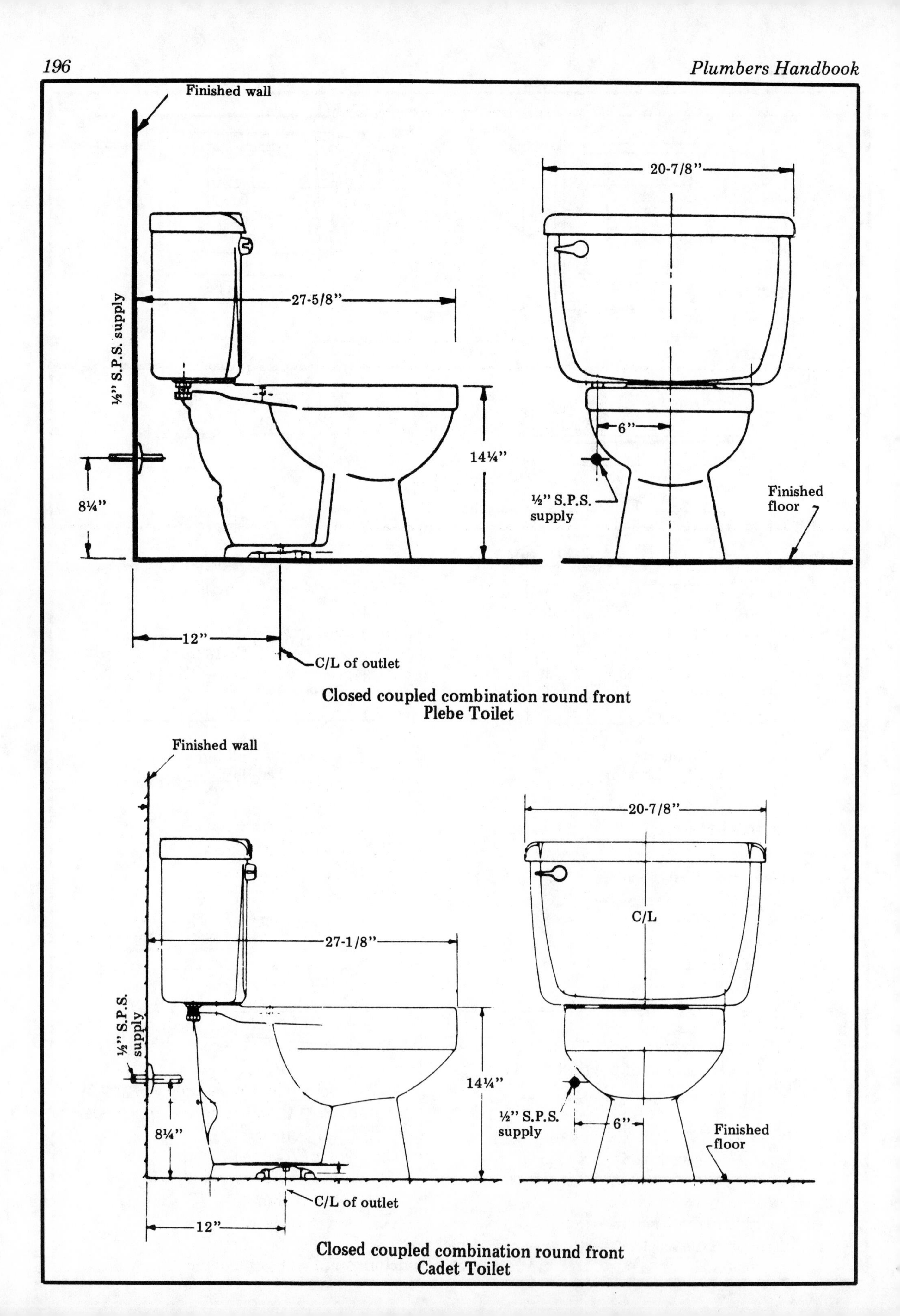

Closed coupled combination round front
Plebe Toilet

Closed coupled combination round front
Cadet Toilet

Examination Day

This chapter is a 210-question test which will help you evaluate your understanding of the plumbing code and better prepare you for the journeyman's or master's examination. The multiple choice questions here are the same type but not the same questions that are on the examination. This chapter should help you locate areas where you need additional study. Answers to the questions are at the end of the chapter.

The topics covered in this chapter include drainage, waste, vent piping and fittings, private disposal systems, trailer park requirements, public and private water systems, swimming pools, fire standpipe systems, solar energy systems, gas systems, fixture requirements and mathematical problems.

You will get the greatest benefit from this chapter by putting your answers on a *separate* sheet of paper and completing the test before checking the answer pages. When you know which answers you missed, review the section of the book that covered that topic. Then review all the questions you missed until you could answer every question correctly.

MULTIPLE CHOICE QUESTIONS

1- The primary function of a relief vent is to
- (A) prevent back siphonage.
- (B) supply fresh air to a bathroom.
- (C) vent the water heater to prevent its explosion.
- (D) provide circulation of air between drainage and vent systems.

2- All piping passing under the footings of a building must have a clearance of at least_____ inches between the top of the pipe and the bottom of the footing.

(A) 2 (B) 4 (C) 1 (D) 6

3- Except when deeper seals are required for interceptors, a fixture trap must have a water seal of between
- (A) one and three inches.
- (B) one and four inches.
- (C) two and four inches.
- (D) three and five inches.

4- A waste stack may receive the discharge from
- (A) lavatory and urinal.
- (B) bed pan washer and lavatory.
- (C) urinal and bed pan washer.
- (D) all of these.

5- Fixture trap inlets measured vertically from the bottom of the fixture to the top of the trap seal shall not exceed
- (A) twelve inches.
- (B) eighteen inches.
- (C) fifteen inches.
- (D) twenty inches.

6- Horizontal wet vents shall

(A) receive discharge from fixture branches only.
(B) exceed fifteen feet.
(C) never connect to a vertical wet vent more than six feet in length.
(D) none of these.

7- The type of non-metallic pipe code approved for use in building sewer installations is
(A) asbestos cement.
(B) bituminous fiber.
(C) vitrified clay.
(D) both B and C.

8- Which one of the following is *not* suitable for use as underground vent piping?
(A) cast iron soil pipe.
(B) lead pipe.
(C) galvanized steel pipe.
(D) brass pipe.

9- All cleanouts must be accessible, and have a clearance of
(A) six inches.
(B) eighteen inches.
(C) twelve inches.
(D) twenty four inches.

10- In a sanitary drainage system, the smallest pipe diameter allowable for soil stack, that carries no waste from urinals or bed pan washers, is
(A) two inches.
(B) two and one-half inches.
(C) three inches.
(D) four inches.

11- Which one of the following may *not* be used in a drainage system?
(A) 45 degree wyes
(B) traps
(C) supports
(D) running threads

12- Materials approved by the code for fire standpipes include
(A) galvanized steel pipe
(B) cast iron pipe
(C) lead pipe
(D) all of these

13- Pneumatic water supply tanks must be equipped with an air volume control valve to
(A) remove excessive air in the tank.
(B) prevent water hammer.
(C) prevent rapid on-off operation of the pump.
(D) prevent air from getting in the pipes of the plumbing system.

14- Rainwater pipes within a building shall be tested with a minimum of______feet of head pressure
(A) 4 (B) 5 (C) 7 (D) 10

15- The suction line from the water supply well to the pump (if less than 40 feet) shall be a minimum of
(A) 1 inch.
(B) 1¼ inches.
(C) 1½ inches.
(D) 2 inches.

16- A boiler outlet to permit emptying or discharging of sediment, is defined as a
(A) safety valve.
(B) blow-off.
(C) drain pipe.
(D) relief valve.

17- The maximum temperature at which waste water can be discharged into a building drainage system is
(A) 90 degrees.
(B) 125 degrees.
(C) 140 degrees.
(D) 180 degrees.

18- The required minimum cover over vitrified clay sewer pipe is
(A) 6 inches.
(B) 12 inches.
(C) 18 inches.
(D) 24 inches.

19- The water service pipe is the pipe from
(A) the water main to the building served.
(B) other source of water supply to the building served.
(C) none of these.
(D) both of these.

20- A sump is a tank or pit located below the normal grade of the gravity system which must be emptied by mechanical means. Its primary function is to receive

(A) waste from floor drains only.
(B) clear water waste only.
(C) waste containing chemicals in solution.
(D) sewage in general.

21- All alterations, repairs, or extensions which include more than_____feet in length of piping and fittings shall be inspected and tested before final approval.
(A) 8 (B) 10 (C) 12 (D) 16

22- Upon the completion of the entire water distribution system, it shall be tested, inspected and proved tight under a water pressure of not less than______working pressure under which it is to be used
(A) the minimum
(B) the maximum
(C) 60 pounds
(D) 75 pounds

23- Water shock or hammer in a water supply system will be the cause of______on final inspection.
(A) too much pressure
(B) a loose washer
(C) condemnation
(D) a defective faucet

24- No drainage or plumbing system or part thereof shall be covered until it has been
(A) tested.
(B) inspected.
(C) approved.
(D) all of these.

25- It shall be the duty and responsibility of the _____to determine if the plumbing has been inspected before it is covered or concealed.
(A) general contractor
(B) permit holder
(C) plumbing inspector
(D) record control

26- If on inspection and tests any plumbing work shows defects, the defective work or materials shall be replaced within_____days and inspected and the test repeated.
(A) two (B) three (C) four (D) five

27- When a house water system is connected to both a well supply and a public water supply the condition should be corrected by
(A) installing a by-pass line.
(B) installing a backflow preventer on the well side.
(C) installing a check valve on the house side.
(D) disconnecting the well supply and capping it off.

28- Indirect waste piping installed below a slab must have a minimum diameter of
(A) ¾ inch.
(B) 1¼ inches.
(C) 1½ inches.
(D) 2 inches.

29- The maximum height of a fire standpipe hose station valve above the finished floor shall be
(A) 60 inches.
(B) 66 inches.
(C) 72 inches.
(D) 76 inches.

30- The minimum size for water supply serving 2½ inch standpipes shall be
(A) 2½ inches.
(B) 3 inches.
(C) 4 inches.
(D) none of these.

31- A "battery of fixtures" is any group of_____ or more similar adjacent fixtures which discharge into a common horizontal waste or soil branch.
(A) two (B) four (C) six (D) all of these

32- A vent connecting one or more individual vents with a vent stack or stack vent is defined as a
(A) relief vent.
(B) branch vent.
(C) wet vent.
(D) common vent.

33- A building drainage system which cannot drain by gravity into the building sewer is defined as a
(A) subdrain.
(B) conductor.
(C) main.
(D) branch.

34- A combined building sewer receives

(A) storm water.
(B) sewage.
(C) liquid waste.
(D) all of these.

35- A common vent is installed to vent
(A) two or more water closets.
(B) a service sink with a three inch trap only.
(C) two fixture drains installed at the same level in a vertical stack.
(D) the last two fixtures on a horizontal drainage system.

36- For the purpose of the code, one fixture unit flow rate shall be deemed to be
(A) one cubic foot per minute.
(B) four and five-tenths gallons per minute.
(C) five and one-half gallons per minute.
(D) seven and one-third gallons per minute.

37- A horizontal waste line may be connected to the vertical section of a waste stack by using
(A) a saddle tee.
(B) a test tee.
(C) a single sanitary tee.
(D) none of these.

38- Piping passing through cast in place concrete shall be protected by
(A) painting pipe with asphaltum paint.
(B) wrapping pipe with air conditioning tape.
(C) sleeving to give one-half inch annular space around the entire circumference of the pipe.
(D) using type K copper only.

39- Air conditioning equipment may discharge directly or indirectly into
(A) rain water leaders which discharge into any surface gutter.
(B) a vent stack having a minimum size of three inches.
(C) a fixture tail piece, provided it is larger than 1¼ inches.
(D) none of these.

40- Plastic pipe and fittings to support the weight of plumbing fixtures is permitted, providing
(A) the fixture does not weigh more than twelve pounds.
(B) the building is a single family residence.
(C) prior approval is obtained from the plumbing inspector.
(D) none of these.

41- All underground soil, waste and vent piping and fittings within a building located over deleterious fill shall be
(A) lead pipe.
(B) centrifugally spun service weight cast iron pipe.
(C) brass pipe.
(D) all of these.

42- Soil, waste and vent piping above ground within buildings over deleterious fill may be
(A) galvanized pipe.
(B) plastic pipe.
(C) copper type K, L, or DWV.
(D) all of these.

43- Plastic piping from a septic tank to a drainfield located under paving shall be
(A) ABS schedule 40.
(B) PVC schedule 40.
(C) PVC schedule 80.
(D) CPVC piping only.

44- A dilution tank is required for corrosive waste which has a PH of
(A) 4.5 (B) 5.0 (C) 6.0 (D) more than 7.0

45- Fixture unit value as load factors for special fixtures are determined by
(A) the size of the fixture trap.
(B) type of fixture.
(C) the manufacturer's suggested load factor.
(D) location of fixture.

46- A building sewer when connected to a septic tank may be considered and sized as a building drain if the developed length does not exceed
(A) 5 feet. (B) 8 feet. (C) 10 feet. (D) 12 feet.

47- No domestic kitchen sink may be installed on a waste stack less than
(A) 1½ inches.
(B) 2 inches.
(C) 2½ inches.
(D) all of these.

48- Sumps and receiving tanks for liquid waste shall be sized to retain a______peak flow.

(A) 20-minute
(B) 30-minute
(C) 45-minute
(D) one hour

49- Ejector pumps shall be provided with a
(A) check valve on discharge side of gate valve.
(B) gate valve only.
(C) check valve only.
(D) check valve located on the pump side of gate valve.

50- Not more than__________water closet(s) shall discharge into a three inch stack at the *same point.*
(A) one (B) two (C) three (D) four

51- Not more than______water closet(s) shall discharge into a three inch stack at the *same level.*
(A) one (B) two (C) three (D) four

52- When a clothes washing machine utilizes one side of a tap cross with a domestic kitchen sink, the waste pipe size shall be no less than
(A) 2 inches.
(B) 2½ inches.
(C) 3 inches.
(D) 4 inches.

53- Sumps receiving body waste from plumbing fixtures must have a minimum
(A) 2 inch vent.
(B) 2½ inch vent.
(C) 3 inch vent.
(D) 4 inch vent.

54- Sump vents may
(A) extend independently to above the roof.
(B) be connected to the plumbing system.
(C) connect to nearest vent stack.
(D) all of these.

55- The minimum size in diameter for pipe used for subsoil drains when placed under the cellar or basement floor is
(A) 2½ inches.
(B) 3 inches.
(C) 4 inches.
(D) 6 inches.

56- Air conditioning condensate drains may connect
(A) directly to a rain water leader pipe.
(B) by indirect means to the building drainage system.
(C) to any waste pipe receiving clear water waste only.
(D) none of these.

57- Drip pipes from walk-in refrigerator floors or storeroom floors where food is stored shall be
(A) carefully installed to drain dry.
(B) installed in a workmanlike manner.
(C) installed as a direct waste.
(D) installed as an indirect waste.

58- Walk-in refrigerator floors or store-room floors where food is stored shall be______above overflow point of receiving fixture.
(A) 1 inch (B) 2 inches (C) 3 inches (D) 4 inches

59- Air conditioning condensate drains for units with not more than 5 tons capacity may discharge
(A) upon a pervious area.
(B) into a soakage pit.
(C) into a building drainage system.
(D) all of these.

60- Air conditioning drains of PVC shall be a minimum of______below bottom of the slab.
(A) 2 inches
(B) 4 inches
(C) 6 inches
(D) 12 inches

61- For concrete sewer pipe, approximately____ percent of the joint at the base of the socket shall be filled with jute or hemp.
(A) 10 (B) 15 (C) 25 (D) 30

62- Mortar for cement joints shall be composed of
(A) two parts cement, one part sand.
(B) one part cement, two parts sand.
(C) two parts cement, two parts sand.
(D) three parts cement, one part sand.

63- A trap depending upon movable parts to retain its seal may be used
(A) for fixtures with clear water waste only.
(B) for fixtures having integral traps only.
(C) for both of these.
(D) none of these.

64- Where commercial food grinders are installed, the waste from those units shall discharge
(A) into a grease interceptor.
(B) into a floor drain.
(C) directly into building drainage system.
(D) all of these.

65- In a building drain or branch drain cleanouts are required every
(A) 25 feet.
(B) 50 feet.
(C) 75 feet.
(D) 100 feet.

66- Cleanouts shall be the same nominal size as the pipe into which they are installed up to_____ inches.
(A) 6 (B) 8 (C) 10 (D) 12

67- Vertical plastic pipe shall be supported at the following intervals, each
(A) four feet.
(B) six feet.
(C) story.
(D) two stories.

68- Horizontal lead joint soil pipe exceeding five feet in length, may be supported at not more than______intervals.
(A) four-foot
(B) six-foot
(C) eight-foot
(D) ten-foot

69- All extensions of soil, waste and vent stacks shall terminate at least_____inches above the roof.
(A) four
(B) six
(C) eight
(D) twelve

70- Where roofs are used for sun decks, solariums or similar purposes all vents shall extend not less than______above the deck.
(A) six inches
(B) thirty-six inches
(C) sixty inches
(D) eighty-four inches

71- Vent pipes shall be graded to:
(A) prevent air lock.
(B) help circulate the air.
(C) drain dry.
(D) prevent noise.

72- Which of the following may not discharge into a horizontal 3 inch wet vent?
(A) bidet
(B) water closet
(C) shower
(D) bathtub

73- Which of the following vents is prohibited?
(A) yoke
(B) crown
(C) common
(D) back

74- The pipe or dry section of a circuit vent may have a diameter of how many pipe sizes less than the diameter of the pipe of the horizontal soil drain it serves?
(A) none
(B) one
(C) one and one-half
(D) two

75- On a loop system a vent shall be installed vertically downstream, and be within______feet of the first fixture branch.
(A) five
(B) seven
(C) eight
(D) ten

76- The diameter of an individual vent shall be not less than______inch(es).
(A) 1 (B) 1¼ (C) 1½ (D) 2

77- A combination waste and vent stack with a diameter of four inches may receive the discharge from
(A) two water closets
(B) four water closets
(C) six water closets
(D) none of these

78- A combination waste and vent stack with a diameter of two inches, that does not exceed 30 feet in length, shall *not* receive the discharge from
(A) four domestic lavatories
(B) two bathtubs
(C) one kitchen sink
(D) three bidets

79- The minimum size vent stack allowable for an accessory building on a building site connected by a common building sewer, when a water closet is installed in the accessory building is
(A) 2 inches.
(B) 2½ inches.
(C) 3 inches.
(D) 4 inches.

80- The vent terminal of a sanitary plumbing system shall not be less than_______feet from the point of any mechanical air intake opening.
(A) six
(B) eight
(C) ten
(D) twelve

81- The vent terminal of any vent pipe if within ten feet of any door, window, or ventilating opening, shall extend not less than_______feet above such door, window, or ventilating opening.
(A) one
(B) two
(C) three
(D) all of these

82- Vertical piping shall be secured at sufficiently close intervals to cause the pipe_______
(A) to stay in alignment.
(B) to carry the pipe's weight.
(C) to carry the weight of the contents.
(D) all of these.

83- Trap cleanouts are prohibited on
(A) lavatory traps.
(B) barber shop sinks.
(C) concealed traps.
(D) kitchen sinks.

84- Rainwater shall be disposed of by
(A) using adjoining pervious property.
(B) the use of a soakage pit.
(C) flowing across a public sidewalk into a gutter.
(D) none of these.

85- Where it is necessary to install a pipe to carry rainwater located under a sidewalk to a street gutter, the minimum concrete cover over such pipe shall be
(A) 1 inch.
(B) 1½ inches.
(C) 2 inches.
(D) 2½ inches.

86- Rain or storm water drains shall be installed
(A) at 1/8 inch fall per foot.
(B) to drain dry.
(C) with a flapper valve.
(D) with a back water valve.

87- When sizing storm sewers, drains, gutters and leaders, where a vertical wall extends above the roof area in such a manner as to drain into the area being considered, then_______of the area of the vertical wall shall be added to the horizontal projection.
(A) 10 percent
(B) 25 percent
(C) 33⅓ percent
(D) 50 percent

88- For the disposition of rainwater on each building site a minimum ratio of one square foot of effective pervious area for each_______square feet of impervious area shall be provided.
(A) five (B) eight (C) ten (D) twelve

89- Which of the following materials shall *not* be used for storm drainage within a building?
(A) bituminized fiber pipe
(B) galvanized steel pipe
(C) plastic pipe
(D) lead pipe

90- Asbestos cement pipe where permitted to be used shall be protected by
(A) encasing it in four inches of concrete.
(B) installing it a minimum of five feet from building's foundation.
(C) using only approved fittings.
(D) having a minimum cover of 12 inches.

91- The use of asbestos cement pipe shall be limited to
(A) storm sewers.
(B) acid waste.
(C) rain leaders.
(D) none of these.

92- The terms "interceptor" and "separator" may be used interchangeably and may be prefaced by a term indicating the
(A) use.

(B) material separated.
(C) location.
(D) all of these.

93- Gasoline, oil and sand interceptors are required
(A) any place where motor vehicles are repaired.
(B) in public storage garages.
(C) in assembly plants.
(D) where floor drainage is to be provided.

94- When, in the opinion of the plumbing inspector, grease can be introduced into the drainage system from a restaurant sink in such quantity as to effect line stoppage or hinder sewage disposal, he will require that the waste line include
(A) a filter.
(B) a bypass.
(C) a sump.
(D) an interceptor.

95- Interceptors for a commercial laundry must be maintained in efficient operating condition by periodic
(A) removal of accumulated contents.
(B) replacement of "limited use" parts.
(C) flushing with chemical mixtures.
(D) specified combination of all the above.

96-The overflow from a fixture shall be connected to
(A) the crown vent.
(B) the fixture branch.
(C) a drip pan.
(D) the house side of the fixture trap.

97- From any finished or stall compartment wall a lavatory shall be spaced a minimum of______ inch(es).
(A) one (B) two (C) three (D) four

98- Where bucket type floor drains are required the minimum diameter of its outlet shall be_____
(A) 2 inches.
(B) 3 inches.
(C) 4 inches.
(D) all of these.

99- When a water closet is installed next to a bath tub the minimum distance from the center of the bowl to the edge of a tub shall be________
(A) 10 inches.
(B) 12 inches.
(C) 14 inches.
(D) 16 inches.

100- Plumbing fixtures shall be constructed of approved materials. Which of the following materials are *not* acceptable?
(A) china.
(B) cultured marble.
(C) materials that are impervious.
(D) materials that are pervious.

101- Water closets installed for public use must have a
(A) regular closed front seat with or without cover.
(B) regular open front seat with or without cover.
(C) elongated closed front seat with or without cover.
(D) elongated open front seat with or without cover.

102- When sheet lead is used for a shower pan it shall weigh not less than__________pounds per square foot.
(A) 2 (B) 3 (C) 4 (D) 8

103- When lead is used for lead bends, lead stubs or lead straps it shall weight not less than ______________pounds per square foot.
(A) 2 (B) 3 (C) 4 (D) 8

104- Lead and copper shower pans shall be protected against the corrosive effects of concrete by__________.
(A) coating the inside with asphaltum paint.
(B) coating the outside with asphaltum paint.
(C) coating the inside and outside with asphaltum paint.
(D) all of these.

105- Built-in tubs with overhead showers shall have________.
(A) an eight inch shower arm.
(B) a waterproof joint between the tub and wall.
(C) a minimum of four foot tile walls.
(D) either a shower door or curtain.

106- No floor drain or other plumbing fixture

shall be installed in a room__________.
(A) containing air handling machinery.
(B) used for sleeping.
(C) used for the storage of food.
(D) used for recreation purposes.

107- Floor drains serving indirect waste pipes from food or drink storage rooms shall *not* be installed in any__________.
(A) toilet room.
(B) unventilated room.
(C) both of these.
(D) none of these.

108- Floor drains sized four inches may not require a vent if installed within________feet of a vented sewer line when measured horizontally.
(A) 5 (B) 10 (C) 15 (D) 20

109- All garbage or trash chutes shall be supplied with__________.
(A) an odor proof door.
(B) a flushing ring.
(C) a roof jack.
(D) a cleanout.

110- Floor drain trap size may be
(A) 2 inches.
(B) 2½ inches.
(C) 3 inches.
(D) all of these.

111- Unless the plumbing inspector rules otherwise, every floor drain trap which is directly connected to the drainage system shall be provided with a permanent water seal which can be fed from
(A) a drinking fountain.
(B) an ice maker.
(C) an automatic priming device.
(D) all of these.

112- When may previously used piping material be reused in a potable water supply system?
(A) at any time.
(B) when its previous use was for potable water supply system.
(C) when approved by a plumbing inspector.
(D) at no time.

113- Pressure rated plastic service piping shall have a minimum working pressure of________.
(A) street level pressure.
(B) 75 pounds per square inch.
(C) 100 pounds per square inch.
(D) 160 pounds per square inch.

114- Lawn sprinkler systems using potable water shall be equipped with an approved
(A) gate valve.
(B) back-flow preventer.
(C) check valve.
(D) ground joint union.

115- Which of the following hose connected faucets are *not* required to have a back-flow preventer installed?
(A) three compartment commercial sink.
(B) service sink.
(C) automatic clothes washing machine.
(D) outside hose faucet.

116- Plastic water service piping shall have a minimum cover of
(A) 6 inches.
(B) 8 inches.
(C) 10 inches.
(D) 12 inches.

117- The minimum liquid capacity of a septic tank serving a two-bedroom, three-bath residence is________.
(A) 600 gallons.
(B) 750 gallons.
(C) 800 gallons.
(D) 900 gallons.

118- Cast-in-place septic tanks located in a parking lot shall be designed to support the anticipated load but not less than that of a_____.
(A) two-ton truck.
(B) four-ton truck.
(C) six-ton truck.
(D) ten-ton truck.

119- The minimum distance from a drainfield to a basement wall shall be________.
(A) five feet.
(B) ten feet.
(C) twelve feet.
(D) none of these.

120- Where a soakage pit is installed in the vicinity of a septic tank drainfield, distance separation__________.

(A) is not important.
(B) shall be a minimum of five feet.
(C) shall be a minimum of ten feet.
(D) shall be a minimum of fifteen feet.

121- Where reservoir-type drainfields are used, the maximum distance between centers of distribution lines shall not exceed________foot (feet).
(A) one (B) two (C) three (D) four

122- The suction line from a potable water supply well serving a single family residence shall have a union installed before
(A) the pump.
(B) the hair strainer.
(C) the check valve.
(D) none of these.

123- Effluent or drainage from a home clothes washing machine may be disposed of by draining__________.
(A) upon a pervious area on private property.
(B) into a 50 gallon drum filled with ¾ rock.
(C) into an adequate sized soakage pit.
(D) through a sanitary system connected to a septic tank.

124- The minimum size hydropneumatic tank for a single family residence shall be
(A) 20 gallons.
(B) 30 gallons.
(C) 42 gallons.
(D) 50 gallons.

125- A disposal well for roof water shall be preceded by a
(A) settling tank.
(B) running trap.
(C) cleanout tee.
(D) overflow fitting.

126- A four inch concrete pad shall be poured around the well casing and shall extend______ inches on all sides.
(A) eight
(B) ten
(C) fourteen
(D) eighteen

127- Maximum length of a single tile drainfield lateral shall be__________feet.
(A) 20 (B) 30 (C) 40 (D) 50

128- Horizontal gas piping of 1¼ inches and larger shall be supported every__________feet.
(A) four (B) six (C) eight (D) ten

129- Which of the following locations is *not* acceptable for installation of gas piping within a building?
(A) under a concrete floor.
(B) a solid partition.
(C) a hollow partition.
(D) an elevator shaft.

130- Branch outlet pipes for a gas distribution system shall not be taken from the__________of horizontal lines.
(A) top
(B) side
(C) bottom
(D) all of these

131- Plastic pipe, tubing and fittings used for the installation of gas piping shall be joined by
(A) heat-fusion method.
(B) solvent cement method.
(C) adhesive method.
(D) all of these.

132- Equipment that requires mobility during operation within a building may use indoor gas hose connectors, providing the length does not exceed__________feet.
(A) six (B) eight (C) ten (D) twelve

133- For fixtures, other than direct flush valves, minimum service pressure at the point of discharge shall be not less than__________.
(A) 4 psi.
(B) 6 psi.
(C) 8 psi.
(D) none of these.

134- Combination pressure and temperature valves shall be installed so that the temperature sensing element is within______________of the top of the tank.
(A) 2 inches
(B) 4 inches
(C) 6 inches
(D) 8 inches

135- The entire pool pressure piping system,

when ready for inspection, shall be tested with a water test of__________.
(A) 10 psi (B) 20 psi (C) 30 psi (D) 40 psi

136- Pressure discharge line from a domestic water heater may *not* discharge into what fixtures?
(A) service sink.
(B) pot sink.
(C) shower.
(D) all of these.

137- The minimum size for a domestic water heater relief valve discharge shall be:
(A) 3/8 inch I.D.
(B) 3/8 inch O.D.
(C) 1/4 inch I.D.
(D) 1/4 inch O.D.

138- Urinal traps and floor drains installed downstream from a water closet or closets in a circuit vent group shall be________inches in size.
(A) 1½ (B) 2 (C) 3 (D) 4

139- All standpipes in buildings 50 feet or more in height shall extend full size above the main roof a minimum of_________inches.
(A) 10 (B) 20 (C) 30 (D) 40

140- The minimum number of fire standpipes which must extend through the roof of a building having four standpipes are
(A) one. (B) two. (C) three. (D) four.

141- Clear liquid waste is the discharge from
(A) urinals.
(B) drinking fountains.
(C) dental chairs.
(D) none of these.

142- The term "sewage" as defined by the code would *not* include
(A) liquid waste containing animal matter in suspension.
(B) liquid waste containing minerals in solution.
(C) rain water.
(D) liquids containing chemicals in solution.

143- Standpipes and fittings above ground, when placed within the exterior walls of a building, shall be of sufficient strength to withstand________pounds per square inch water pressure at the topmost outlet.
(A) 55 (B) 75 (C) 100 (D) 150

144- Outside rain leaders, when exposed to contact with vehicles, must have cast iron which extends______________feet above grade.
(A) two (B) three (C) four (D) five

145- Indirect waste pipes when not installed below the floor must have a minimum clearance above such floor of________inch(es).
(A) 1 (B) 2 (C) 3 (D) 4

146- Air conditioning condensate drains above floor shall be a minimum of ¾ inch in diameter for one or more units totaling not over________ tons of refrigeration capacity.
(A) 4 (B) 6 (C) 8 (D) 10

147- Horizontal copper tubing shall be supported at approximately______________ intervals.
(A) four foot
(B) six foot
(C) eight foot
(D) ten foot

148- Residential septic tanks:
(A) shall have a minimum capacity of 500 gallons.
(B) shall not receive kitchen sink waste.
(C) may be constructed of masonry block parged with ½ inch portland cement grout.
(D) shall be rectangular in shape.

149- The "invert" of a 4 inch drain inlet to an oil interceptor basin must be at least_________ above the waste line.
(A) 1 inch
(B) 2 inches
(C) 3 inches
(D) 4 inches

150- The minimum size water service pipe from the meter to the house shall be:
(A) ½ inch minimum.
(B) ¾ inch minimum.
(C) 1 inch minimum.
(D) determined by meter size.

Using Figure 17-1, identify each section of piping as specified in questions 151-154 below.

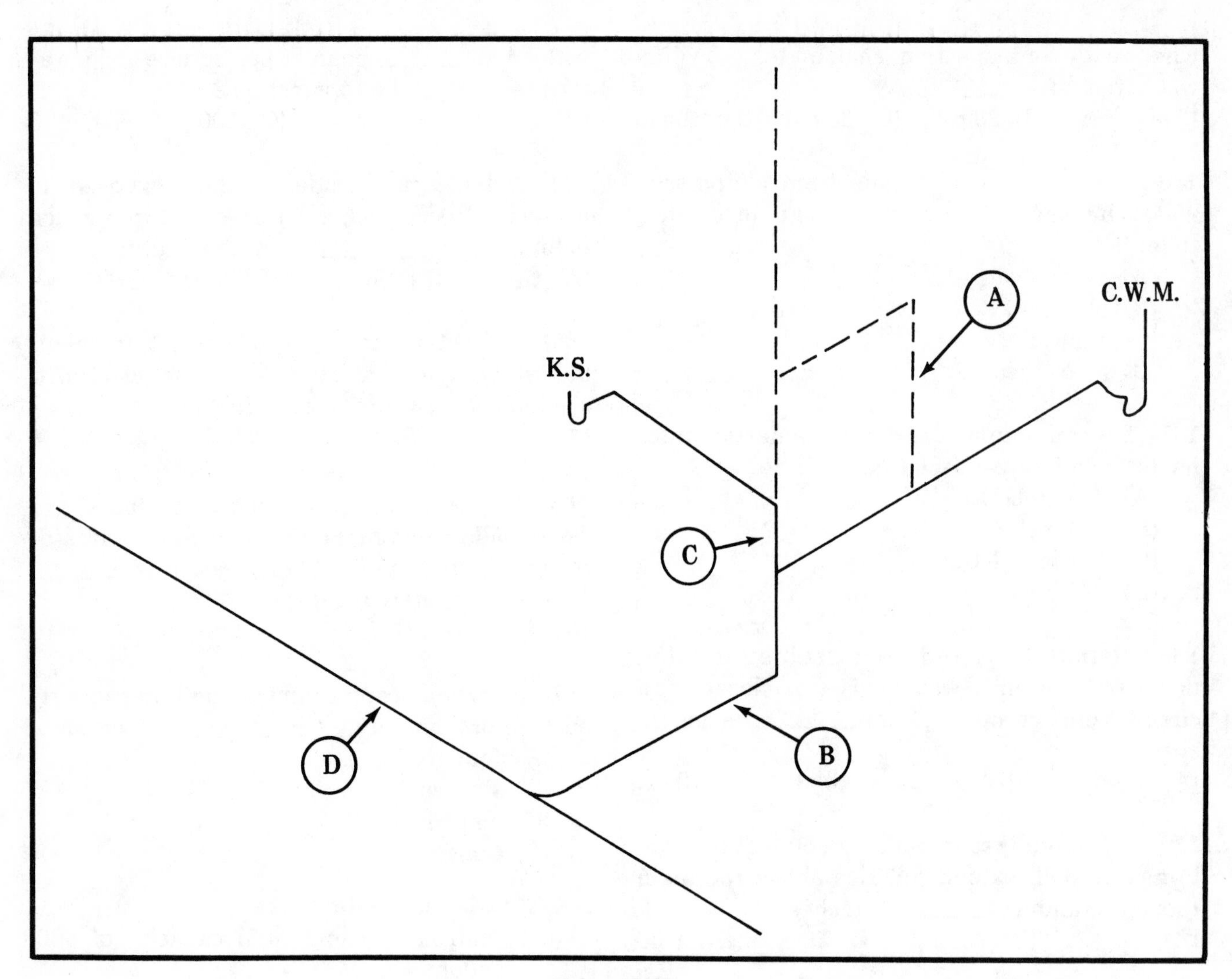

Figure 17-1

151- Section A
(A) island vent
(B) loop vent
(C) relief vent
(D) circuit vent

152- Section B
(A) fixture drain
(B) wet vent
(C) building drain
(D) fixture branch

153- Section C
(A) fixture branch
(B) wet vent
(C) combination waste and vent pipe
(D) dirty waste pipe

154- Section D
(A) soil pipe
(B) waste pipe
(C) building sewer
(D) building drain

Using Figure 17-2, give the fixture units and size the pipes as specified in questions 155-161 below.

155- The combined fixture units at pipe section "A" total is
(A) 12 (B) 21 (C) 25 (D) 28

156- The minimum pipe size at Section "A" is
(A) 2 inches
(B) 3 inches
(C) 4 inches
(D) 6 inches

157- The minimum pipe size at Section "B" is
(A) 1¼ inches

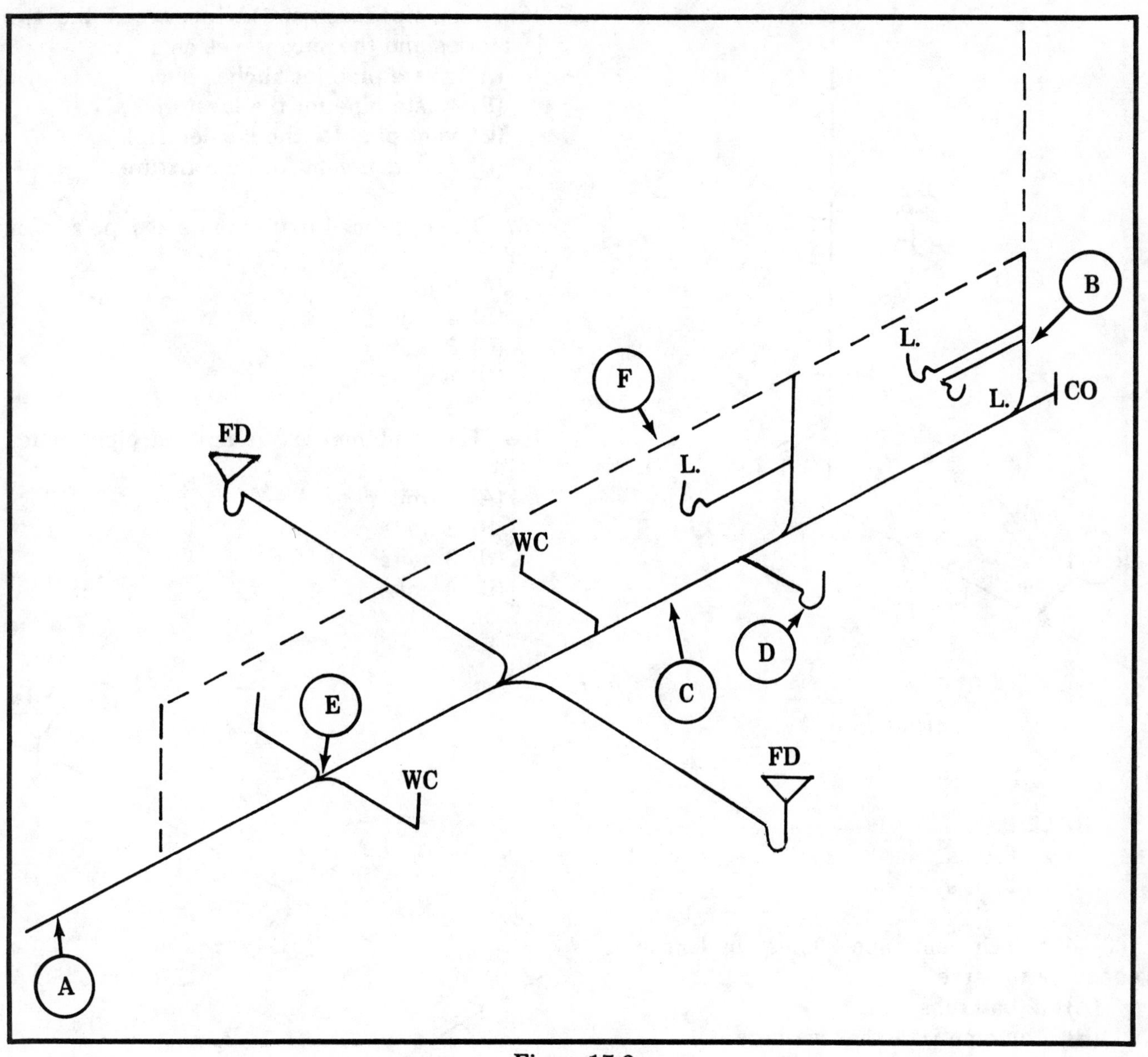

Figure 17-2

(B) 1½ inches
(C) 2 inches
(D) 3 inches

158- The combined fixture units at pipe section "C" total is
(A) 4 (B) 7 (C) 9 (D) 11

159- The minimum size of urinal trap "D" is
(A) 1½ inches
(B) 2 inches
(C) 2½ inches
(D) 3 inches

160- Fitting "E" is known in the trade as a
(A) double combination Y and 1/8 bend
(B) double combination Y and 1/5 bend
(C) double combination Y and 1/16 bend
(D) double drainage wye

161- The minimum size of horizontal vent pipe "F" is
(A) 2 inches
(B) 2½ inches
(C) 3 inches
(D) 4 inches

162- A 3-inch building drain installed at a maximum fall (pitch) of 1/8 inch per foot may receive the discharge from fixtures having a total fixture unit load of
(A) 18 F.U.'s

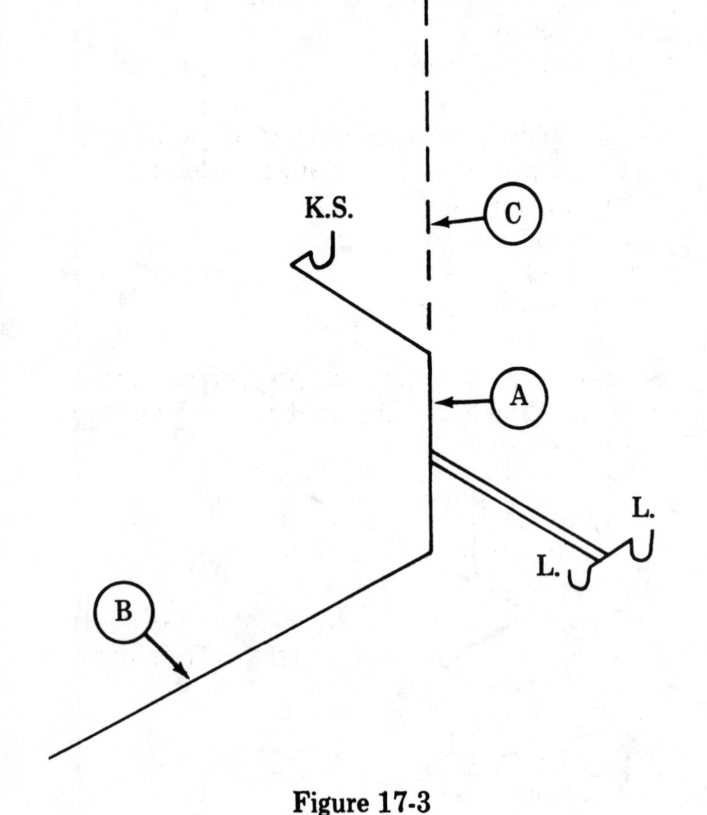

Figure 17-3

(B) 22 F.U.'s
(C) 25 F.U.'s
(D) 28 F.U.'s

163- A 2 inch vent pipe 20 feet in length is adequate to serve
(A) 12 bathtubs
(B) 7 automatic clothes washers
(C) 26 lavatories
(D) 13 showers

164- Identify the one piece of equipment below which may *not* be drained by indirect means.
(A) drinking fountain
(B) hand sink
(C) three compartment glass sink
(D) beverage cooler

165- A clothes washing machine standpipe must have a minimum length of
(A) 10 inches
(B) 14 inches
(C) 18 inches
(D) 28 inches

Use Figure 17-3 to answer questions 166-168.

166- The pipe section "A" between the two lavatories and the sink serves as a
(A) waste pipe for kitchen sink
(B) waste pipe for the lavatories
(C) vent pipe for the kitchen sink
(D) common vent for both fixtures

167- The combined fixture units at pipe section "B" is
(A) 1 unit
(B) 2 units
(C) 3 units
(D) 4 units

168- The combined fixture units at pipe section "C" is
(A) 1 unit
(B) 2 units
(C) 3 units
(D) 4 units

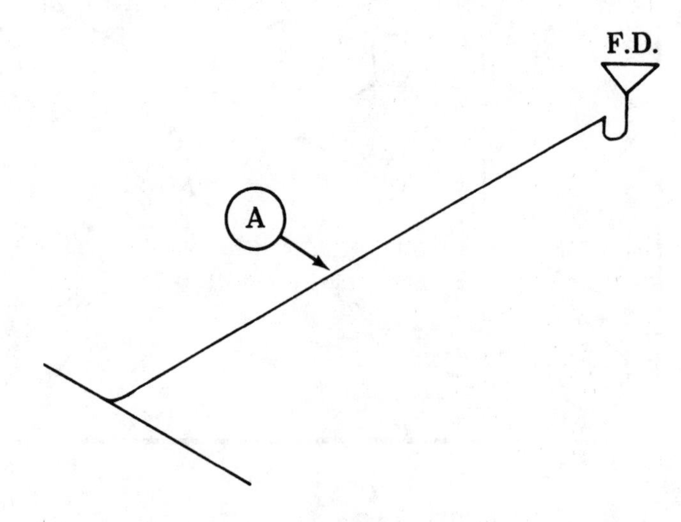

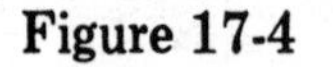
Figure 17-4

169- Using Figure 17-4, what is the maximum length of pipe "A" serving a floor drain from a vented building drain?
(A) 3 feet
(B) 10 feet
(C) 15 feet
(D) 20 feet

170- Lead pipe is permitted to be fully wiped and joined to one of the following materials:
(A) plastic
(B) brass
(C) stainless steel
(D) wrought iron

171- A flaring tool is used to make flare joints in
(A) type K copper pipe
(B) type L copper pipe
(C) type L copper tubing
(D) type M copper pipe

172- Which of the following materials is considered semi-rigid?
(A) type K copper pipe
(B) type L copper pipe
(C) type L copper tubing
(D) type M copper pipe

173- Piping used to convey corrosive gases may be
(A) brass
(B) galvanized steel
(C) type K copper pipe
(D) schedule 80 plastic

174- Vent piping serving a gas water heater that penetrates the roof near a 2 foot parapet wall should be a minimum of
(A) one foot high
(B) two feet high
(C) three feet high
(D) four feet high

175- A water closet using a flushometer valve must have a vacuum breaker located above the rim of the bowl a minimum of
(A) 1 inch
(B) 3 inches
(C) 6 inches
(D) 8 inches

176- A single flushometer valve may be used to serve
(A) 1 water closet
(B) 2 water closets
(C) 1 water closet and 1 urinal
(D) 1 water closet and 2 urinals

177- The code prohibits the use of one of the following materials in a water piping system:
(A) CPVC plastic
(B) brass
(C) type M copper pipe
(D) aluminum pipe

178- Steel fittings used in a water piping system within a building must
(A) be beaded
(B) have standard pipe threads
(C) be plain
(D) be galvanized

179- A building sewer constructed of bituminized fiber pipe may be installed under a
(A) family room
(B) driveway
(C) porch
(D) screened patio

180- The size of a gas supply piping outlet serving any gas appliance must not be less than
(A) 3/8 inch
(B) 1/2 inch
(C) 3/4 inch
(D) the appliance inlet pipe

181- The combustion chamber of a gas fired water heater installed in a private garage must be above the floor level at least
(A) 12 inches
(B) 18 inches
(C) 24 inches
(D) 28 inches

182- Rainwater drains conveying contents from leaders and discharging directly into soakage pits must be provided with
(A) a cleanout
(B) a test tee
(C) a backflow valve
(D) an overflow fitting

183- A sanitary tee may be used to change the direction from
(A) horizontal to horizontal
(B) vertical to horizontal
(C) horizontal to vertical
(D) none of these

184- The minimum air space within a septic tank is
(A) 4 inches
(B) 8 inches
(C) 10 inches
(D) 12 inches

185- Each drain unit for a reservoir-type drainfield serving a 900 gallon septic tank equals
(A) 2 square feet
(B) 2 linear feet

(C) 4 square feet
(D) 4 linear feet

186- Trailer parks must provide minimum laundry facilities of one automatic washer and one 2-compartment laundry tray for each
(A) 15 trailer coach spaces
(B) 25 trailer coach spaces
(C) 35 trailer coach spaces
(D) 45 trailer coach spaces

187- A 4 inch sewer installed at 1/8'' slope per foot, provided it is individually vented, may serve__________trailers.
(A) 10 (B) 20 (C) 30 (D) 40

188- Vent pipes penetrating a roof must be made water-tight by the use of
(A) approved flashing material
(B) approved roofing cement
(C) 30 lb. felt sealed with hot tar
(D) approved caps

189- The fittings for galvanized drainage, waste and vent pipes must be
(A) galvanized inside only
(B) galvanized outside only
(C) straight type
(D) recessed type

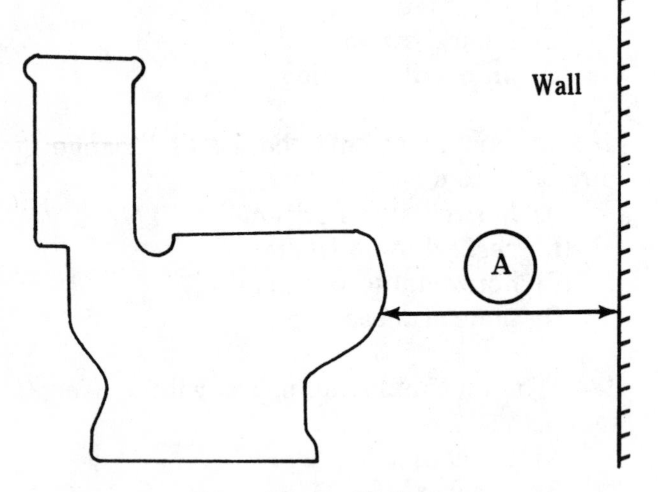

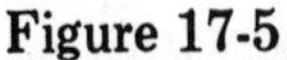
Figure 17-5

190- Using Figure 17-5, distance A for a water closet must be a minimum of
(A) 12 inches
(B) 15 inches
(C) 21 inches
(D) 30 inches

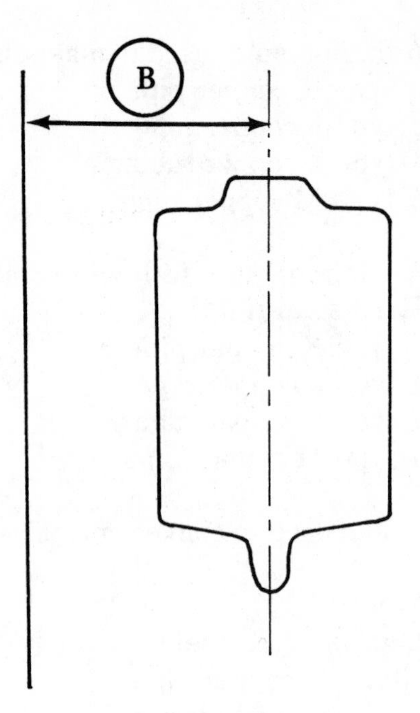

Figure 17-6

191- Using Figure 17-6, distance B for a wall hung urinal must be a minimum of
(A) 10 inches
(B) 12 inches
(C) 15 inches
(D) 16 inches

192- Again, using Figure 17-6, distance B for a stall urinal must be a minimum of
(A) 10 inches
(B) 12 inches
(C) 15 inches
(D) 16 inches

193- The minimum clearance from an insulated gas fired water heater to any door or walls constructed of combustible materials is
(A) 1 inch
(B) 2 inches
(C) 3 inches
(D) 4 inches

194- A thermosyphon solar system requires
(A) a circulating pump
(B) a thermostat sensor
(C) a 60 gallon storage tank
(D) none of these

195- One of the following materials is *not* recommended for a solar heat collector deck:
(A) black plastic

(B) aluminum
(C) copper
(D) steel

196- A gas appliance consumption of 220,000 B.t.u.'s per hour is equal to
(A) 22 cubic feet
(B) 220 cubic feet
(C) 2,220 cubic feet
(D) 22,200 cubic feet

197- The cleanout fitting for an 8 inch building sewer must not be smaller than
(A) 2 inches
(B) 4 inches
(C) 6 inches
(D) 8 inches

198- A commercial food waste disposer unit must be connected to waste directly into
(A) the sanitary system
(B) the greasy waste system
(C) a bucket type trap
(D) a grease trap

199- A gallon of water is considered to weigh most nearly
(A) 8.34 lb.
(B) 9.5 lb.
(C) 10.3 lb.
(D) 12.4 lb.

200- A cubic foot of water contains
(A) 6.50 gallons
(B) 7.48 gallons
(C) 7.85 gallons
(D) 8.48 gallons

201- A square foot contains
(A) 54 sq. in.
(B) 76 sq. in.
(C) 120 sq. in.
(D) 144 sq. in.

202- An indirect waste pipe installed above a floor must have a clearance above such floor of at least
(A) 1 inch
(B) 2 inches
(C) 3 inches
(D) 4 inches

203- The local vent of a sump receiving waste from plumbing fixtures having a total of 7 fixture units may be
(A) 1¼ inches
(B) 2 inches
(C) 2½ inches
(D) 3 inches

204- A check valve for such sump may be located
(A) in basement drain
(B) on sump side of gate valve
(C) on building drain side of gate valve
(D) B and C only

205- Fixtures connected directly to the sanitary drainage system must be equipped with
(A) a water seal trap
(B) a vent
(C) a backwater valve
(D) a waste pipe

206- A rectangular tank 4 feet long, 2 feet wide and 2 feet deep would contain________cubic feet.
(A) 12 (B) 14 (C) 16 (D) 18

207- A rectangular tank 4 feet long, 2 feet wide and 2 feet deep would contain approximately___ gallons of water.
(A) 105.0 (B) 119.7 (C) 126.2 (D) 132.5

208- A 2 inch pipe is equivalent to______one-inch pipes.
(A) 3.2 (B) 4.0 (C) 5.6 (D) 9.8

209- The minimum floor area of a shower stall must be
(A) 400 sq. in.
(B) 600 sq. in.
(C) 860 sq. in.
(D) 1024 sq. in.

210- One of the following materials used in a gas system is *not* permitted outside a building, underground:
(A) plastic pipe
(B) aluminum pipe
(C) copper pipe
(D) brass pipe

MULTIPLE CHOICE ANSWERS

1. D
2. A
3. C
4. A
5. B
6. A
7. D
8. C
9. B
10. C
11. D
12. A
13. C
14. B
15. A
16. B
17. C
18. B
19. D
20. D
21. B
22. B
23. C
24. D
25. B
26. B
27. D
28. B
29. C
30. C
31. D
32. B
33. A
34. D
35. C
36. A
37. C
38. C
39. D
40. D
41. B
42. D
43. C
44. A
45. A
46. C
47. B
48. B
49. D
50. B
51. D
52. B
53. C
54. A
55. C
56. B
57. D
58. B
59. D
60. A
61. C
62. B
63. D
64. C
65. C
66. A
67. C
68. D
69. B
70. D
71. C
72. B
73. B
74. B
75. A
76. B
77. D
78. C
79. A
80. C
81. C
82. D
83. C
84. B
85. C
86. B
87. C
88. C
89. A
90. D
91. A
92. D
93. D
94. D
95. A
96. D
97. D
98. C
99. B
100. D
101. D
102. C
103. D
104. C
105. B
106. A
107. C
108. C
109. B
110. C
111. C
112. B
113. D
114. B
115. C
116. D
117. B
118. D
119. B
120. C
121. D
122. A
123. D
124. C
125. A
126. D
127. C
128. D
129. D
130. C
131. D
132. A
133. C
134. C
135. D
136. D
137. A
138. C
139. C
140. D
141. D
142. C
143. C
144. D
145. C
146. D
147. C
148. D
149. A
150. B
151. C
152. A
153. B
154. D
155. C
156. C
157. D
158. B
159. B
160. A
161. C
162. D
163. A
164. B
165. C
166. A
167. D
168. D
169. C
170. B
171. C
172. C
173. B
174. C
175. C
176. A
177. D
178. D
179. B
180. D
181. B
182. D
183. C
184. B
185. C
186. B
187. B
188. A
189. D
190. C
191. C
192. C
193. B
194. D
195. A
196. B
197. C
198. A
199. A
200. B
201. D
202. C
203. D
204. B
205. A
206. C
207. B
208. C
209. D
210. B

Definitions, Abbreviations, And Symbols

Definitions

The terms included here are found in most plumbing codes. Some words in the code have become so descriptive and specialized that their meaning is different from what a dictionary might give.

Two words that appear repeatly in every code are *shall* and *may*. To be able to comply with the code, you must clearly understand the specialized meanings of these words.

Shall is mandatory. It requires compliance without deviation. For example, part of the requirements for waste disposal is that "sewage and liquid waste *shall* be treated and disposed of as hereinafter provided." *May* is a permissive term. When used in the code, it means allowable or optional, but not required. For example, "Drinking fountains *may* be installed with indirect waste only for the purpose of resealing required traps of floor drains."

"Building drain" and "building sewer" are two terms often used improperly. Many professionals assume that both terms apply to the same part of the drainage system. However, note the code definition for each term: Building drain is the main horizontal collection system within the walls of a building which extends to five feet beyond the building line. (This distance may vary in some codes.) A building sewer is defined as that part of the horizontal drainage system outside the building line which connects to the building drain and conveys the liquid waste to a legal point of disposal.

Effective and constructive code interpretation is possible only when the words and terms used in the code are understood. Many definitions can be illustrated through isometric drawings. It is possible to have some variation and still remain within the intent of the code definitions. The alert professional will discover that most isometric drawings fit into these definitions. Isometric illustrations provide a better understanding of code definitions.

DEFINITIONS

Absorption Drainfield absorption area.

Air gap (in a water supply system) The unobstructed vertical distance through the free atmosphere between the lowest opening from any pipe or faucet supplying water to a tank, plumbing fixture, or other device, and the flood level rim of the receptacle.

Anaerobic living without free oxygen. Anaerobic bacteria found in septic tanks are beneficial in digesting organic matter.

Approved Approved by the plumbing official or other authority given jurisdiction by the code.

Area drain A receptacle designed to collect

surface or rain water from an open area.

Backfill That portion of the trench excavation up to the original earth line which is replaced after the sewer or other piping has been laid.

Backflow The flow of water or other liquids, mixtures, or substances into the distributing pipes of a potable supply of water, and any other fixture or appliance from any source other than its intended course.

Backflow connection Any arrangement whereby backflow can occur.

Backflow preventer A device or means to prevent backflow into the potable water system.

Back Siphonage The flow of water or other liquids, mixtures or substances into the distributing pipes of a potable supply of water or any other fixture, device, or appliance from any sources other than its intended course, due to a negative pressure in such pipe.

Base The lowest point of any vertical pipe.

Battery of fixtures Any group of two or more similar adjacent plumbing fixtures which discharge into a common horizontal waste or soil branch.

Boiler blow-off An outlet on a boiler to permit emptying or discharging of water or sediment in the boiler.

Branch Any part of the piping system other than a main, riser or stack.

Branch interval A length of soil or waste stack (vertical pipe) generally one story in height (approximately nine feet, but not less than eight feet) into which the horizontal branches from one floor or story of a building are connected to the stack.

Branch vent The vent that connects one or more individual vents with a vent stack or stack vent.

Building drain The main horizontal sanitary collection system, inside the wall line of the building, which conveys sewage to the building sewer beginning five feet (more or less in some codes) outside the building wall. The building drain excludes the waste and vent stacks which receive the discharge from soil, waste and other drainage pipes, including storm water.

Building sewer That part of the horizontal piping of a drainage system which connects to the end of the building drain and conveys the contents to a public sewer, private sewer, or individual sewage disposal system.

Building storm drain A drain used to receive and convey rain water, surface water, ground water, subsurface water and other clear water waste, and discharge these waste products into a building storm sewer or a combined building sewer beginning five feet outside the building wall.

Building storm sewer Connects to the end of the building storm drain to receive and convey the contents to a public storm sewer, combined sewer, or other approved point of disposal.

Building subdrain Any portion of a drainage system which cannot drain by gravity into the building sewer.

Caulking Any approved method for rendering a joint water and gas tight. For cast iron pipe and fittings with hub joints, the term refers to caulking the joint with lead and oakum.

Code Regulations and their subsequent amendments or any emergency rule or regulation lawfully adopted to control the plumbing work by the administrative authority having jurisdiction.

Combined building sewer A building sewer which receives storm water, sewage and other liquid waste.

Common vent The vertical vent portion serving to vent two fixture drains which are installed at the same level in a vertical stack.

Conductor See "leader."

Continuous waste A drain connecting a single fixture with more than one compartment or other permitted fixtures to a common trap.

Cross connection Any physical connection or arrangement between two separate piping systems, one containing potable water and the other water of unknown or questionable safety.

Dead end A branch leading from a soil, waste or vent pipe, building drain or building sewer which is terminated by a plug or other closed fitting at a developed distance of two feet or more. A dead end is also classified as an extension for future connection, or as an extension of a cleanout for accessibility.

Developed length The length as measured along the center line of the pipe and fittings.

Diameter The nominal diameter of a pipe or fitting as designed commercially, unless specifically stated otherwise.

Downspout See "leader."

Drain Any pipe which carries liquid, waste water or other water-borne wastes in a building drainage system to an approved point of disposal.

Drainage system All the piping within public or private premises which conveys sewage, rain water, or other types of liquid wastes to a legal point of disposal.

Drainage well Any drilled, driven or natural cavity which taps the underground water and into which surface waters, waste waters, industrial waste or sewage is placed.

Durham system An all-threaded pipe system of rigid construction, using recessed drainage fittings to correspond to the types of piping being used.

Effective opening The minimum cross-sectional area of the diameter of a circle at the point of water supply discharge.

Effluent The liquid waste as it flows from the septic tank and into the drainfield.

Fixture branch The drain from the trap of a fixture to the junction of that drain with a vent. Some codes refer to a fixture branch as a "fixture drain."

Fixture drain The drain from the fixture branch to the junction of any other drain pipe, referred to in some codes as a "fixture branch."

Fixture unit A design factor to determine the load-producing value of the different plumbing fixtures. For instance, the unit flow rate from fixtures is determined to be one cubic foot, or 7.5 gallons of water per minute.

Fire lines The complete wet standpipe system of the building, including the water service, standpipe, roof manifold, Siamese connections and pumps.

Flood level rim The top edge of a plumbing fixture or other receptacle from which water or other liquids will overflow.

Floor drain An opening or receptacle located at approximately floor level connected to a trap to receive the discharge from indirect waste and floor drainage.

Floor sink An opening or receptacle usually made of enameled cast iron located at approximately floor level which is connected to a trap, to receive the discharge from indirect waste and floor drainage. A floor sink is more sanitary and easier to clean than a regular floor drain, and is usually used for restaurant and hospital installations.

Flushometer valve A device actuated by direct water pressure which discharges a predetermined quantity of water to fixtures for flushing purposes.

Grade The slope or pitch, known as "the fall," usually expressed in drainage piping as a fraction of an inch per foot.

Horizontal pipe Any pipe or fitting which makes an angle of more than 45 degrees with the vertical.

Horizontal branch A drain pipe extending laterally from a soil or waste stack or building drain. May or may not have vertical sections or branches.

Indirect wastes A waste pipe charged to convey liquid wastes (other than body wastes) by discharging them into an open plumbing fixture or receptacle such as floor drain or floor sink. The overflow point of such fixture or receptacle is at a lower elevation than the item drained.

Industrial waste Liquid waste, free of body waste, resulting from the processes used in industrial establishments.

Insanitary Contrary to sanitary principles, injurious to health.

Interceptor A device designed and installed to separate and retain deleterious, hazardous, or undesirable matter from normal wastes and permit normal sewage or liquid wastes to discharge by gravity into the disposal terminal or sewer.

Leader The vertical water conductor or downspout from the roof to the building storm drain, combined building sewer, or other approved means of disposal.

Liquid waste The discharge from any fixture, appliance or appurtenance that connects to a plumbing system which does not receive body waste.

Load factor The percentage of the total connected fixture unit flow rate which is likely to occur at any point with the probability factor of simultaneous use. It varies with the type of occupancy, the total flow unit above this point being considered.

Loop or circuit waste and vent A combination of plumbing fixtures on the same floor level in the same or adjacent rooms connected to a common horizontal branch soil or waste pipe.

Main The principal artery of *any system* of continuous piping, to which branches may be connected.

Main vent The principal artery of the venting system, to which vent branches may be connected.

May The word "may" as used in the code book is a permissive term.

Mezzanine An intermediate floor placed in any story or room. When the total area of any such mezzanine floor exceeds 33⅓ percent of the total floor area in that room or story, it is considered as constituting an additional story rather than a mezzanine.

Pitch "Grade," also referred to as "slope."

Plumbing Includes any or all of the following: (1) the materials including pipe, fittings, valves, fixtures and appliances attached to and a part of a system for the purpose of creating and maintaining sanitary conditions in buildings, camps and swimming pools on private property where people live, work, play, assemble or travel; (2) that part of a water supply and sewage and drainage system extending from either the public water supply mains or private water supply to the public sanitary, storm or combined sanitary and storm sewers, or to a private sewage disposal plant, septic tank, disposal field, pit, box filter bed or any other receptacle or into any natural or artificial body of water, water course upon public or private property; (3) the design, installation or contracting for installation, removal and replacement, repair or remodeling, of all or any part of the materials, appurtenances or devices attached to and forming a part of a plumbing system, including the installation of any fixture, appurtenance or devices used for cooking, washing, drinking, cleaning, fire fighting, mechanical or manufacturing purposes.

Plumbing fixtures Receptacles, devices, or appliances which are supplied with water or which receive or discharge liquids or liquid borne waste, with or without discharge, into the drainage system with which they may be directly or indirectly connected.

Plumbing official inspector The chief administrative officer charged with the administration, enforcement and application of the plumbing code and all amendments thereto.

Plumbing system The drainage system, water supply, water supply distribution pipes, plumbing fixtures, traps, soil pipes, waste pipes, vent pipes, building drains, building sewers, building storm drain, building storm sewer, liquid waste piping, water treating, water using equipment, sewerage treatment, sewerage treatment equipment, fire standpipes, fire sprinklers, and relative appliances and appurtenances, including their respective connections and devices, within the private property lines of a premises.

Potable water Water which is satisfactory for drinking, culinary and domestic purposes and meets the requirements of the health authority having jurisdiction.

Private property For the purposes of the code, all property except streets or roads dedicated to the public, and easements (excluding easements between private parties).

Private or *private use* In relation to plumbing fixtures: in residences and apartments, and in private bathrooms of hotels and similar installations where the fixtures are intended for the use of a family or an individual.

Private sewer A sewer privately owned and not directly controlled by public authority.

Public or *public use* In relation to plumbing fixtures: in commercial and industrial establishments, in restaurants, bars, public buildings, comfort stations, schools, gymnasiums, railroad stations or places to which the public is invited or which are frequented by the public without special permission or special invitation, and other installations (whether paid or free) where a number of fixtures are installed so that their use is similarly unrestricted.

Public sewer A common sewer directly controlled by public authority.

Public swimming pool A pool together with its buildings and appurtenances where the public is allowed to bathe or which is open to the public for bathing purposes by consent of the owner.

Relief vent A vent, the primary function of which is to provide circulation of air between drainage and vent systems.

Rim In code usage, an unobstructed open edge at the overflow point of a fixture.

Rock drainfield Three-quarter inch drainfield rock 100 percent passing a one inch screen and a maximum of ten percent passing a one-half inch screen.

Roof drain An outlet installed to receive water collecting on the surface of a roof which discharges into the leader or downspout.

Roughing-in The installation of all parts of the plumbing system which can be completed prior to the installation of plumbing fixtures; includes drainage, water supply, and vent piping, and the necessary fixture supports.

Sanitary sewer A pipe which carries sewage and excludes storm, surface and ground water.

Second hand A term applied to material or plumbing equipment which has been installed and used, or removed.

Septic tank A watertight receptacle which receives the discharge of a drainage system or part thereof, so designed and constructed as to separate solids from liquid, digest organic matter through a period of detention, and allow the liquids to discharge into the soil outside the tank through a sub-surface system of open-joint or perforated piping, or other approved methods.

Sewage Any liquid waste containing animal, mineral or vegetable matter in suspension or solution. May include liquids containing chemicals in solution.

Shall A mandatory term, as used in the code.

Slope See "grade."

Soil pipe Any pipe which conveys the discharge of water closets or fixtures having similar functions, with or without the discharge from other fixtures, to the building drain or building sewer.

Stack The vertical pipe of a system of soil, waste or vent piping.

Stack vent (Sometimes called a waste vent or soil vent) the extension of a soil or waste stack above the highest horizontal drain connected to the stack.

Standpipe system A system of piping installed for fire protection purposes having a primary water supply constantly or automatically available at each hose outlet.

Storm sewer A sewer used for conveying rain water and/or surface water.

Subsurface drain A drain which receives only subsurface or seepage water and conveys it to a place of disposal.

Sump A tank or pit which receives sewage or liquid waste, located below the normal grade of the gravity system and which must be emptied by mechanical means.

Supports (Also known as "hangers" or "anchors.") Devices for supporting and securing pipe and fixtures to walls, ceilings, floors or structural members.

Supply well Any artificial opening in the ground designed to conduct water from a source bed through the surface when water from such well is used for public, semi-public or private use.

Trap A fitting or device so designed and constructed as to provide a liquid seal which will prevent the back passage of air without materially affecting the flow of sewage or waste water through it.

Trap seal The maximum vertical depth of liquid that a trap will retain, measured between the crown weir and the top of the dip of the trap.

Vent stack A vertical vent pipe installed primarily for the purpose of providing circulation of air to and from any part of the drainage system.

Vent system A pipe or pipes installed to provide a flow of air to or from a drainage system or to provide a circulation of air within such system.

Vertical pipe Any pipe or fitting which is installed in a vertical position or which makes an angle of not more than 45 degrees with the vertical.

Waste pipe Any pipe which receives the discharge of any fixture, except water closets or fixtures having similar functions, and conveys it to the building drain or to the soil or waste stack.

Water-distributing pipe A pipe which conveys water from the water service pipe to the plumbing fixtures, appliances and other water outlets.

Water main A water supply pipe for public or community use.

Water outlet As used in connection with the water-distributing system, the discharge opening for the water (1) to a fixture, (2) to atmospheric pressure (except into an open tank which is part of the water supply system), (3) to a boiler or heating system, (4) to any water-operated device or equipment requiring water to operate, but not a part of the plumbing system.

Water service pipe The pipe from the water main or other source of water supply to the building served.

Water supply system Consists of the water service pipe, the water-distributing pipes, standpipe system and the necessary connecting pipes, fittings, control valves and all appurtenances in or on private property.

Wet vent A waste pipe which serves to vent and convey waste from fixtures other than water closets.

Yoke vent A pipe connecting upward from a soil or waste stack for the purpose of preventing pressure changes in the stacks.

ABBREVIATIONS

The abbreviations here are often found on blueprints (building plans) and in plumbing reference books (including the code) to identify plumbing fixtures, pipes, valves and nationally-recognized associations.

A	area
AD	area drain
AGA	American Gas Association
AISI	American Iron and Steel Institute
ASA	American Standard Association
ASCE	American Society of Civil Engineering
ASHRAE	American Society of Heating, Refrigeration and Air Conditioning Engineers
ASME	American Society of Mechanical Engineers
ASSE	American Society of Sanitary Engineering
ASTM	American Society for Testing Materials
AWWA	American Water Works Association
B.S.	bar sink
B	bidet
B. T.	bathtub
B.t.u.	British Thermal Unit
C to C	center to center
CI	cast iron
CISPI	Cast Iron Soil Pipe Institute
C	condensate line
C. O.	cleanout
C. W.	cold water
cu. ft.	cubic feet
cu. in.	cubic inches
C. W. M.	clothes washing machine
C. V.	check valve
D. F.	drinking fountain
D. W.	dish washer
E to C	end to center
EWC	electric water cooler
°F	degrees Fahrenheit
F	Fahrenheit
F. B.	foot bath
F. F.	finish floor
F. C. O.	floor cleanout
F. D.	floor drain
F. D. C.	fire department connection
F. E. C.	fire extinguisher cabinet
F. G.	finish grade
F. H. C.	fire hose cabinet
F. L.	fire line
F. P.	fire plug
F. S. P.	fire standpipe
F. U.	fixture unit
GAL.	gallons
gpm or Gal. per Min.	gallons per minute
Galv.	galvanized
G. S.	glass sink
G. V.	gate valve
G. P. D.	gallons per day
H. B.	hose bib
Hd or H.D.	head
H. W.	hot water
H. W. R.	hot water return
HWT	hot water tank
in.	inch
I. D.	inside diameter
I W	indirect waste
IPS	iron pipe size
K. S.	kitchen sink
L or LAV.	lavatory
L. T.	laundry tray
L	length
lb.	pound
Max.	maximum
Mfr.	manufacturer
Min.	minimum
M. H.	manhole
NAPHCC	National Association of Plumbing Heating and Cooling Contractors
NBFU	National Board of Fire Underwriters
NBS	National Bureau of Standards
NFPA	National Fire Protection Association
NPS	nominal pipe size
O	oxygen
O. D.	outside diameter
Oz.	ounce
P. D.	planter drain
P. P.	pool piping
PSI	pounds per square inch
Rad.	radius
R. D.	roof drain
red.	reducer
R. L.	roof leader
San.	sanitary
Sh	Shower
Spec.	specification
Sq.	square
S. B.	Sitz bath

Sq. Ft.	square feet
S. P	swimming pool
SS	service sink
Std.	Standard
SV	service
SW	service weight
S & W	soil and waste
T	temperature
U or Urn	urinal
V	volume
Vtr	vent through roof
W	waste
WC	water closet
WH	water heater
XH	extra heavy

RECOMMENDED SYMBOLS FOR PLUMBING FIXTURES

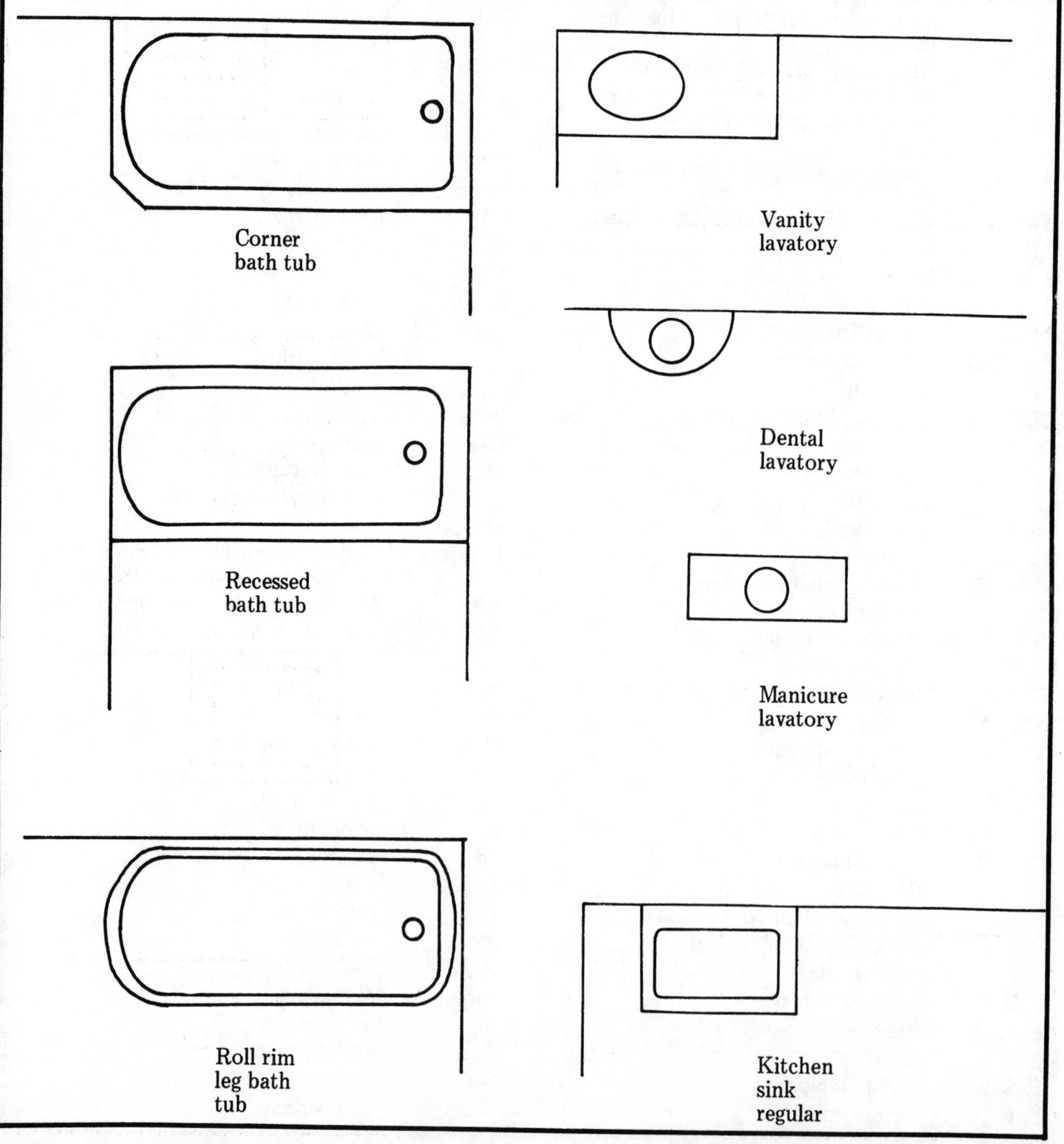

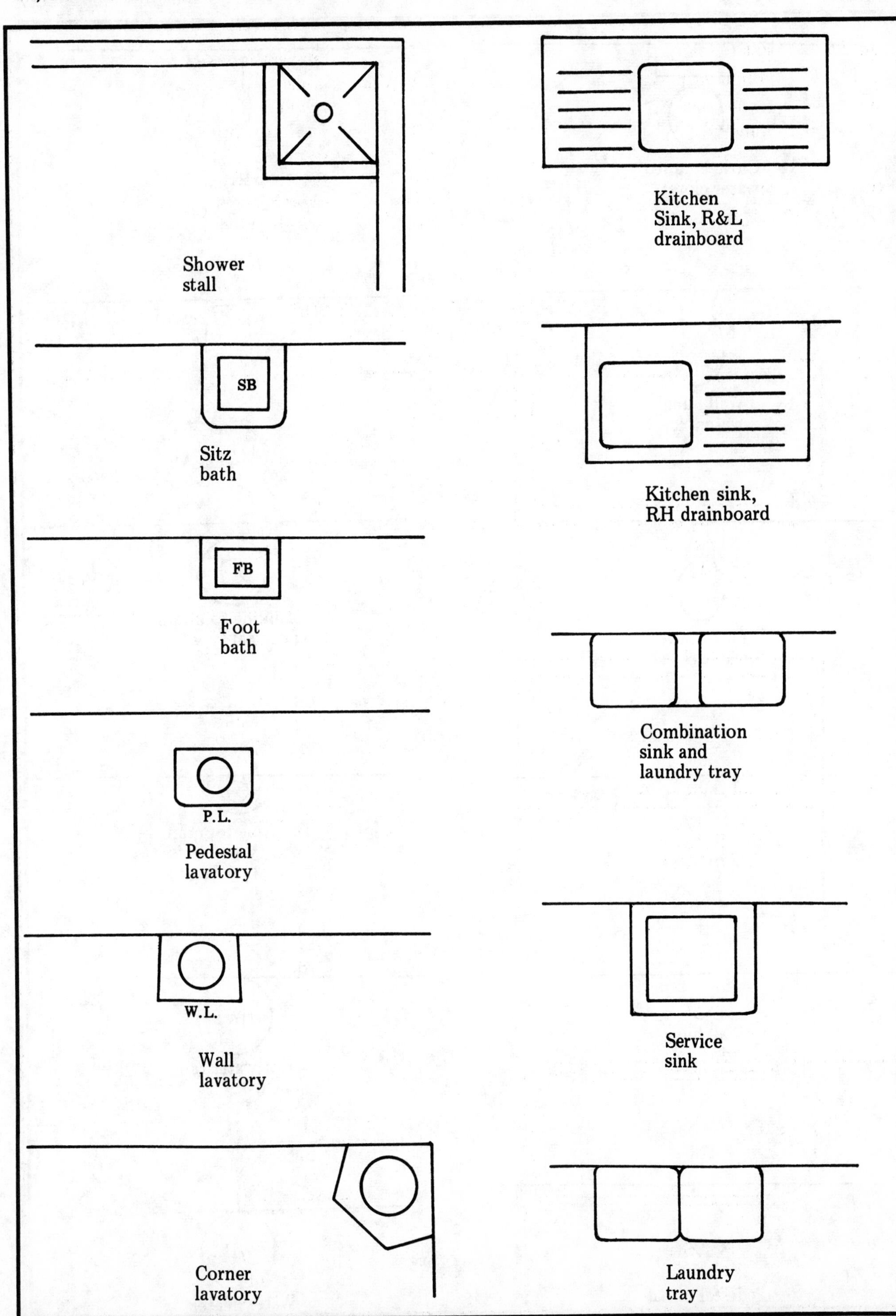
Shower
stall
Kitchen
Sink, R&L
drainboard
SB
Sitz
bath
Kitchen sink,
RH drainboard
FB
Foot
bath
Combination
sink and
laundry tray
P.L.
Pedestal
lavatory
W.L.
Wall
lavatory
Service
sink
Corner
lavatory
Laundry
tray

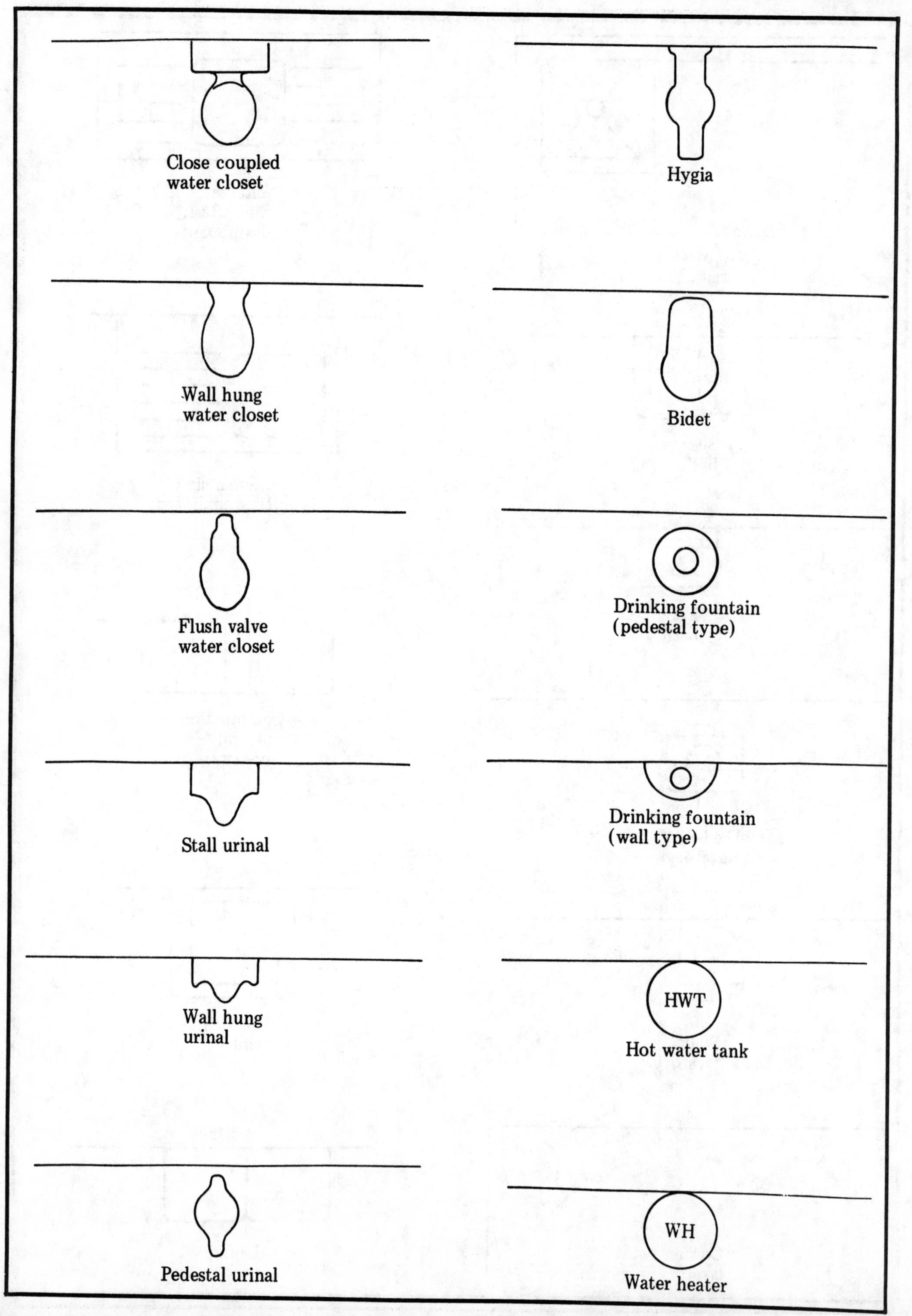
Close coupled
water closet
Hygia
Wall hung
water closet
Bidet
Flush valve
water closet
Drinking fountain
(pedestal type)
Stall urinal
Drinking fountain
(wall type)
Wall hung
urinal
HWT
Hot water tank
Pedestal urinal
WH
Water heater

PLUMBING SYMBOL LEGEND

Symbol	Description
C.W.	Cold water
H.W.	Hot water
H.W.R.	Hot water return
W.L.	Waste line
V.L.	Vent line
S S	Sanitary sewer
C	Condensate line
S D	Storm drain
R.W.L.	Rain water leader
I.W.	Indirect waste
F	Fire line
G	Gas line
	Gate valve
	Globe valve
	Check valve
R	Relief line
	P&T relief valve
F.C.O.	Floor cleanout
F.D.	Floor drain
P.D.	Planter drain
R.D.	Roof drain
H.B.	Hose bibb
A.D.	Area drain

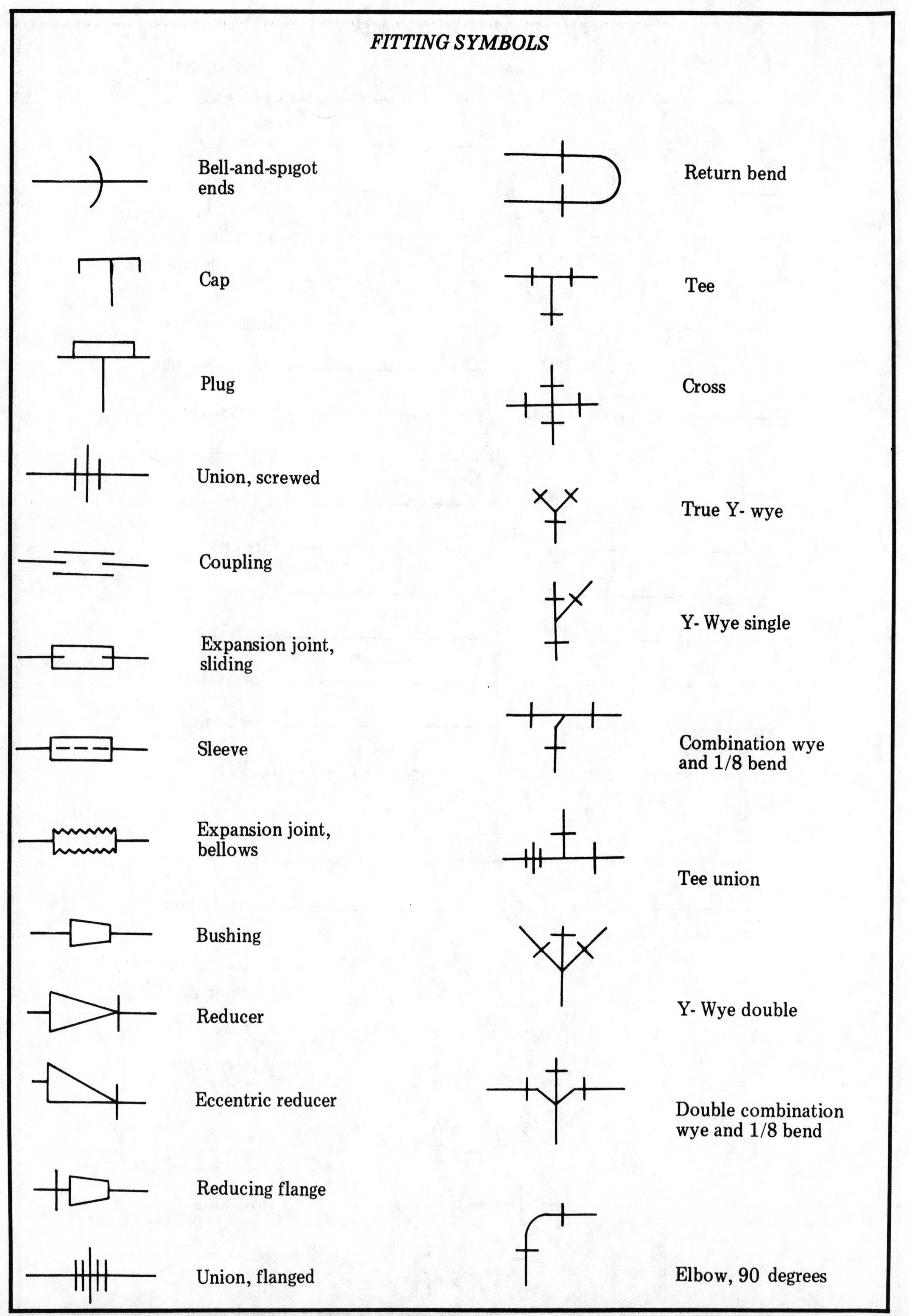
FITTING SYMBOLS
Bell-and-spigot ends
Cap
Plug
Union, screwed
Coupling
Expansion joint, sliding
Sleeve
Expansion joint, bellows
Bushing
Reducer
Eccentric reducer
Reducing flange
Union, flanged
Return bend
Tee
Cross
True Y- wye
Y- Wye single
Combination wye and 1/8 bend
Tee union
Y- Wye double
Double combination wye and 1/8 bend
Elbow, 90 degrees

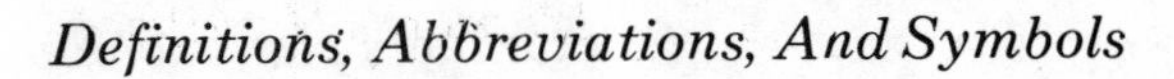

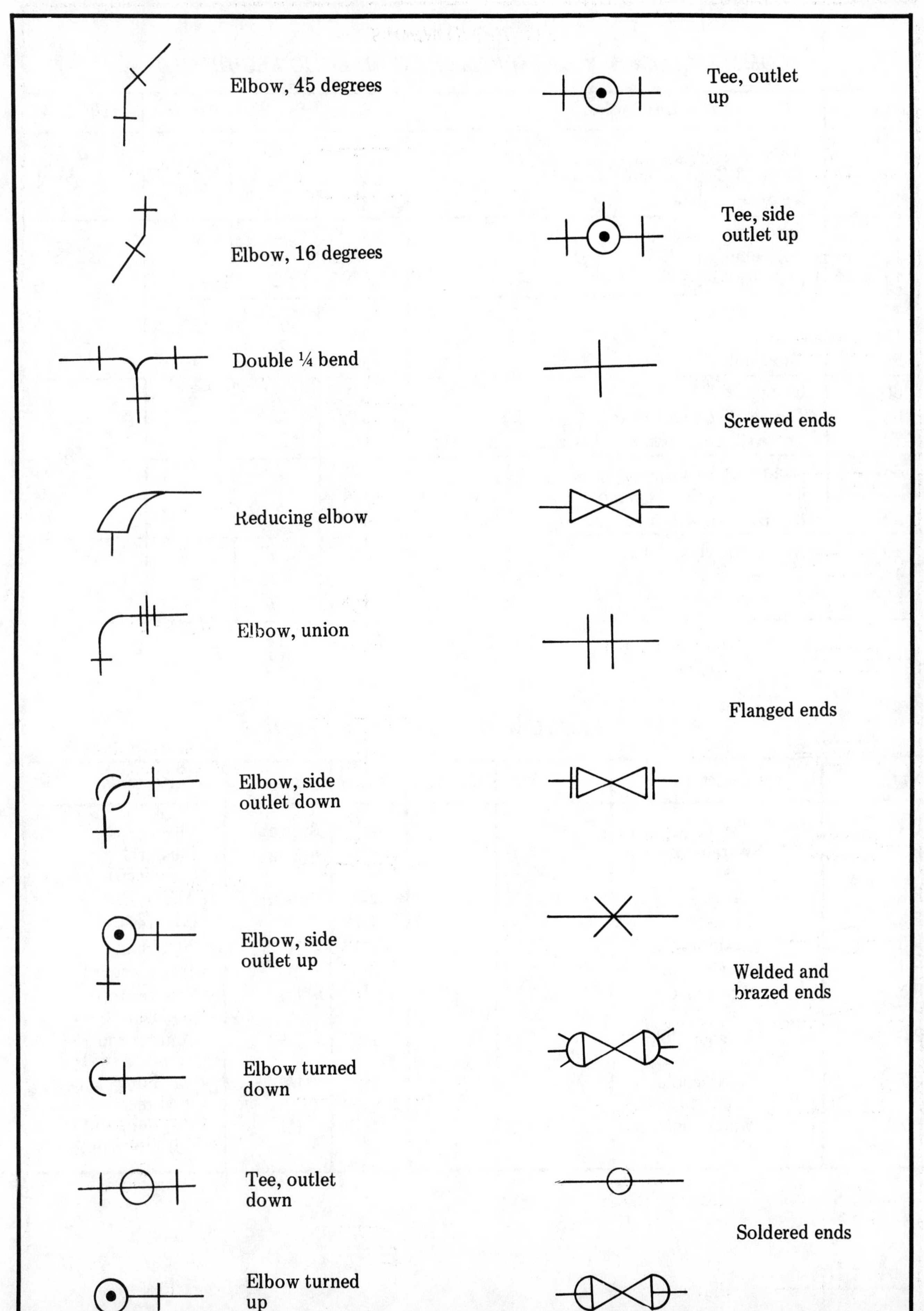
Elbow, 45 degrees
Elbow, 16 degrees
Double ¼ bend
Reducing elbow
Elbow, union
Elbow, side outlet down
Elbow, side outlet up
Elbow turned down
Tee, outlet down
Elbow turned up
Tee, outlet up
Tee, side outlet up
Screwed ends
Flanged ends
Welded and brazed ends
Soldered ends

COMMERCIAL KITCHEN EQUIPMENT CONNECTION SCHEDULE

Item	Description	Waste	C. W.	H. W.	180° H. W.
4	Water station	1"	½"	--	--
7	Three (3) compartment sink	2"	½"	½"	½"
10	Ice cream cabinet	1"	--	--	--
11	Sandwich refrigerator	1"	--	--	--
13	Coffee maker	--	½"	--	--
14	Milk dispenser	1"	--	--	--
17	Refrigerator	1"	--	--	--
18	Freezer	1"	--	--	--
24	Hand sink	1½"	½"	½"	--
30	Ice maker	1"	½"	--	--
31	Three (3) compartment bottle cooler	1"	--	--	--
34	Three (3) compartment sink w/2 drainboards	2"	½"	½"	½"
36	Hand gun, beverage spray	½"	--	--	--
37	Ice chest, cocktail combo	1"	½"	--	--
39	Refrigerated wall base	1"	--	--	--
40	Automatic glass cleaner	1"	½"	½"	--
44	Waitress station	1"	½"	--	--
53	24" x 24" sink	2"	½"	½"	--

PLUMBING FIXTURE CONNECTION SCHEDULE

Mark	Description	C. W.	H. W.	Waste	Trap	Remarks
P-1	Water closet	1"	--	4"	Integral	Flush valve
P-1A	Water closet	1"	--	4"	Integral	Flush valve (handicap)
P-2	Urinal	¾"	--	2"	Integral	Wall hung
P-3	Lavatory	½"	½"	1¼"	1¼"	Counter top
P-3	Lavatory	½"	½"	1¼"	1¼"	Wall hung
P-4	Service sink	½"	½"	3"	3"	Trap standard
P-5	Work sink	½"	½"	1½"	1½"	Counter top (stainless steel)
P-6	Kitchen sink	½"	½"	1½"	1½"	Counter top (stainless steel)
P-7	Water cooler	½"	--	1¼"	1¼"	Wall hung simu-lated recess
P-7A	Water cooler	½"	--	1¼"	1¼"	Wall hung (30" A.B.F.F. handicap)

Index

T

U

V

W

Other Practical References

Other Plumbing References

Estimating Plumbing Costs
Offers a basic procedure for estimating materials, labor, and direct and indirect costs for residential and commercial plumbing jobs. Explains how to interpret and understand plot plans, design drainage, waste, and vent systems, meet code requirements, and make an accurate take-off for materials and labor. Includes sample cost sheets, manhour production tables, complete illustrations, and all the practical information you need to accurately estimate plumbing costs. **224 pages, 8½ x 11, $17.25**

Process Plant and Equipment Cost Estimation
Current cost data and estimating methods for process plant construction. Includes nearly 100 pages of cost data from a broad sample of U.S., European and Asian projects: typical equipment and plant costs, labor cost and productivity comparisons for the entire project duration, escalation indexes for both plant and equipment, manpower distributions, cost adjustments based on the selection of alternate materials, typical project durations and cost overruns, the cost of chemicals, construction materials and utilities, location cost indexes, and operating costs. **240 pages, 8½ x 11, $19.00**

Planning and Designing Plumbing Systems
Explains in clear language, with detailed illustrations, basic drafting principles for plumbing construction needs. Covers basic drafting fundamentals: isometric pipe drawing, sectional drawings and details, how to use a plot plan, and how to convert it into a working drawing. Gives instructions and examples for water supply systems, drainage and venting, pipe, valves and fixtures, and has a special section covering heating systems, refrigeration, gas, oil, and compressed air piping, storm, roof and building drains, fire hydrants, and more. **224 pages, 8½ x 11, $13.00**

Plumber's Exam Preparation Guide
Lists questions like those asked on most plumber's exams. Gives the correct answer to each question, under both the Uniform Plumbing Code and the Standard Plumbing Code — and explains why that answer is correct. Includes questions on system design and layout where a plan drawing is required. Covers plumbing systems (both standard and specialized), gas systems, plumbing isometrics, piping diagrams, and as much plumber's math as the examination requires. Suggests the best ways to prepare for the exam, how and what to study and describes what you can expect on exam day. At the end of the book is a complete sample exam that can predict how you'll do on the real tests. **320 pages, 8½ x 11, $21.00**

Process & Industrial Pipe Estimating
A clear, concise guide to estimating costs of fabricating and installing underground and above ground piping. Includes types of pipes and fittings, valves, filters, strainers, and other in-line equipment commonly specified, and their installation methods. Shows how a take-off is consolidated on the estimate form and the bid estimate derived using the complete set of manhour tables provided in this complete manual of pipe estimating. **240 pages, 8½ x 11, $18.25**

Basic Plumbing with Illustrations
The journeyman's and apprentice's guide to installing plumbing, piping and fixtures in residential and light commercial buildings: how to select the right materials, lay out the job and do professional quality plumbing work. Explains the use of essential tools and materials, how to make repairs, maintain plumbing systems, install fixtures and add to existing systems. **320 pages, 8½ x 11, $17.50**

Remodeling References

Manual of Professional Remodeling
This is the practical manual of professional remodeling written by an experienced and successful remodeling contractor. Shows how to evaluate a job and avoid 30-minute jobs that take all day, what to fix and what to leave alone, and what to watch for in dealing with subcontractors. Includes chapters on calculating space requirements, repairing structural defects, remodeling kitchens, baths, walls and ceilings, doors and windows, floors, roofs, installing fireplaces and chimneys (including built-ins), skylights, and exterior siding. Includes blank forms, checklists, sample contracts, and proposals you can copy and use. **400 pages, 8½ x 11, $18.75**

How to Sell Remodeling
Proven, effective sales methods for repair and remodeling contractors: finding qualified leads, making the sales call, identifying what your prospects really need, pricing the job, arranging financing, and closing the sale. Explains how to organize and staff a sales team, how to bring in the work to keep your crews busy and your business growing, and much more. Includes blank forms, tables, and charts. **240 pages, 8½ x 11, $17.50**

Paint Contractor's Manual
How to start and run a profitable paint contracting company: getting set up and organized to handle volume work, avoiding the mistakes most painters make, getting top production from your crews and the most value from your advertising dollar. Shows how to estimate all prep and painting. Loaded with manhour estimates, sample forms, contracts, charts, tables and examples you can use. **224 pages, 8½ x 11, $19.25**

Remodeler's Handbook
The complete manual of home improvement contracting: Planning the job, estimating costs, doing the work, running your company and making profits. Pages of sample forms, contracts, documents, clear illustrations and examples. Chapters on evaluating the work, rehabilitation, kitchens, bathrooms, adding living area, re-flooring, re-siding, re-roofing, replacing windows and doors, installing new wall and ceiling cover, repainting, upgrading insulation, combating moisture damage, estimating, selling your services, and bookkeeping for remodelers. **416 pages, 8½ x 11, $18.50**

Electrical References

Home Wiring: Improvement, Extension, Repairs
How to repair electrical wiring in older homes, extend or expand an existing electrical system in homes being remodeled, and bring the electrical system up to modern standards in any residence. Shows how to use the anticipated loads and demand factors to figure the amperage and number of new circuits needed, and how to size and install wiring, conduit, switches, and auxiliary panels and fixtures. Explains how to test and troubleshoot fixtures, circuit wiring, and switches, as well as how to service or replace low voltage systems. **224 pages, 5½ x 8½, $15.00**

Residential Wiring
Shows how to install rough and finish wiring in both new construction and alterations and additions. Complete instructions are included on troubleshooting and repairs. Every subject is referenced to the 1987 National Electrical Code, and over 24 pages of the most needed NEC tables are included to help you avoid errors so your wiring passes inspection — the first time. **352 pages, 5½ x 8½, $18.25**

Electrical Blueprint Reading
Shows how to read and interpret electrical drawings, wiring diagrams and specifications for construction of electrical systems in buildings. Shows how a typical lighting plan and power layout would appear on the plans and explains what the contractor would do to execute this plan. Describes how to use a panelboard or heating schedule and includes typical electrical specifications. **128 pages, 8½ x 11, $13.75**

Estimating Electrical Construction
A practical approach to estimating materials and labor for residential and commercial electrical construction. Written by the A.S.P.E. National Estimator of the Year, it explains how to use labor units, the plan take-off and the bid summary to establish an accurate estimate. Covers dealing with suppliers, pricing sheets, and how to modify labor units. Provides extensive labor unit tables, and blank forms for use in estimating your next electrical job. **272 pages, 8½ x 11, $19.00**

Construction Estimating Guides

Estimating Home Building Costs
Estimate every phase of residential construction from site costs to the profit margin you should include in your bid. Shows how to keep track of manhours and make accurate labor cost estimates for footings, foundations, framing and sheathing finishes, electrical, plumbing and more. Explains the work being estimated and provides sample cost estimate worksheets with complete instructions for each job phase. **320 pages, 5½ x 8½, $14.00**

Construction Estimating Reference Data
Collected in this single volume are the building estimator's 300 most useful estimating reference tables. Labor requirements for nearly every type of construction are included: site work, concrete work, masonry, steel, carpentry, thermal & moisture protection, doors and windows, finishes, mechanical and electrical. Each section explains in detail the work being estimated and gives the appropriate crew size and equipment needed. **368 pages, 11 x 8½, $20.00**

Cost Records for Construction Estimating
How to organize and use cost information from jobs just completed to make more accurate estimates in the future. Explains how to keep the cost records you need to reflect the time spent on each part of the job. Shows the best way to track costs for sitework, footing, foundations, framing, interior finish, siding and trim, masonry, and subcontract expense. Provides sample forms. **208 pages, 8½ x 11, $15.75**

Electrical Construction Estimator
If you estimate electrical jobs, this is your guide to current material costs, reliable manhour estimates per unit, and the total installed cost for all common electrical work: conduit, wire, boxes, fixtures, switches, outlets, loadcenters, panelboards, raceway, duct, signal systems, and more. Explains what every estimator should know before estimating each part of an electrical system. **416 pages, 8½ x 11, $25.00**

Estimating Tables for Home Building
Produce accurate estimates in minutes for nearly any home or multi-family dwelling. This handy manual has the tables you need to find the quantity of materials and labor for most residential construction. Includes overhead and profit, how to develop unit costs for labor and materials and how to be sure you've considered every cost in the job. **336 pages, 8½ x 11, $21.50**

National Construction Estimator
Current building costs in dollars and cents for residential, commercial and industrial construction. Prices for every commonly used building material, and the proper labor cost associated with installation of the material. Everything figured out to give you the "in place" cost in seconds. Many timesaving rules of thumb, waste and coverage factors and estimating tables are included. **528 pages, 8½ x 11, $18.50. Revised annually.**

Building Cost Manual
Square foot costs for residential, commercial, industrial, and farm buildings. In a few minutes you work up a reliable budget estimate based on the actual materials and design features, area, shape, wall height, number of floors and support requirements. Most important, you include all the important variables that can make any building unique from a cost standpoint. **240 pages, 8½ x 11, $14.00. Revised annually**

Berger Building Cost File
Labor and material costs needed to estimate major projects: shopping centers and stores, hospitals, educational facilities, office complexes, industrial and institutional buildings, and housing projects. All cost estimates show both the manhours required and the typical crew needed so you can figure the price and schedule the work quickly and easily. **344 pages, 8½ x 11, $30.00**

Carpentry References

Stair Builders Handbook
If you know the floor to floor rise, this handbook will give you everything else: the number and dimension of treads and risers, the total run, the correct well hole opening, the angle of incline, the quantity of materials and settings for your framing square for over 3,500 code approved rise and run combinations—several for every 1/8 inch interval from a 3 foot to a 12 foot floor to floor rise. **416 pages, 8½ x 5½, $13.75**

Carpentry Estimating
Simple, clear instructions show you how to take off quantities and figure costs for all rough and finish carpentry. Shows how much overhead and profit to include, how to convert piece prices to MBF prices or linear foot prices, and how to use the tables included to quickly estimate manhours. All carpentry is covered: floor joists, exterior and interior walls and finishes, ceiling joists and rafters, stairs, trim, windows, doors, and much more. Includes sample forms, checklists, and the author's factor worksheets to save you time and help prevent errors. **320 pages, 8½ x 11, $25.50**

Building Layout
Shows how to use a transit to locate the building on the lot correctly, plan proper grades with minimum excavation, find utility lines and easements, establish correct elevations, lay out accurate foundations and set correct floor heights. Explains planning sewer connections, leveling a foundation out of level, using a story pole and batterboards, working on steep sites, and minimizing excavation costs. **240 pages, 5½ x 8½, $11.75**

Rafter Length Manual
Complete rafter length tables and the "how to" of roof framing. Shows how to use the tables to find the actual length of common, hip, valley and jack rafters. Shows how to measure, mark, cut and erect the rafters, find the drop of the hip, shorten jack rafters, mark the ridge and much more. Has the tables, explanations and illustrations every professional roof framer needs. **369 pages, 8½ x 5½, $12.25**

Roof Framing
Frame any type of roof in common use today, even if you've never framed a roof before. Shows how to use a pocket calculator to figure any common, hip, valley, and jack rafter length in seconds. Over 400 illustrations take you through every measurement and every cut on each type of roof: gable, hip, Dutch, Tudor, gambrel, shed, gazebo and more. **480 pages, 5½ x 8½, $19.50**

Wood-Frame House Construction
From the layout of the outer walls, excavation and formwork, to finish carpentry, and painting, every step of construction is covered in detail with clear illustrations and explanations. Everything the builder needs to know about framing, roofing, siding, insulation and vapor barrier, interior finishing, floor coverings, and stairs. . . complete step by step "how to" information on what goes into building a frame house. **240 pages, 8½ x 11, $11.25. Revised edition**

Contractor's Guide to the Building Code
Explains in plain English exactly what the Uniform Buildi Code requires and shows how to design and constru residential and light commercial buildings that will pass i spection the first time. Suggests how to work with the inspe tor to minimize construction costs, what common buildi short cuts are likely to be cited, and where exceptions a granted. **312 pages, 5½ x 8½, $16.25**

Carpentry for Residential Construction
How to do professional quality carpentry work in homes a apartments. Illustrated instructions show you everything fro setting batter boards to framing floors and walls, installi floor, wall and roof sheathing, and applying roofing. Cove finish carpentry, also: How to install each type of cornic frieze, lookout, ledger, fascia and soffit; how to hang windo and doors; how to install siding, drywall and trim. Each j description includes the tools and materials needed, t estimated manhours required, and a step-by-step guide each part of the task. **400 pages, 5½ x 8½, $19.75**

Reducing Home Building Costs
Explains where significant cost savings are possible a shows how to take advantage of these opportunities. S chapters show how to reduce foundation, floor, exterior wa roof, interior and finishing costs. Three chapters show effe tive ways to avoid problems usually associated with b weather at the jobsite. Explains how to increase labor produ tivity. **224 pages, 8½ x 11, $10.25**

Rough Carpentry
All rough carpentry is covered in detail: sills, girders, column joists, sheathing, ceiling, roof and wall framing, roof trusse dormers, bay windows, furring and grounds, stairs and insul tion. Many of the 24 chapters explain practical code approv methods for saving lumber and time without sacrificing qua ty. Chapters on columns, headers, rafters, joists and girde show how to use simple engineering principles to select t right lumber dimension for whatever species and grade y are using. **288 pages, 8½ x 11, $16.00**

Finish Carpentry
The time-saving methods and proven shortcuts you need to d first class finish work on any job: cornices and rakes, gutte and downspouts, wood shingle roofing, asphalt, asbestos a built-up roofing, prefabricated windows, door bucks a frames, door trim, siding, wallboard, lath and plaster, stai and railings, cabinets, joinery, and wood flooring. **192 page 8½ x 11, $10.50**

Roofers Handbook
The journeyman roofer's complete guide to wood and aspha shingle application on both new construction and reroofi jobs: How professional roofers make smooth tie-ins on a job, the right way to cover valleys and ridges, how to hand and prevent leaks, how to set up and run your own roofi business and sell your services as a professional roofer. Ov 250 illustrations and hundreds of trade tips. **192 pages, 8½ 11, $9.25**

Builder's Office References

Contractor's Survival Manual
How to survive hard times in construction and take full advantage of the profitable cycles. Shows what to do when the bills can't be paid, finding money and buying time, transferring debt, and all the alternatives to bankruptcy. Explains how to build profits, avoid problems in zoning and permits, taxes, time-keeping, and payroll. Unconventional advice includes how to invest in inflation, get high appraisals, trade and postpone income, and how to stay hip-deep in profitable work. **160 pages, 8½ x 11, $16.75**

Contractor's Year-Round Tax Guide
How to set up and run your construction business to minimize taxes: corporate tax strategy and how to use it to your advantage, and what you should be aware of in contracts with others. Covers tax shelters for builders, write-offs and investments that will reduce your taxes, accounting methods that are best for contractors, and what the I.R.S. allows and what it often questions. **192 pages, 8½ x 11, $16.50**

Builder's Guide to Accounting Revised
Step-by-step, easy to follow guidelines for setting up and maintaining an efficient record keeping system for your building business. Not a book of theory, this practical, newly-revised guide to all accounting methods shows how to meet state and federal accounting requirements, including new depreciation rules, and explains what the tax reform act of 1986 can mean to your business. Full of charts, diagrams, blank forms, simple directions and examples. **304 pages, 8½ x 11, $17.25**

Computers: The Builder's New Tool
Shows how to avoid costly mistakes and find the right computer system for your needs. Takes you step-by-step through each important decision, from selecting the software to getting your equipment set up and operating. Filled with examples, checklists and illustrations, including case histories describing experiences other contractors have had. If you're thinking about putting a computer in your construction office, you should read this book before buying anything. **192 pages, 8½ x 11, $17.75**

Builder's Guide to Construction Financing
Explains how and where to borrow the money to buy land and build homes and apartments: conventional loan sources, loan brokers, private lenders, purchase money loans, and federally insured loans. How to shop for financing, get the valuation you need, comply with lending requirements, and handle liens. **304 pages, 5½ x 8½, $11.00**

Builder's Office Manual, Revised
Explains how to create routine ways of doing all the things that must be done in every construction office — in the minimum time, at the lowest cost, and with the least supervision possible: Organizing the office space, establishing effective procedures and forms, setting priorities and goals, finding and keeping an effective staff, getting the most from your record-keeping system (whether manual or computerized). Loaded with practical tips, charts and sample forms for your use. **192 pages, 8½ x 11, $15.50**

Craftsman

Craftsman Book Company
6058 Corte del Cedro
P.O. Box 6500
Carlsbad, CA 92008

In a Hurry?
We accept charge card phone orders
Call (619) 438-7828

(Please print clearly)

any

ss

State/Zip

d check or money order
al enclosed ________ (In California add 6% tax)
ou prefer, use your ☐ Visa or ☐ MasterCard
rd no. ________
xpiration date ________ Initials ________

e ship within 48 hours of receiving your order. If you prefer, e can ship U.P.S. For the fastest service available call in your der with your Visa or MasterCard number. We will ship within hours.

ese books are tax deductible when used to improve or main- n your professional skill.

10 Day Money Back GUARANTEE

- ☐ 17.50 Basic Plumbing with Illustrations
- ☐ 30.00 Berger Building Cost File
- ☐ 17.25 Builder's Guide to Accounting Revised
- ☐ 11.00 Builder's Guide to Const. Financing
- ☐ 15.50 Builder's Office Manual Revised
- ☐ 14.00 Building Cost Manual
- ☐ 11.75 Building Layout
- ☐ 25.50 Carpentry Estimating
- ☐ 19.75 Carpentry for Residential Construction
- ☐ 17.75 Computers: The Builder's New Tool
- ☐ 20.00 Const. Estimating Ref. Data
- ☐ 16.25 Contractor's Guide to the Building Code
- ☐ 16.75 Contractor's Survival Manual
- ☐ 16.50 Contractor's Year-Round Tax Guide
- ☐ 15.75 Cost Records for Const. Estimating
- ☐ 13.75 Electrical Blueprint Reading
- ☐ 25.00 Electrical Construction Estimator
- ☐ 19.00 Estimating Electrical Construction
- ☐ 14.00 Estimating Home Building Costs
- ☐ 17.25 Estimating Plumbing Costs
- ☐ 21.50 Estimating Tables for Home Building
- ☐ 10.50 Finish Carpentry
- ☐ 15.00 Home Wiring: Improvement, Extension, Repairs
- ☐ 17.50 How to Sell Remodeling
- ☐ 18.75 Manual of Professional Remodeling
- ☐ 18.50 National Construction Estimator
- ☐ 19.25 Paint Contractor's Manual
- ☐ 13.00 Planning and Designing Plumbing Systems
- ☐ 21.00 Plumber's Exam Preparation Guide
- ☐ 18.25 Process & Industrial Pipe Estimating
- ☐ 19.00 Process Plant & Equip. Cost Est.
- ☐ 12.25 Rafter Length Manual
- ☐ 10.25 Reducing Home Building Costs
- ☐ 18.50 Remodeler's Handbook
- ☐ 18.25 Residential Wiring
- ☐ 19.50 Roof Framing
- ☐ 9.25 Roofers Handbook
- ☐ 16.00 Rough Carpentry
- ☐ 13.75 Stair Builder's Handbook
- ☐ 11.25 Wood-Frame House Construction
- ☐ 18.00 Plumbers Handbook Revised